Creation and Transfer of Knowledge

Institutions and Incentives

Springer
Berlin
Heidelberg
New York
Barcelona
Budapest
Hong Kong
London
Milan
Paris
Singapore
Tokyo

G. Barba Navaretti · P. Dasgupta
K.-G. Mäler · D. Siniscalco
Editors

Creation and Transfer of Knowledge

Institutions and Incentives

With 28 Figures
and 16 Tables

Springer

Prof. Giorgio Barba Navaretti
University of Milan
Department of Economics
Faculty of Political Science
Via Conservatorio 7
I-20122 Milan, Italy

Prof. Partha Dasgupta
University of Cambridge
Faculty of Economics
Sidgwick Ave.
CB3 9DE Cambridge, England

Prof. Karl-Göran Mäler
The Royal Swedish Academy of Sciences
The Beijer International Institute of Ecological Economics
P.O. Box 50005
S-10405 Stockholm, Sweden

Prof. Domenico Siniscalco
University of Turin
Department of Economics
Faculty of Economics and Commerce
Corso Unione Sovietica 218 bis
I-10134 Turin, Italy

ISBN 978-3-642-08408-9

Library of Congress Cataloging-in-Publication Data
Die Deutsche Bibliothek – CIP-Einheitsaufnahme
Barba Navaretti, G.; Dasgupta, P.; Mäler, K.-G.; Siniscalco, D. (eds.): Creation and Transfer of Knowledge: Institutions and Incentives; with 28 figures and 16 tables / Giorgio Barba Navaretti et al. – Berlin; Heidelberg; New York; Barcelona; Budapest; Hong Kong; London; Milan; Singapore; Paris; Tokyo: Springer, 1998

Springer-Verlag Berlin Heidelberg New York
a member of BertelsmannSpringer Science+Business Media GmbH

Printed in Germany

Hardcover-Design: Erich Kirchner, Heidelberg

Preface

This volume collects a set of papers that were presented at a conference organised by Fondazione Eni Enrico Mattei and the Beijer International Institute of Ecological Economics within their joint research programme on 'Knowledge as an Economic Good' The editors wish to thank Roberta Ranzini and Silvia Simoni for their invaluable effort in putting the manuscript together. They also wish to thank Christina Leijonhufvud and Astrid Auraldsson for their help in organising the conference and dealing with the Beijer's network. Finally, the editors are grateful to Springer's staff and particularly to Martina Bihn, the Economics Editor, for supporting this project.

TABLE OF CONTENTS

PART II: TRANSMISSION AND DIFFUSION OF KNOWLEDGE

1. Production and Transmission of Knowledge: Institutions and Economic Incentives. An Introduction

Giorgio Barba Navaretti, Partha Dasgupta, Karl Göran Mäler and Domenico Siniscalco*

Is knowledge an economic good? Which are the characteristics of the institutions regulating the production, the transfer and the difffusion of knowledge? This volume adopts a multidisciplinary approach to bring knowledge at the forefront, as a key economic issue.

No economic endeavor can be undertaken without learning. Indeed, knowledge has recently moved at the core of most new developments of economic analysis. It is an explicit variable in an increasing number of theoretical and empirical contributions dealing with fundamental issues such as: economic growth and accumulation of human capital; education, social segmentation and income distribution; location and clustering of economic activities; R&D contracts and networks; technological choices within the firm; North-South transfers of technology; fertility and population. The methodology and the results developed by this recent literature also bear profound insights for events like scientific progress, artistic and craft development which have rarely been addressed as socio-economic institutions and which have generally been the domain of sociologists and historians rather than economists.

For the most part we have a picture of each of these issues in isolation. It has been uncommon to relate them, to see them at once as endeavors in the production, dissemination and use of knowledge. The present collection, which includes some of the papers presented at a conference jointly organised by Fondazione Eni Enrico Mattei and the Beijer International Institute of Ecological Economics, is aimed precisely at studying the similarities and differences in these endeavors.

The unifying framework is the characteristic of knowledge as a semi-public good, with non easily enforceable property rights. Its diffusion, in principle, is good for social well being, but bad for private returns: no one wants to invest in the creation of new knowledge, if the rents generated are not, at least partly, appropriable. Institutions that govern the creation and the diffusion of knowledge have invariably been moulded by this trade off.

This volume explores resource allocation mechanisms, more generally, socio-economic institutions, that can be relied upon to produce knowledge in an efficient

* Università degli Studi di Milano, Fondazione Eni Enrico Mattei and Centro Studi Luca d'Agliano; University of Cambridge, U.K.; The Beijer International Institute of Ecological Economics; Università di Torino and Fondazione Eni Enrico Mattei.

manner, that is institutions that can sustain an efficient production and allocation of knowledge of all kinds. In principle, there may be several resource allocation mechanisms that are equally efficient as regards the production of knowledge. In these situations, it is a matter of social indifference which institution is established. But in the world as we know it, matters are different. It is socially desirable to institute not one, but simultaneously more than one resource allocation mechanism in the "knowledge sector". There may be social advantages in encouraging these institutions in part to assume adversarial roles.

Of all possible resource allocation mechanisms, the one that has been most studied in economics is the market mechanism. As is now well known, if the market mechanism is not aided by further social contrivances, such as, for example, intellectual property rights, it can be relied upon *not* to sustain an efficient production of knowledge. More generally, the market mechanism has a tendency to discourage the production of public goods because of an inability on the part of producers to appropriate fully the value of the fruits of their activity.[1]

Three schemes are available for overcoming the market's deficiency in this regard. The *first* scheme consists in the government engaging itself directly in the production of knowledge, allowing free use of it, and financing the production cost from general taxation) A *second* possible scheme is for society to grant intellectual property rights to private producers for their discoveries, and permit them to charge (possibly differential) fees for their use by others A *third* possible scheme is for society to encourage private production of knowledge by offering public subsidies for its production, and by relying upon general taxation to finance these subsidies. This collection considers all these allocation schemes in completely different settings and explores how market and non market based institutions interact.

Institutions can be classified broadly into those ruling production and those ruling transmission of knowledge. The system of patronage in arts, research and development in firms and universities mostly concern creation; schools, art academies, transfer of technology through multinationals mostly concern transmission. We follow the same organising principle: the first section deals with production and the second one with transfer and diffusion.

This said, it needs emphasis that the distinction between production, transmission, and diffusion of knowledge is blurred. As in most service industries, producers and customers often work together to make use of the knowledge that is produced: rewards for the discovery of a new piece of knowledge depend upon the mechanisms of disclosure of this very piece of knowledge. Thus, institutions for the creation of new knowledge often develop the rules of transmission. Compare

---

[1] This is the well-known "free rider" problem in the supply of public goods. The earliest writers on this were Knut Wicksell and Eric Lindahl. See Musgrave and Peacock (1958) for abridged versions of the Wicksell-Lindahl analyses. Modern statements of the allocation problem associated with public goods are in Samuelson (1954) and Arrow (1971). The basic economics of public goods can be found in any textbook on public finance. See e.g. Musgrave (1968) and Stiglitz (1988).

speculative, theoretical knowledge to applied, practical knowledge. The first one is characterised by a reward system where the rule of priority applies. Whoever discloses first has the prize and the honour. Thereon, though, the new piece of knowledge turns into a public good (shared by anyone interested, at no cost). The second one has patenting as its reward system. Patenting preserves property righs on the new piece of knowledge (which is therefore a private good) and associates diffusion with the exploitation of a rent.

Equally, institutions for the transmission and dissemination of knowledge can affect the pattern of future knowledge production. Consider, for example, convents in Europe in the Middle Ages, music schools in Northern India, and Ph.D programs in modern universities. These are institutions where production and transmission of knowledge go hand in hand. One may think of a continuum of institutions, which range from pure production to pure diffusion mechanisms, with the most frequent and interesting cases lying in the middle.

Even though the creation and dissemination of knowledge are closely interrelated, there are two reasons why we should consider them separately. First, the technological, institutional and financial characteristics of the institutions that produce and disseminate knowledge depend on exogenous shocks and investments, which are very often external to these institutions, but which create feed-backs on the world of ideas as well as on social cohesion and segmentation. Secondly, despite their heterogeneous genesis and nature, the mechanisms for the dissemination of knowledge are characterized by important regularities and lend themselves to a common analysis.

The papers included are not comprehensive surveys, but sketches, examples of how different disciplines deal with the same subject. This approach discloses the problem of language. Indeed, multidisciplinarity implies bridging jargons not at all meant to communicate between themselves. The development of a codified language is a central element of the mechanics of knowledge diffusion. Codes speed up interactions between peers and keep outsiders at bay. For this reason, we decided to render the problem of language explicit. To enhance intelligibility, the editors' introduction unfolds the threads between the different approaches and issues developed in the papers. Furthermore, papers with a stronger technical flavour (partcularly the economic ones) are preceded by introductions which put the technical work in its broader context.

1.1 Production of knowledge

Patronage, property rights and their impact on the creation of a new style in architecture is the object of *Mary Hollingsworth*'s paper, 'Patronage and Innovation in Architecture'. The most fascinating aspect of Hollingsworth's paper, which was focused on the emergence of the renaissance style in Florence in the 15th century, is the property rights issue of architectural innovation. The patron, and not the artist, was generally considered as the 'creator' at those times. The identification between the work of art and the artist was therefore weak; the latter was deprived of part of the rents generated by his innovative work, particularly in terms of future glory. This outcome had partly to do with construction techniques and partly with bargaining power. In early architecture, the profession of the architect was ill defined. The patron, rather than the designer of the building, was indeed the co-ordinator of the works. He was therefore entitled to take the credit for the building. When construction techniques became more sophisticated, the role of the architect was finally recognised in public (Brunelleschi's dome of Florence Duomo is a clear example). But also simpler works of arts, like sculptures and paintings, were rarely identified with the author. This system was probably inefficient. The inability to fully exploit the rents generated by his knowledge and his ability, reduced the artist's incentive to maximise his effort. This inefficiency had to be counterbalanced by the contracts of the time. Often, the price of the work of art was to be defined by an independent commission, after the work was accomplished. The composition of the commission showed that this was a sheer matter of bargaining power. Only when the artist was powerful enough, he was allowed to name a commissioner.

Perhaps Renaissance patrons would have opted for a more equitable sharing of property rights, if contracts were designed under the same conditions assumed by modern contract theory. Indeed, the paper by *Philippe Aghion* and *Jean Tirole*, 'Opening the Black Box of Innovation' raises very similar issues, although in the context of industrial research and development (R&D). Aghion and Tirole underline the importance of introducing organisational considerations for understanding the relationship between technological innovation and growth. They argue that the relationship between institutions carrying out R&D is often hierarchical, of a customer-supplier (or patron-artist) type. The distribution of property rights on the innovation between the different parties has a fundamental impact on the efficiency of the innovative activity. In this context, the customer provides the investment and the research unit (the supplier) the research effort. The optimal allocation of property rights depends on the relative importance of the investment *vis à vis* the research effort. But if the customer (who has the money) has a strong relative bargaining power, property rights could well be allocated in an inefficient way.

Agreements for R&D cooperation between firms are not always hierarchical (of the customer-supplier type). Two or more firms, sometimes even competitors, may form agreements to develop together new products, components or production processes. As documented by *Giorgio Barba Navaretti* and *Andrea Bigano* ('R&D Inter-firm Agreements in Developing Countries: Where? Why? How?)' R&D joint ventures also take place with firms based in developing countries. Interesting, because these firms are generally endowed with asymmetric stocks of knowledge and North-South technology flows are tipically hierarchical, whereby southern firms are passive recipients of northern technologies. As extensively discussed in the section on technology transfer, multinational activity in developing countries generates knowledge spillovers, which are larger when local firms are particularly dynamic and competitive. As knowledge spillovers are inevitable, under given circumstances, multinationals may decide to internalise them by forming agreements with local competitors. Extending the agreement to joint R&D allows multinationals to accede cheaper talent and technological skills, to be closer to the market or to the manufacturers of final products and to free scarce resources at home and use them in more sophisticated activities. Finally, R&D cooperation favours the development of a relationship of trust between the partners and make arm-length agreements more stable. This argument is supported by the empirical analysis developed in the paper, based on a sample of more than 600 technological agreements between firms based in advanced and developing countries.

Even if we discard organisational considerations, there are other reasons why the technological choices made by firms may not necessarily maximise growth in productivity. *Yaw Nyarko* and *Bogdan Jovanovic*'s paper, 'Research and Productivity' examine the way knowledge stocks cumulated by a firm affects its productivity and the consequent choice of its R&D effort. What the firm does now depends on the technologies it has used and on the past research it has carried out. But conversion of technological experience into stocks of knowledge and into increases in productivity is not straightforward. The authors emphasise the 'creative destruction' feature of research. The faster a firm upgrades its technology, the more quickly knowledge that is specific to the abandoned technology depreciates. As a consequence, firms which are in the same industry may choose completely different research strategies. Some firms may focus on applied research (improvement of the existing technology), instead of carrying out basic research (which completely innovates on technology and allows firms to achieve larger productivity increases in the longer term), because they are so familiar with the existing technologies that the short term costs of innovation are too high. This is an example of how short term returns from the adoption of a given body of knowledge in the past may obstacle deeper innovation and technological upgrading in the future.

Business and academic research are, in principle, ruled by different institutional mechanisms. The first one characterised by patenting and the second one by full disclosure of the results of the innovation and peer group recognition. Yet the borderline is not always neat. *Rebecca Henderson, Adam Jaffe* and *Manuel Trejtenberg* present an empirical analysis of US Universities patenting between 1965 and 1988 ("Universities as a source of Commercial Technology: a Detailed Analysis of University patenting 1965-88). They show that University patenting increased 15 fold over the period and research spending almost tripled. A probably central cause for this development, (along with increased industry funding) is a 1980 change in the patenting federal law, the Bayh-Dole act. The new law allowed universities to patent the results of publicly funded research. The old legislation (which was aimed at preserving the public good nature of the new knowledge created with public funds) hampered any incentive to develop basic research into commercial applications. However, the quality of patents, in terms of their importance (measured by the number of future invention each patent generates) and generality (measured by the range of fields where the patent is applied) did not increase. This result is somehow related to Nyarko and Jovanovic's finding that history or changes in the reward system may shift the focus of research from wide ranging basic research to focused and short term oriented applied research.

Whatever their nature, research networks have common features in their dynamics. This the object of the last paper of this section, by *Paul David* ('Communications, Creativity and Research Network Dynamics: an Elementary Economic Model of Scientific Communities'). David and Flemming analyse the dynamics of scientific research communities and how ideas and statements circulate, acquire validity as reliable knowledge and are recombined to generate new ideas. In other words, they analyse how scientific communities achieve 'knowledge consensus'. They show how both the expected time for achieving the consensus and the expected rate of creativity of the network depend upon specific endogenous features of the network, like communication behaviours and size. At the same time, they argue that local networks are part of much larger global networks. The interaction between global and local networks also affect creativity and diffusion of consensus.

1.2 Transmission and diffusion of knowledge

Jack Goody, with his paper on 'Literacy and the Diffusion of Knowledge Across Cultures and Times', delivers a fascinating historical account on how the written word gradually became the permanent repository and the major vehicle for the transmission of human knowledge. Goody's paper is an example of how the evolution of institutions in times and space can partly be explained in terms of economic incentives. Literacy followed precisely that ambiguous pattern between secrecy and disclosure that is so common for any form of knowledge. Its diffusion was driven by the economic returns that it was able to generate for that very society at that very time. Learning is a time consuming endeavour, and early societies could not afford to withdraw a large share of the labour force from manual activities. At the same time, the secrets of literacy were for long preserved by small groups of individuals, who could enjoy very large private rents in return. Only when the needs of new economic activities and social structures and new sets of values prevailed on these private interests (with the diffusion of industry, the spread of religions and democracy) literacy could finally take off.

The diffusion of education, though, (and consequently its impact on economic growth) does not follow new economic incentives in a straight forward manner. *Giuseppe Bertola* and *Daniele Coen Pirani,* in their paper 'Market Failures Education and Macroeconomics', argue hat the role of human capital in fostering economic growth is now well established in the growth literature. However, the microeconomic foundations of the provision of education are not as simple as the macroeconomic literature makes them. The two authors surveyed these microeconomic foundations and discussed their implications in terms of income distribution and growth. Education is characterised by imperfections, particularly related to uncertainty, information asymmetries and social and human capital spillovers. In advanced societies with deep specialisation in skills, education also gives information on the absolute and comparative advantage of each individual within a heterogeneous work force. The authors show that the possibility of setting up an efficient screening mechanism through education is strictly related to financial market imperfections. If financial markets are imperfect, and if education is self financed, schools are unlikely to deliver efficient outcomes, as the poor cannot finance their education even when they are capable. *Jaques Thisse and Ives Zenou*'s contribution 'How to Finance Education when the Labour Force is Heterogeneous' focus precisely on this point. They develop a job matching model and they analyse how the heterogeneity of both workers and firms affect the formation of wages on the labour market and the way in which the costs of education are shared between firms and workers. In this setting, they are able to contrast the two cases where education is a pure private (firm specific training) and a pure public (general education) good.

Similar issues emerge if we look at education in Arts. Both the papers by *Ilaria Bignamini* and *Enrico Castelnuovo* on 'The Role of the Educational Institutions in the Art Sector: from Academies to the Grand Tour', and by *Alaknanda Patel*, 'Oral Transmission in Indian Classical Music: the Gharana system', show how educational institutions are also tools for the generation and the transfer of knowledge related rents. Castelnuovo and Bignamini analysed Art Academies: institutions moulded by the ruling powers to define a predominant national artistic style. Centralised education had a fundamental role in the development of the arts. Early Academies like the Florentine 'Accademia del Disegno' gave young artists access to living models or to the study of theory, geometry, perspective and human anatomy, means that were not provided by workshop training. At the same time, the objective of defining a national style led to great achievements, like the diffusion of renaissance in 15th century Florence. However, the strict institutionalisation of learning and of the artistic carrier were detrimental to innovation. Only unregulated and self supporting private academies like the early English Academy were more likely to promote innovations and changes in the major stylistic patterns. The Gharana system of music provides an alternative example. As described in Alaknanda Patel's paper, Gharana music was never written and was transmitted through a very strict, secretive and life-long master-pupil relationship. Rents were preserved by secrecy rather than by institutional patents. Secrecy had very clear economic reasons: authorship and intellectual property rights are very shaky when the music is transmitted orally. The whole system developed into a social structure with well precise family lines, rules of conduct and collusive behaviours that allowed Gharana masters to preserve their share of secrecy. The oral tradition and the multiplicity of patrons forced masters to define their identity by innovating. But the extent of innovation is difficult to assess, given that masters were faced by the very difficult task of showing off their abilities without disclosing the secret of their art.

North-South transfers of technology are examined by the last two papers of the book. This setting is very useful to understand transmission of knowledge across groups rooted in different cultures and in societies at different levels of economic development. *Jim Markusen* and *Tony Venables'* paper 'The International Transmission of Knowledge by Multinational Firms: Impact on Source and Host Country Skilled Labour', addresses the fundamental question of how country differences in skill and knowledge capital may affect location decisions of multinationals, real wages and national standards of living. Which is the link between technology and skilled labour? Skilled labour may be required for the production or the use of firm level knowledge capital or plant level knowledge capital. The former refers to the body of centralised knowledge (e.g. blueprints) which is owned by a multinational and can be diffused across plants. The latter refers to knowledge specifically related to individual plants. Generally, multinational headquarters locate in skilled labour abundant countries and set up unskilled labour intensive production plants in the host (skill scarce) country. They, therefore, substitute skill-intensive producer services for commodity exports. In skill scarce countries, imports of skilled intensive services through multinationals compete with local skilled labour that will witness a decline in its relative wages. In contrast, if skilled labour is required for plant specific activities, the multinational subsidiary will demand local skilled labour and skilled wages will therefore rise.

Magnus Blomstrom and Ari Kokko ('Multinational Corporations and Spillovers: a Review of the Evidence') produce a very broad survey of the empirical evidence available on knowledge externalities generated by multinationals towards local firms in developing countries. Spillovers result because of forward or backward linkages with multinational subsidiaries, imitation of multinationals' technologies, hiring of workers trained by multinationals and access to foreign markets. Blomstrom and Kokko's conclusion is that spillovers may indeed be substantial, both within and between industries, but there is no strong evidence on their exact nature and magnitude. Spillovers, though, are likely to increase with the level of local capabilities and competition. As discussed in the previous section, increasing spillovers, which are impossible to internalise, and dynamic local firms may gradually generate non hierarchical technological flows and the setting up of North-South R&D agreements.

1.3 On the dynamics of institutions and policy implications

Institutions for the creation and transmission of knowledge emerge and evolve endogenously. They change according to the the type of knowledge they rule, the interests they serve and the returns they generate. There is probably a mechanism of selection, which primarily takes place directly and indirectly, through competition.

Yet, the dynamics of institutions (being them market or non market based) has inherent market failures and it is not necessarily optimal in terms of social welfare. There are many cases reported, where institutions have been negatively affected by vested interests both related to knowledge itself, or related more generally to the regulation of society.

Exclusive control over a given body of knowledge and vehicles for transmission creates rents for insiders. This can happen in many different ways. Peer groups devise languages and codes which speed up internal interaction. But codes raise entry costs and keep outsiders at bay. No one can become a peer, unless she learns the code. The code can strictly be a language, a technique or a set of specific notions. Preserving that very set of notions as the dominant one also preserves the rents of the insiders, but does not necessarily favours the creation of new, useful knowledge. Examples abound. We have discussed above the role of art academies in teaching and defining one style and therefore preventing stylistic innovation (see the paper by Bignamini and Castelnuovo); or the role of early literacy in preserving rents of small groups of *learned* individuals (See the paper by Jack Goody).

Inaccessibility of knowledge could be due to secrecy or simply technological complexities. The technology of different media may have a tremendous impact on society, in terms of cohesion and segmentation, strictly related to learning patterns. Initial diffusion of the press increased social cohesion among those who were able to read but also strengthened social exclusion of the illiterates. Only later, with growing litteracy, the press became a vehicle for global diffusion. The same thing is happening for the Internet. In the short run, computer technology appears to widen the gap among "ins" and "outs", as well as generations,with repercussions on the segmentation of the labor market.

At the opposite level, when knowledge is easily reproducible, the inability to define and enforce effciently intellectual property rights may well jeopardise private incentives to invest in knowledge creation. The volume discusses at length the functioning of property rights as incentive schemes.

Understanding the mechanisms of institutions is extremly important when we move to policy design. This is now well established in regulation, but perhaps less so for science, technology and education policies. In other words, policies should be designed as tools to improve the working of institutions which have their own endogenous dynamics.

The problem, though, is that the lack or malfunctionig of institutions that effectively promote creation and diffusion of knowledge must sometimes be addressed at the broader level of the overall values of society. In other words, payoffs from the creation and diffusion of knowledge (and thus the threshold between private and public good) also depend on the overall sets of values of the society.

William Baumol's analysis of the allocation of entrepreneurship between productive and unproductive activities applies very well here. Baumol argues that such an allocation is influenced by the relative payoffs society offers to such activities. To discuss the unfavourable reward system of Ancient Rome, among other evidence, he reports a beautiful anedoct drawn from Finley:

> '(..) a man invented unbreakable glass and demonstrated it to Tiberius in anticipation of a great reward. The emperor asked the inventor whether anyone shared his secret and he was assured that there was no one else; whereupon his head was promptly removed, lest, said Tiberius, gold be reduced to the value of mud'[2]

---

[2] Baumol, 1990, p. 900 reported from Finley, 1985, p. 147

We do not want to push the stroy of Tiberius and the unbreakable glass any further, as we would move to the realm of the political economy of institutional design, which is well beyond the scope of this work. Yet, once we acknowledge that institutions have an endogenous dynamics, we must make sure that we understand where their roots lie.

References

Arrow, K.J. (1971) "Political and Economic Evaluation of Social Effects and Externalities" in M.Intrigator (ed.), *Frontiers of Qualitative Economics*, (Contributions to Economic Analysis vol. 71, Amsterdam: North-Holland Publishing Co.)

Baumol, William, J., 1990, 'Entrepreneurship: Productive, Unproductive and Distructive', *Journal of Political Economy*, vol. 98, pp. 893-921

Dasgupta, P.S. and P.A. David (1987), "Information Disclosure and the Economics of Science and Technology", Ch. 16 in G.R. Feiwel, ed., *Arrow and the Accent of Modern Economics Theory*, New York: New York University Press.

Finley, Moses, I., 1985, '*The Ancient Economy*', 2nd ed. London, Hogarth

Musgrave, R.A. and A.T. Peacock, eds. (1958) *Classics in the Theory of Public Finance*, London and New York: MacMillan.

Musgrave, R.A. (1968) *The Theory of Public Finance*, Second Edition, New York: McGraw-Hill.

Samuelson, P. and Stiglitz, J. (1954) "The Pure Theory of Public Expenditure", *Review of Economics and Statistics*, vol. 36: pp. 387-89.

PART I

Production of Knowledge

2. Patronage and Innovation in Architecture

Mary Hollingsworth*

The Renaissance was a period of dramatic upheaval: changes to the political, religious, economic, social and cultural structures in Europe marked the end of traditional medieval certainties and the beginning of the modern world. And the period is justly famous for its art. The new order was given visual expression in radical stylistic change that saw the revival of the artistic language of ancient Rome to provide the basis for the emergence of new styles in painting, sculpture and architecture. But the process by which these innovations were created and transferred is not as straightforward as is usually thought. The Renaissance is rich in myths, and perhaps none is more compelling than the idea that artists in fifteenth- and sixteenth-century Italy were free to explore their own ideas and create their masterpieces for enlightened patrons. This myth is reinforced by the methodology of many art historians who study the Renaissance via its artists, an approach that disguises the fact that it was the patron who was the real initiator of the art of the period, and that he played a significant role in determining the final appearance of his commissions.[1]

Renaissance patrons were not passive connoisseurs; they were active consumers. For them, art was not a statement of their aesthetic sensibilities: it was the prime vehicle for the display of their achievements, their social status, political ambitions or commercial prowess, their religious beliefs and their civic pride. The magnificent palaces and lavish decoration commissioned by governments, guilds and individuals were designed to demonstrate the wealth and power of their owners. Rich Florentine bankers, Venetian guilds or cardinals in Rome commissioned chapels, embellished with altarpieces, fresco cycles and tombs, to provide settings for the prayers needed for their souls after death; and they endowed these chapels with substantial revenues to fund the cost of regular prayers. This could be expensive. After the death of Giovanni di Bicci de' Medici his sons gave 900 florins to San Lorenzo to pay for a service every Monday for his soul;[2] the cost of an altarpiece could range from 25 to 250 florins, for a really expensive one, while a skilled mason in full time employment earned only 60

* University of East Anglia, Norwick, U.K.

[1] The role of the patron in Renaissance art forms the basis of my two books: *Patronage in Renaissance Italy: from 1400 to the Early Sixteenth Century*, London 1994 (hereinafter Hollingsworth 1994) and *Patronage in Sixteenth-Century Italy*, London 1996.

[2] R.Gaston, 'Liturgy and Patronage in San Lorenzo, Florence, 1350-1650', in F.W.Kent & P.Simons (eds.), *Patronage, Art and Society in Renaissance Italy*, Oxford 1987, 122.

florins a year.[3] Architecture, above all, was visible, permanent and expensive. They understood its value as propaganda. Pope Nicholas V insisted that magnificent buildings were essential to convince ordinary people of the supreme power of the Church.[4] In 1496 the Venetian diarist, Domenico Malipiero, reported that his government had begun to build an costly clock tower at a time of economic instability in order to demonstrate that the state was not bankrupt.[5] Renaissance patrons spent conspicuously. Grand scale, costly materials and lavish decoration provided recognisable evidence of their wealth and importance. Their choice of style provided visual evidence of the image they wished to present to the world. And, unlike their modern counterparts, they kept close control not only of the financial aspects of their projects, but also of the design and its execution through a system that was significantly different from that current today.

I propose first to outline the process of design and construction of Renaissance architecture and then to discuss the projects of several patrons in Florence and Venice, both individual and corporate, to examine their motives, to show how they gave visual expression to their particular aims and ambitions, and the extent to which they took control of the final appearance of their buildings. The fifteenth century was a period of major stylistic change in both centres as the Gothic pointed arch was rejected in favour of the classical language of ancient Rome. The process was not uniform, nor was it propelled by artists. I hope to show you how patrons acted as the catalyst for change.

One important difference between modern architectural patronage and the system current in the Renaissance was the position of the patron.[6] After choosing an architect and approving the design, a modern patron's involvement is largely limited to finance. This was not the case during the Renaissance. Fifteenth and sixteenth-century patrons were central to their projects. It was the patron who hired the workforce, drawing up contracts with the stone masons, wallers and unskilled labourers who formed the bulk of the workforce and the self-employed masters, such as smiths and foundations experts, who were contracted for specific jobs. He also drew up contracts with suppliers of all building materials, sometimes even buying his own quarries to ensure a regular supply of stone and marble. The patron also hired a supervisor to take overall charge of construction; this supervisor, significantly, took his orders directly from the patron. When technical problems arose that the supervisor could not solve, it was the patron who sought advice from experts. But the most striking difference between the two periods was in the field of design. Our concept of the architect as the creative force behind the design of a building hardly existed in the fifteenth century. Nor was there any concept of the training and profession of an architect. In the account books of

---

3 On comparative costs, see Hollingsworth 1994, op.cit., 4, 20-1, 28-9, 34, 38.

4 C.W.Westfall, *In This Most Perfect Paradise*, University Park PA & London 1974, 33.

5 Howard Burns, *Andrea Palladio 1508-1580*, London 1975, 113.

6 The following discussion is based on my article, 'The Architect in Fifteenth-Century Florence', *Art History*, 7 (1984), 385-410 (hereinafter Hollingsworth 1984).

building projects, many of which survive, it is usually impossible to isolate a single designer, for the simple reason that design was not the exclusive field of one individual. Our desire to identify a single artistic personality behind the form and style of a building disguises the reality of the creative forces that produced the architecture of the Renaissance. Anyone could design buildings in fifteenth- and sixteenth-century Italy. The range of talents involved in design was huge: stonemasons, carpenters, goldsmiths, painters, intellectuals with no craft training and civil servants all made significant contributions to the stylistic development of Renaissance architecture. Moreover, designs were not blueprints: the supervisor on any project was closely involved with interpreting the plans. But the person with the strongest interest in the appearance of a building was its patron; his position holding the purse strings at the centre of the system gave him ample opportunity to control the design. In the context of innovation he played a key role.

Florence in 1400 was a prosperous commercial centre, home to bankers who dominated the European credit market and entrepreneurs who operated throughout Europe and the Middle East; with a population of 50,000, this was one of the largest cities in Europe.[7] The merchants who were responsible for her economic prowess were also in charge of government. Armed with the potent weapon of money, they had challenged and expelled the aristocratic ruling class in the 1250s and set up their own representative government. Mercantile wealth replaced inheritance as the basis of power and guild membership, which had long been obligatory for trade, now became the prerequisite for political office. The new order gave visual expression to its authority and prestige in two major architectural projects: a new town hall (the Palazzo della Signoria) (begun 1299) and a new cathedral (begun 1294).

Responsibility for the cathedral was allocated to one of the leading guilds, the Wool Merchants, but their financial resources were inadequate for this huge project and the state deducted 3 lire from every will to supplement the building fund.[8] The committee, or *Opera*, in charge of the project included both guild members and government appointees, a deliberate reflection of the financial arrangements. These were hard-headed businessmen; many also served on the government committees that dealt with political issues, such as finance or defence, and their presence on the *Opera* of the cathedral was part of their civic duties, not an indication of artistic sensibility. The *Opera* was responsible for the day-to-day running of the Cathedral as well as for its construction and embellishment; it appointed both the organist and the master mason, bought candle wax and vestments and also organised the supply of building materials, employed the workforce and commissioned independent sculptors to carve statues for the

---

[7] For a basic outline, and further bibliography, see Hollingsworth 1994, op.cit., 11-16.

[8] Margaret Haines, 'Brunelleschi and Bureaucracy: The Tradition of Public Patronage at the Florentine Cathedral', *I Tatti Studies*, 3 (1989), 92-3; for further bibliography, see Hollingsworth 1994, op.cit., 29-32.

elaborate facades. But major issues required a broader consensus. When decisions had to be taken that affected the visual appearance of the building, they were made by a wider cross-section of the Florentine electorate.

Work had started in 1294 at the west end of the long nave but it proceeded slowly, hampered by war, economic slump, and civic unrest that afflicted the early fourteenth century. By 1366 first two bays of the nave had been constructed and the need to decide on the final form of the east end had become paramount. If there were only to be three bays, then the design of the third set of piers would depend on decisions still to be made about the dome; this issue was delayed by the decision to build a fourth bay. During the sixteen months between July 1366 and October 1367 the *Opera* held meetings with technical experts and with large numbers of citizens to discuss the options.[9] Finally two designs were shortlisted: the design by the two supervisors in charge of work on the Cathedral was rejected and it was decided to go ahead with a design that had been submitted by a group of four stone cutters and four painters, but significantly altered by a small committee of citizens and monks appointed by the *Opera* to modify it. The supervisors continued in their posts as foremen of the project. Designed on a scale unparalleled since the end of the Roman Empire, the ribbed octagonal dome was intended to be the largest in Christendom; it was also distinctive and bore no resemblance to the cathedral domes in other Italian cities. This unambiguous statement of the unique character of Florentine power was only accepted after a referendum of the city's electorate. But construction was again hampered by war. By 1400 it was still unfinished; the completion of this statement of national pride now became an urgent priority.

The innovative choice posed major problems for the craftsmen in charge of constructing it and a competition was held in 1418 to find a solution.[10] Typical of the wide variety of people involved in architectural design, projects were received from all types of craftsmen as well as from Giovanni di Gherardo da Prato, a lecturer on Dante at the University and chief administrator on the Cathedral. The project chosen was by Brunelleschi, a sculptor and goldsmith by training. His intention to construct a double-shelled dome with the ingenious use of herringbone brickwork and without the traditional and extremely expensive use of a wooden support structure, caused so much controversy that he was appointed to supervise its execution. This appointment, which might seem normal to us, was relatively unusual in the fifteenth century.[11] Design might have been open to a wide range of talent, but construction required technical knowledge of building

[9] The impact of these meetings on the cupola design are discussed in Howard Saalman, 'Santa Maria del Fiore: 1294-1418', *Art Bulletin*, 46 (1964), 471-500.

[10] For a contemporary account of the building of the dome, see Antonio di Tuccio Manetti, *The Life of Brunelleschi*, H. Saalman (ed.), University Park PA & London 1970, 62-94; see also Howard Saalman, 'Giovanni di Gherardo da Prato's Designs Concerning the Cupola of Santa Maria del Fiore in Florence', *Journal of the Society of Architectural Historians*, 18 (1959), 11-20.

[11] Hollingsworth 1984, op.cit., 399-401.

and it was invariably undertaken by the supervisor (or supervisors) employed on the site. We may put a higher value on a designer and his creative ideas than an engineer and his technical expertise, but this was not the case in the Renaissance. Brunelleschi was extensively praised by his contemporaries for his engineering achievements more than for his architectural style. And when it met to discuss Brunelleschi's memorial, the *Opera* of the Cathedral praised him for saving them so much money.[12]

Florence was transformed in the second half of the fifteenth century in a spectacular building boom, financed by the wealthy elite of bankers and merchants, who gave visual expression to their commercial success, their status and, above all, their civic pride, in projects to rebuild churches and embellish them with new chapels, cloisters, tabernacles, altarpieces and fresco cycles, all proclaiming the identity of their patrons in coats-of-arms, portraits or inscriptions. They also spent conspicuously building or remodelling over a hundred palaces.[13] These palaces were built by men who had made fortunes in shrewd commercial dealing and cost a large proportion of their patron's capital. Surprisingly, they were investments that brought no financial return and, moreover, that this boom coincided with a general lack of confidence in the city's economy. Resale and rental values bore little relationship to the cost of construction. The largest palace in fifteenth-century Florence cost its patron, Filippo Strozzi, 40,000 florins, almost half his estimated income over twelve years.[14] Giovanni Rucellai spent 1,500 florins just acquiring the site for his palace.[15] Some patrons bankrupted themselves in the process. But these palaces were a very real investment in family honour, an issue of prime importance in fifteenth-century Florence.

In this context, Filippo Strozzi and Giovanni Rucellai deserve closer attention. Both were exceptionally rich bankers and also, through their links with the exile Palla Strozzi, politically suspect. Both had returned to favour through the influence of the powerful Medici family and embarked on major building programmes that were manifestly designed to give visual expression to their return to power in Florence. In addition to their palaces, Strozzi built a chapel in Santa Maria Novella and villas on his country estates, while Rucellai financed an

---

[12] Haines, op.cit., 123; see also Hollingsworth 1994, 31.

[13] On Florentine palaces, see R.Goldthwaite, 'The Florentine Palace as Domestic Architecture', *American Historical Review*, 77 (1972), 977-1012; R. Goldthwaite, *The Building of Renaissance Florence*, Baltimore MD & London 1980; F.W.Kent, 'Palaces, Politics and Society in Fifteenth-Century Florence', *I Tatti Studies*, 2 (1987), 41-70; see also Hollingsworth 1994, op.cit., 36-7, 57-9 for further bibliography.

[14] On Strozzi and his palace, see R.Goldthwaite, 'The Building of the Strozzi Palace: the Construction Industry in Renaissance Florence', *Studies in Medieval & Renaissance History*, 10 (1973), 99-194 (hereinafter Goldthwaite 1973); see also Hollingsworth 1994, op.cit., 59, 62-3.

[15] On Rucellai and his palace, see F.W.Kent, 'The Making of a Renaissance Patron of the Arts', in *Giovanni Rucellai ed il suo Zibaldone*, (Studies of the Warburg Institute 24), London 1981, vol.2, 9-95; see also Hollingsworth 1994, op.cit., 60-3.

expensive marble facade for the same church, emblazoned with his name and his coat-of-arms. Rucellai, looking back over his life, listed with pride his health (he was 70 years old) and his spectacular commercial success; as the prime indicators of his fortune he boasted that he had paid 60,000 florins in taxes, 10,000 florins in dowries and huge sums on buildings, and reflected that the making and spending of money were two of the greatest pleasures that man has but he could not decide which he preferred![16] Their palaces were both designed for effect but they were strikingly different. Strozzi chose to build on a large scale but to a traditional style. Rucellai on the other hand opted for an innovative design. The first, and the only fifteenth-century Florentine palace to display the classical orders on its facade, it has an important place in architectural history. Its impressive use of classical detail has led art historians traditionally to attribute the design to Alberti, but there is no positive evidence of this in Rucellai's copious records. Likewise the ledgers of the Palazzo Strozzi, which have survived intact, offer little help to the art historian in search of the architect.[17] Payments were made to Giuliano da Sangallo for a wooden model, which still exists, but its lack of detail made it clearly inappropriate as a blueprint for building. The supervisor of the project was Cronaca, who started work six months after the building had been begun; his main task was supervising the quarrying, transport and cutting of stone from Filippo's quarries, and he also designed details such as capitals. But there is no evidence that he was responsible for the design of the whole palace. What the ledgers do reveal is Filippo's central position in the construction of his project and, in the light of the evidence it is likely that both patrons played significant roles in the design of their palaces.

The evidence for Filippo Strozzi and Giovanni Rucellai may be somewhat circumstantial, but it is more convincing for Lorenzo de' Medici. By the middle of the fifteenth century, the Medici family had become the *de facto* rulers of Florence. The city still retained its old representative government, but it was exclusively dominated by the Medici family and their supporters (their manipulation of the system was understandably not popular). Lorenzo de' Medici was leader of the Medici faction from 1469 to his death in 1492; and he dominated not only the political but also the cultural life of the city.[18] The family may have lacked the status of their aristocratic contemporaries who ruled in Urbino, Mantua and Milan, but they had the financial resources to acquire the expensive trappings of princely prestige and, under Lorenzo, to buy the political clout to enable them to act as heads of state. They were extravagant patrons: they amassed an

[16] *Giovanni Rucellai ed il suo Zibaldone* (Studies of the Warburg Institute), London 1981, vol.1, 120-1.

[17] Goldthwaite 1973, op.cit., 123-35 & *passim*; Hollingsworth 1984, op.cit., 387-90.

[18] The literature on Lorenzo is immense; see, for example, the contributions to G.C.Garfagnini (ed.), *Lorenzo il Magnifico e il suo mondo* (Convegno internazionale di studi, Florence 1992), Florence 1994; see also Hollingsworth 1994, op.cit., Ch.5, for further bibliography.

ostentatiously expensive collection of antique cameos, vases of semi-precious stones, gems, medals and coins, including the Tazza Farnese, a Hellenistic cameo that was valued at 10,000 florins.[19]

Lorenzo also made a major contribution to the development of villa design. Rural property was an important feature of the investment portfolio of the Florentine merchant; it balanced assets in more speculative commercial ventures, but also produced wine, oil, fruit and vegetables for consumption in town. Florentine businessmen, including Lorenzo, read classical texts describing villas as a retreat from the demands of urban life, where one could appreciate the beauties of nature and indulge in intellectual and cultural pursuits; and they were inspired with ideas of ancient patrician grandeur. Traditionally these villas were plain and unadorned, unless fortified; but this was not an appropriately classical image. The design of Lorenzo's villa at Poggio a Caiano reflected the need for change.[20] A major landmark in architectural history, the villa's facade incorporated an Ionic portico elegantly decorated with a frieze inspired by classical reliefs. Lorenzo's interest in architecture is well-documented.[21] He owned a copy of the ancient Roman treatise on architecture by Vitruvius and had Alberti's treatise on the subject read to him (apparently while he was in his bath); he also requested drawings of new palaces and churches being built elsewhere in Italy. He used his political position to influence the design of several public buildings in Florence, recommended artists as favours to foreign rulers and sent his own design for a villa to the King of Naples. The original and innovative idea of applying a temple front to the villa is far more likely to have been the result of his knowledge of classical and modern design than that of his works foreman, Giuliano da Sangallo.

Venice was also a prosperous commercial city, larger than Florence, and capital of a trading empire that extended down the ex-Yugoslavian coast of the Adriatic into the Aegean: at one stage she even owned Athens.[22] Venetian society was highly stratified; at its top was the patrician class, members of mercantile families who had seized power around 1300 and restricted government office to themselves and their descendants. The Doge, the head of state, was a patrician elected for life who acted as chairman of the government committees rather than an absolute ruler. Unlike Florence, birth was the prerequisite for political power. The middle

---

[19] The inventory is published in E.Müntz, *Les Collections des Medicis au XVe siècle*, Paris 1888.

[20] Hollingsworth 1994, op.cit., 72-3.

[21] On Lorenzo's architectural expertise, see E.Gombrich, 'The Early Medici as Patrons of Art', reprinted in E.Gombrich, *Norm and Form*, London 1971, 54; Mario Martelli, 'I pensieri architettonici del Magnifico', *Commentarii*, 17 (1966), 107-11; F.W.Kent, 'Lorenzo de' Medici's Acquisition of Poggio a Caiano in 1474 and an Early Reference to his Architectural Expertise', *Journal of the Warburg & Courtauld Institutes*, 42 (1979), 250-7; Hollingsworth 1984, op.cit., 402-4; B.L.Brown, 'An Enthusiastic Amateur: Lorenzo de' Medici as Architect', *Renaissance Quarterly*, 46 (1993), 1-22.

[22] For an outline of Venetian history, society and government, see Hollingsworth 1994, op.cit., 97-104, with bibliography.

classes, *cittadini*, were also controlled by law: their status relied on proving long-term residency and at least 2 generations of non-manual trade (foreigners could acquire *cittadini* status after 25 years residence). Ruthless self-interest and devious diplomacy during the Fourth Crusade of 1202-4 had given Venice the chance to establish herself as a major power in the eastern Mediterranean. Having agreed to supply fully-armed galleys and ships for transporting the crusaders, she made an alliance with the deposed Byzantine Emperor who promised rich rewards in return for help to restore him to his throne in Constantinople. He was restored in 1203 but failed to pay up; the following year the angry and dissatisfied armies, egged on by Doge Enrico Dandolo, ransacked the city and installed their own Emperor. By 1400 Venice had become the leading maritime power in the eastern Mediterranean; traditionally independent from mainland Italy, her commercial and cultural links were primarily with the Byzantine East. This was abundantly evident in churches and palaces throughout the city, above all, in San Marco, the prime image of Venetian power. Designed (c.1050) on a Greek cross plan, and crowned with huge hemispherical domes, it deliberately recalled Emperor Constantine's church of the Apostles in Constantinople. The church bristled with imagery that testified to her wealth and power: one of the statues embellishing the side was a red porphry group of the four rulers of the Roman Empire, looted, like the famous bronze horses on the facade, during the sack of Constantinople (1204).

It is hard to exaggerate the importance of art as state propaganda in Venice. The complex created around San Marco, its huge piazza, the Doge's Palace and the Piazzetta leading down to the lagoon, provided an impressive focus for the display of Venetian prestige and one that had been carefully designed for maximum impact on visitors and Venetians alike.[23] Not surprisingly it was the patrician elite which was responsible for these images of state power. Decisions concerning the construction or decoration of these projects were political and were decided in exactly the same way as other political issues. The initial impetus emerged in discussions of the small but powerful committees, and were then ratified by the Grand Council, the base of Venetian government consisting of all male patricians over the age of 25 who met every Sunday to vote through legislation. San Marco, the building and its substantial investments in property throughout the empire, were the responsibility of the Procurators of St Marks, elected for life and, after the Doge, the most prestigious men in Venice. Most other state projects were financed and supervised by the magistrates in charge of the Salt Office, who also controlled the manufacture, purchase and sale of salt, a major source of state revenue. Like the *Opera* of the cathedral in Florence, these committees handled all aspects of their projects, choosing a design, employing craftsmen to execute it, paying for supplies and labour and keeping account of their expenditure of public funds.

In the 1420s Venice decided on a major change of policy: reversing her traditional preference of isolation from mainland Italy, the government voted to extend Venetian control into the Po plain to counter the threat of a powerful and

---

[23] On the Venetian government as patrons, see Hollingsworth 1994, op.cit., 105-19.

expansionist Milan. This new policy was given powerful visual expression in the Doge's Palace with the decision to rebuild the old Byzantine structure on the Piazzetta. The new facade, begun in 1424, with its cusped ogee arches, quatrefoils and pointed arches, was elegant and Gothic.[24] This was the language of European power, largely rejected in Florence, and it was a significant visual statement of the new political policy. In 1438 an overtly Gothic entrance portal, the Porta della Carta, was commissioned by the Salt Office from Giovanni and Bartolomeo Bon for 1700 ducats.[25] But, preoccupied with carving out an Italian empire, the Venetians had ignored the growing threat from the powerful Ottoman Turks, who took Constantinople in 1453 and began to conquer Venice's Mediterranean possessions. With her authority challenged by the superior Turkish navies, Venice became increasingly dependent on her Italian possessions. At this time of crisis Venetian patricians again opted for a radically new style for the display of state power, turning back to their city's Byzantine roots to assert her distinctive identity in the context of mainland Italy.[26] This is what art historians call the Venetian Renaissance, but it took a very different form from elsewhere in Italy. They did not opt for the language of ancient Rome, but chose to revive links with Constantinople. A new entrance portal for the Arsenale, the state shipyards, was ornamented with Byzantine filigree capitals, quite unlike Roman prototypes.[27] When the east wing of the Doge's Palace was badly damaged by fire in 1483, plans to rebuild it were debated in the Senate and approved by the Grand Council before work could begin.[28] The stylistic complexity of the new wing poses problems for art historians. With round arches on the ground floor, pointed arches on the main floor and two elaborately decorated storeys above, it does not conform to the modern notion of stylistic uniformity. But there can be little doubt that the complexity was deliberate. The Gothic arches recall the facade of the Doge's Palace, while many other details were taken from San Marco. Financial problems dominated the project from the start.[29] The government allocated 500 ducats a month to the project from funds normally scheduled for the repair of the city's sea defences and work started under the Salt Office's new supervisor, Antonio Rizzo. Rizzo claimed his salary of 100 ducats a year was inadequate, and it was raised to 125 ducats and then to 200 ducats a year. Expenditure on the project got out of hand and in 1493 work had to be suspended through lack of funds. The government was becoming suspicious and ordered two senators to audit the account books. Rizzo was accused of embezzling 10,000 ducats, and he was sacked. The new wing was only completed in the sixteenth century.

---

[24] Hollingsworth 1994, op.cit., 110-12.

[25] For the contract, see D.Chambers, *Patrons and Artists in the Italian Renaissance*, London 1970, 66-9

[26] Hollingsworth 1994, op.cit., 112-15.

[27] D.Howard, *The Architectural History of Venice*, London 1980, 104-6.

[28] For a contemporary account of the discussion, see J.J.Norwich (ed.), *Venice: A Traveller's Companion*, London 1990, 79-81; see also Hollingsworth 1994, op.cit., 117-19.

[29] A.M.Schulz, *Antonio Rizzo, Sculptor and Architect*, Princeton NJ 1983, 128-35.

A similar pattern of stylistic change was evident in the grand palaces built by patricians on the Grand Canal, the principal thoroughfare of Venice. There are over 40 fifteenth-century palaces remaining on the Grand Canal.[30] Coats-of-arms proclaimed ownership but the extensive use of heraldic emblems, so common in Florence, was rare in Venice. Up until c.1475 all palaces displayed a strikingly uniform feature: the windows and balconies of their main rooms were decorated with the same distinctive Gothic style of tracery that was clearly visible on the Doge's Palace. The choice of style and uniformity was clearly intentional. It not only proclaimed the owner's wealth but, more importantly, the family's membership of the city's ruling elite. One of the best known palaces, the Ca' d'Oro, literally the House of Gold, was so called after the gilding that originally decorated the tracery.[31] Its patron, Marin Contarini, was a successful merchant and a Procurator of St Mark. The palace is one of the few for which detailed building accounts survive and it is clear that Contarini, like Rucellai and Strozzi, was closely involved in both design and construction. The books make no mention of an overall designer and it was he who co-ordinated the contributions of his craftsmen. He ordered the six-bay window on the first floor from the workshop of Matteo Raverti. Ornamented with the cusped pointed arches and quatrefoil roundels of the Doge's Palace, this window was an unambiguous statement of his patrician status. One of the first classical palaces to appear on the Grand Canal, now the Palazzo Vendramin-Calergi, was built by another patrician, Andrea Loredan. Overtly classical, the facade contained free-standing classical columns and it was conspicuously different from other Venetian palaces of the period, and was not imitated for nearly a century. We know little about its commission. But we do know that Loredan had doubts about his innovative choice. On the ground facade, and easily visible by anyone passing on the canal, was an inscription *Non nobis Domine, non nobis*, the first words of Psalm 115 'Not unto us, O Lord, not unto us, but unto Thy name give glory'.[32]

---

[30] On Venetian palaces, see Howard, op.cit., 85-97; Hollingsworth 1994, op.cit., 148-52.

[31] Howard, op.cit., 92-4; Hollingsworth 1994, op.cit., 149-50.

[32] J.B.Onians, *Bearers of Meaning*, Princeton NJ 1988, 128.

3. Opening the Black Box of Innovation

Philippe Aghion, Jean Tirole*

A common feature of the existing patent race and endogenous growth literatures is their representation of R&D activities as being performed by a collection of individuals playing simultaneously the roles of financier, creator, owner, and user of the innovation. In practice, however, R&D takes place either within firms where researchers-inventors are subject to employment contracts, or through contractual agreements between independent research units and users of their innovations. In both cases, the contractual provisions on how to finance the research activities, how to share property rights on innovations, and how to structure the monetary compensations to the inventors, are complex and suggest that the aggregated view of the R&D process may be too limited.

Opening the black box of R&D activities can serve both *micro* and *macro* purposes. On the micro side, it can provide theoretical foundations: (a) to interpret existing findings by the empirical R&D literature; (b) to better understand the "law and economics" of R&D contracting and to help build more efficient organization of research; (c) to guide government intervention in the R&D sector. On the macro side, a theory of the organization of R&D can provide new insights into the relationship between the volume and nature of R&D activities and the business cycle; finally, it can help introduce organizational considerations into the analysis of the process of economic growth.[1]

The purpose of this note is to summarize some preliminary work in that direction by Aghion and Tirole (1993). This work relies primarily upon the incomplete contracts approach to property rights and vertical integration developed by Grossman and Hart (1986).

* University College, London, and European Bank for Reconstruction and Development, London; IDEI, Université des Sciences Sociales, Toulouse Cedex, CERAS, Paris, France and MIT, Boston, USA.

[1] For example, the recent 'neo-Schumpeterian' literature points out that the long-run rate of productivity growth depends on the *frequency* and the *size* of innovations. As shown in Aghion and Tirole (1993), both turn out to be affected by the organizational form of research.

3.1 The model

The basic research contracting relationship to be analyzed involves a research unit or researcher (RU) and a customer (C), who for the moment is the only potential user of the innovation. The customer directly benefits from the innovation: He may be either a manufacturer who 'develops' or commercializes the innovation, or a user who purchases the resulting product, or the supplier of complementary products, or a combination of the three. The research unit contributes its ideas and knowledge but has no independent resources to pay for equipment or labor. It must therefore look for outside financing investment, which can be supplied by the customer in exchange for a contract specifying how to share property rights on using the innovation and royalties to be paid to the research unit.

The value of the innovation for the customer is $V>0$. Let e denote the noncontractible (research) effort supplied by RU and E denote the investment provided by C, both at a linear cost. We allow E to be either contractible (monetary investment) or noncontractible (as is often the case for the supply of proprietary technology or demand information). For the moment we also assume that the customer faces no cash constraint. The probability of making the innovation is assumed to be increasing, strictly concave, and separable in (e, E), namely: $p(e, E) = q(e)+r(E)$. Both parties are income risk-neutral and have reservation utility 0. The socially optimal (or first-best) effort and investment are defined by

(1) $$\max_{\{e, E\}} \{p(e, E)\, V - e - E\}$$

or

$$q'(e^*(V))\, V = r'(E^*(V))V = 1$$

However, the first best cannot be implemented in general, due to the incompleteness of the research contract. More specifically, we assume that the innovation cannot be described ex ante, so that the two parties cannot contract for delivery of a specific innovation. All what the contract can specify, therefore, is the allocation of *property rights* on any forthcoming innovation, a *sharing rule* on the revenue (license fee) obtained by the research unit, and (if it is contractible) the customer's investment E.

If property rights on the innovation are allocated to C, then C can freely use the innovation. This we call the *integrated case*. In that case RU receives no reward for innovation.[2]

[2] In practice successful research employees are rewarded ex post through salary increases, cash awards, fringe benefits, stocks or promotions. Such rewards are generally not commensurate with the value of the innovation.

If RU owns the innovation, C and RU bargain over the licensing fee once the innovation has been made. This we call the *nonintegrated case*. For simplicity, we assume that the total pie V is then equally split ex post between the owner (RU) and the customer (C), so that RU gets a licence fee equal to $V/2$.[3]

As we will see in Section 3, C's taking equity in RU has no economic effect. Either C has property right and RU receives no licensing fee. Or RU has the property right and the two parties necessarily bargain over the *net* licensing fee, namely the formal licensing fee minus the share that goes back to the customer through the equity participation. For this reason, we can ignore this contracting dimension.

3.2 Who should own the innovation?

The answer to this qucstion hinges on *three* considerations: (a) the *marginal efficiency* of RU's effort compared with the marginal efficiency of C's investment; (b) the *ex ante bargaining pouwer* of the two parties, which reflects the extent to which the research unit is the only candidate to perform the research; and (c) the *cash constraint* faced by C.

3.2.1 Marginal efficiency

In this subsection and the next, we assume for expositional simplicity that E is noncontractible. Under *C-ownership*, RU receives no reward for innovating and therefore should supply no effort: $e=0$. On the other hand, C is residual claimant for V and therefore supplies the first-best investment level $E^*(V)$. Aggregate utility is then given by: $U_{RU} + U_C = 0 + [p\,(0, E^*(V))\,V - E^*(V)]$.

Under RU-ownership, each party receives $V/2$ once the innovation occurs. Thus one ends up with second-best effort and investment equal to $e^*(V/2)$ and $E^*(V/2)$ respectively. The corresponding aggregate utility level is:

$$\text{(2)} \qquad \tilde{U}_{RU} + \tilde{U}_C = p\left(e^*\left(\frac{V}{2}\right), E^*\left(\frac{V}{2}\right)\right)\frac{V}{2} \times 2 - e^*\left(\frac{V}{2}\right) - E^*\left(\frac{V}{2}\right)$$

Clearly, $\tilde{U}_{RU} + \tilde{U}_C > U_{RU} + U_C$ whenever the marginal efficiency of the research unit's effort is sufficiently large relative to that of the customer's investment: in this case non-integration yields a higher total surplus than integration (otherwise, integration may be more efficient).

---

[3] The equal split outcome could result from a Rubinstein bargaining process with alternative offers by the two parties and no time delay between two successive offers. Focusing on the equal split case involves no loss of insights in this model where utility in the ex post bargaining game is transferable in the relevant range (where RU's income is positive).

3.2.2 Ex ante bargaining power

Utility is not always transferable in the *ex ante* bargaining over property rights between RU and C because the cash constrained RU is unable to compensate C for a transfer of ownership to RU, even if such a transfer results in a higher total surplus *(pV-e-E)*. Whether the equilibrium allocation of property rights is efficient therefore depends on the *ex ante* bargaining power of the two parties. In particular, one can easily see that the equilibrium allocation of property rights is always efficient when RU has the *ex ante* bargaining power: indeed, if $\tilde{U}_{RU}+\tilde{U}_C \geq U_{RU}+U_C$, RU allocates the property right to itself; otherwise, it allocates the property right to C in exchange for a cash transfer equal to the customer's subsequent surplus.

The same reasoning, however, does not carry over to the case where C has the *ex ante* bargaining power; for, RU cannot make the necessary cash transfer that would induce C to give up its property right on the innovation even when *nonintegration* is the efficient organizational form.

3.2.3 Cash constraints

Suppose that *E* is contractible and consider the situation where *C*-ownership would be selected provided the customer could finance *E*(V)* without going on the capital market, but assume now that the customer has less than *E*(V)* and must borrow in a market that is imperfect for informational reasons. External financing is then more costly than internal financing and a move from *C*-ownership back to RU-ownership may become attractive since it reduces the customer's required (monetary) investment from *E*(V)* to *E*(V/2)*, say, and thus reduces the amount of cash C needs to borrow on the capital market. Financial constraints therefore bias the organizational form toward the use of creative inputs and away from capital expenditures. An interesting implication of this analysis is the prediction that new firms or firms which have experienced hard times will tend to farm out their research activities more than established, healthy firms. This is nothing but the well-known 'long-purse' hypothesis enunciated by Schumpeter, according to which a firm's R&D investment should be positively correlated with its assets.

3.2.4 Schumpeterian hypotheses

This 'Grossman-Hart approach' applied to innovation can be used to assess the 'Schumpeterian hypotheses'. These hypotheses state that a firm's incentives for R&D are higher, the larger the scale, scope, initial monopoly power (and financial resources) of the firm. In terms of our model, increasing the scale, scope or market power of the customer firm may under some assumptions all be thought of as increasing the value of the innovation *V*. How do R&D inputs *e* and *E* and the R&D output *p (e, E)* react to such an increase in *V*? The answer is obvious *when one takes the organizational form* (C or RU-ownership) *as given*: Then *e*, *E* and *p* all increase with V[4] as enunciated by Schumpeter and his disciples. However, the conclusions may change once the organizational form or ownership structure is endogenized.

Let us use the case in which the customer has the *ex ante* bargaining power to convey the intuition for why an increase in the value of innovation may reduce the probability of discovery (the output of the R&D process) or the R&D expenditures (the measurable input). An increase in V may either induce the customer to insist on having property rights (it becomes important to capture the rent) or have him voluntarily relinquish property rights to the research unit (it also becomes important to give proper incentives to RU). A switch to C-ownership increases the customer's expenditures, but reduces the creative input and may thus reduce the probability of discovery, in which case the input and output measures are negatively correlated. Conversely, a switch to RU-ownership may decrease expenditures.

3.3 Who should finance the innovation?

3.3.1 Cofinancing by private outside investors

Our analysis has overlooked the fact that private investors that do not directly benefit from the innovation, namely banks, venture capitalists or parent companies do often contribute to the financing of research units.

A first question is, why can't the customer itself provide cash to the research unit in exchange for equity shares in RU? The answer to this question turns out to be quite simple and relies on the following irrelevance argument: Suppose that C is given $(1-\alpha)$ shares in RU. The real license fee paid by C when both parties agree on a nominal license fee l is: $\bar{l} = l - (1-\alpha)\, l = \alpha l$. The equilibrium level of $\bar{l}$ is, as before with l, driven by time impatience only and not by the sharing rule $(\alpha/2,\ 1-\alpha/2)$ implied by the initial contract.[5] Therefore C's equity shares in RU have no effect on the net license fee and are thus irrelevant.

On the other hand, introducing third parties as co-owners of RU can help promote (efficient) RU-ownership in a situation where C has the bargaining power *ex ante*. More specifically, C may demand cofinancing E_l from a competitive investor in exchange of a claim of a fraction $(1-\alpha)$ of RU's profits. As before RU and C bargain over the license fee after the innovation occurs and C must still pay a license fee equal to $V/2$ to RU's owners. However, the researchers themselves now receive a return of $\alpha V/2$ for the innovation whereas the outside investor receives $(1-\alpha)V/2$. Assuming *ex ante* competition among outside investors leads to

---

[4] Indeed, under C-ownership, $E=E^*(V)$ and $p= r(E^*(V))$ with $dE^*/dV>0$ and $r'>0$. Under RU-ownership, $e = e^*(V/2)$, $E = E^*(V/2)$ and $p = r(E^*(V/2)) + q(e^*(V/2))$ with $de^*/d V > 0$ and $q'> 0$.

[5] Put differently, the initial sharing rule cannot influence the ex post bargaining game between RU and C since in contrast to other incomplete contracts models with renegotiation, the object (innovation) to be traded ex post is not contractible.

the following free-entry condition, where E_I (resp. E_C) denotes the outside investor's (resp. the customer's) investment:

$$(3) \quad E_I = p\left(e^*\left(\alpha \frac{V}{2}\right), E_I + E_C\right)(1-\alpha)\frac{V}{2}$$

C's profit is then

$$(4) \quad p\left(e^*\left(\alpha \frac{V}{2}\right), E\right)\frac{V}{2} - E_C = p\left(e^*\left(\alpha \frac{V}{2}\right), E\right)\left(1 - \frac{\alpha}{2}\right)V - E,$$

where E is total investment. The optimal choice of a results from the following trade-off: First, a lower α increases cofinancing and thereby the rent that the customer can extract from RU under RU ownership [$(1-\alpha/2)$ decreases with α]; second, the dilution of RU's shares implied by a lower α reduces RU's incentives and therefore reduces the probability of discovery [$e^*(\alpha V/2)$ and therefore p increases with α]. The trade-off between these 'rent extraction' and 'incentive' effects determines an optimal sharing coefficient α^* that is less than 1, except when either the RU has *ex ante* bargaining power or the research unit's effort decreases rapidly with a dilution of its equity. Otherwise RU-ownership with cofinancing dominates pure RU-ownership as well as C-ownership, when the customer's investment is contractible.

3.3.2 The scope for public R&D subsidies

The traditional approach to government intervention in the R&D sector (e.g. followed by the patent race or the endogenous growth literatures) was formulated in purely *quantitative* terms: Namely, from the point of view of allocative efficiency or optimal growth, should aggregate R&D investments be subsidized or taxed? The typical answer then would be that R&D investments should be subsidized whenever positive externality effects dominate and as a result growth under laissez-faire is suboptimal [as in Romer (1986, 1987)]. On the other hand, R&D investments should be taxed if too much 'business-stealing' or creative destruction take place under laissez-faire [see Aghion and Howitt (1992)].

In contrast, our analysis points at the possibility of a more *targeted* or *qualitative* approach to public R&D financing. Should R&D subsidies be primarily targeted at integrated (C-owned) or non-integrated (RU-owned) research undertakings? We have done some *preliminary* work on this issue. Suppose for instance that C has the bargaining power *ex ante* and farms out its R&D to one of several potential research units;[6] that C has no cash constraint; that there is a shadow cost of public funds so that the government wants to economize on R&D subsidies; and that participation in a government-sponsored R&D program is

[6] As we showed in Section 2.2, C may inefficiently perform R&D internally. We will not discuss this case: let us just mention that the government may then want to subsidize the farming out of R&D.

voluntary (so the customer is not hurt by the R&D program). Then it is optimal for the government to subsidize R&D if the shadow cost of public funds is under some threshold. This governmental intervention can take several forms. One consists in raising RU's incentives by reducing investor's share in the research unit. This takes the form of the government's buying up some of investor's shares in RU and turning them back for free to the research unit, in exchange for the customer also contributing to this increase in RU's shareholding. Another form of intervention consists in raising the customer's contractible investment by subsidizing it (again, in exchange of the customer's also contributing to the increase in investment).[7]

3.4 On the dynamic management of innovations

3.4.1 Breakaways and trailer clauses

Many research contracts between a customer firm (employer) and its (integrated) research employees specify that an innovation made by the latter *after breakaway* belongs to the employer if made shortly after quitting the firm and to the research employees otherwise. An efficiency rationale for such 'trailer clauses' can be easily obtained through the following extension of our basic framework, where *several* property rights (namely on the current innovation V_0, *and* on future innovations V_t, $t \geq 1$) are to be simultaneously allocated *ex ante* between C and RU.

Suppose that the effort *e* and the investment *E* produce a time sequence of innovations with value V_t in which the customer's marginal efficiency decreases over time:[8] $p(e, E, t) = q(e) + h(t)\, r(E)$, $h'(t) < 0$. Let α_t denote RU's share of the date *t* return. Because both parties never have an incentive to overinvest,[9] an optimal contract maximizes the stake of one party (say, the customer) for a given value of the other party's stake:

$$(5) \qquad \max_{(\alpha_t)} \left\{ \sum_t (1-\alpha_t)\, h(t)\, V_t \right\}$$

$$\text{s.t.} \quad \sum_t \alpha_t V_t \geq \bar{V}_{RU}$$

[7] These are not the only possible interventions. For example, a government that lacks technological information may take a share in profit. Indeed, in France, the Agence Nationale de Valorisation de la Recherche has two types of policies: 'subventions'(subsidies targeted mainly to the purchase of R&D inputs) and 'avances remboursables' (interest free loans which are paid back in cases of success. The reimbursement rate is about 50 percent).

[8] E.g., because the RU has left the firm at date $t = 1$.

[9] Individual stakes indeed never exceed $\sum_t V_t$ (we assume away discounting).

The solution to this program satisfies

$$t < t_0 \Rightarrow \alpha_t = 0 \quad \text{(C - ownership)}$$

$$t > t_0 \Rightarrow \alpha_t = 1 \quad \text{(RU - ownership)}$$

The optimal policy is thus very similar to the trailer clauses observed in practice.[10] Similar rationales can be round in Aghion and Tirole (1993) for the (perhaps more interesting) institutions of 'shop rights' and 'hired for doctrine'.

3.4.2 Comparing research joint ventures and fully integrated structures

A joint venture is typically defined as a *temporary* association among several parties to carry out a specific (research) program. Our analysis in 5.1 suggests that research joint ventures between a research unit (RU) and a development unit (C) may dominate fully (i.e. *durable)* integrated structures; for, RJVs do presumably better at preserving the RU's entrepreneurial spirit and incentives aimed at *future innovations.*[11] The idea [see Aghion and Tirole (1993) for details] is that the research unit has more incentives to invest in the next product generation if it has the ability to develop it. This idea seems to be consistent with industry reasoning (for example, with the desire of pharmaceutical companies to keep an arm's-length relationship with biotechnological firms).

What are the costs of RJVs? As before we consider the relationship between RU and C, where C contributes capital and RU contributes 'its knowledge' to the production of a current innovation aimed at a third party.[12] RU's decision whether to transfer *(e=1)* or not to transfer *(e=0)* its knowledge is noncontractible and must therefore be incentive-compatible from RU's perspective. This technology transfers affects the probability of occurrence of the current innovation, that is:

(6) $$p = q_0 e + r\,(E), \quad q_0 > 0 \text{ (where E denotes C's capital investment)}$$

If V_1 denotes the price at which the current innovation is sold *ex post* to the third party and α_1 (resp., $1\text{-}\alpha_1$) denotes RU's (resp., C's) share of the resulting income,

---

10 For example in research contracts proposed by Polaroid or Gulf-Oil (see Neumeyer, 1971)

11 In which the current customer's contribution is implicitly assumed to be negligible.

12 That a third party rather than C uses the innovation allows us to cleanly identify the dynamic considerations behind the choice of ownership, since from a static viewpoint incentives are entirely provided by the rule for sharing the income from this third party and ownership is irrelevant. One could dearly reintroduce the static dimension of ownership by assuming, as in Sections 1-2, that C is the only user of the innovation. However, this would only restrict the set of feasible sharing rules for the first innovation without adding any insights as to the basic trade-offs governing the dynamic management of innovations.

then technological diffusion yields a net benefit of $\alpha_1 q_0 V_1$ to RU in the absence of subsequent innovation.

Now let us introduce the possibility of a costless future innovation entirely based on RU's knowledge and with a value of V_2 to a third party. *Consider first an RJV*: If RU does not transfer its technology to C in the first period, it remains full and exclusive owner of the future innovation and thus appropriates the whole value V_2; otherwise, if RU transfers its technology to C, it will have to compete with C on the product market. Assuming Bertrand competition *ex post*, RU's second-period payoff shrinks from V_2 to zero in the latter case. A technology transfer from RU to C in period 1 is thus incentive-compatible if

$$\alpha_1 q_0 V_1 \geq V_2 \qquad \text{or} \qquad \alpha_1 \geq \frac{V_2}{q_0 V_1}$$

Consider now a fully integrated structure where only C can commercialize the future innovation: RU gets $V_2/2$ if it has not transferred its knowledge to C in period 1, since in spite of being the owner C cannot dispense with RU's private knowledge *ex post*. A technology transfer from RU to C is now incentive-compatible if

$$\alpha_1 q_0 V_1 \geq \frac{V_2}{2} \qquad \text{or} \qquad \alpha_1 \geq \frac{V_2}{2 q_0 V_1}$$

Thus, *the cost of technological diffusion* (both, in terms of α_1, that is of C's reduced incentives to invest in the current innovation, *and* of future rent dissipation due to *ex post* competition between RU and C) *is lower in the integrated case*; it is all the more so when *ex post* competition cannot be prevented by an adequate patent system.

References

Aghion, P. and P. Howitt, 1992, A model of growth through creative destruction, Econometrica 60, 323-351.

Aghion. P. and J. Tirole, 1993, On the management of innovation, Quarterly Journal of Economics, forthcoming.

Grossman, S. and O. Hart, 1986, The costs and benefits of ownership: A theory of laleral and vertical integration, Journal of Political Economy 94, 691-719.

Neumeyer. F., 1971, The employed inventor in the United States: R&D policies, law and practice (MJT Press, Cambridge, MA).

Romer, P.M., 1986, Increasing returns and long-run growth, Journal of Political Economy, 94, 1002-1037.

Romer, P.M., 1987, Growth based on increasing returns due to specialization, American Econonic Review, Papers and Proceedings 77, 56-72.

4. R&D Inter-firm Agreements in Developing Countries. Where? Why? How?

Giorgio Barba Navaretti, Andrea Bigano*

Foreign investments and inter-firm contractual agreements are the main channels for the diffusion of technologies towards Developing Countries (LDCs). The literature on the determinants of multinationals is very broad and mostly revolves around Dunning's '*eclectic theory*' (Dunning, 1981). Multinational activity is the combined result of the OLIs triade: ownership, localisation and internalisation factors.

Knowledge is an important determinant of the decision of starting production abroad. Both the empirical and the theoretical literature emphasise its role in creating ownership advantages. Indeed, it is a firm specific intangible asset, which can be transferred at a lower cost than other capital assets (Horstman and Markusen 1987). At the same time, its public good nature influences the effectiveness and, consequently the form, of its transfer. Many empirical studies have now gathered a quite impressive body of evidence about knowledge spillovers from multinational activities to host countries' firms (see Blomström and Kokko's chapter in this volume for a comprehensive survey of this literature). Moreover, spillovers are incurred by any innovating firm, in some proportion, when information about the innovation can be gathered at no cost or at a low cost by competing firms. Fears of spillovers may induce multinationals to choose internalised international transactions (subsidiaries), instead of arm-length ones (e.g. licensing) (Ethier 1986, Ethier and Markusen 1991).

The literature which has analysed these issues (well summarised in Markusen 1995) generally adopts a straightforward hierarchical characterisation of the transferring process: host countries are passive recipient of technologies developed in the home country. This assumption, which is even more stringent when applied to the North-South space, is a correct representation of reality in most cases. All the same, it is not consistent with some recent evidence about multinational activity. *First*, R&D activity carried out by multinationals is often

---

* Università degli Studi di Milano, Fondazione Eni Enrico Mattei and Centro Studi Luca d'Agliano; Université Catholique de Louvain.
Comments from all the participants to the International Conference on 'Creation and Transfer of Knowledge: Incentives and Institutions', organised by the Beijer International Institute of Ecological Economics and by Fondazione Eni Enrico Mattei, Castelgandolfo, September 1995 and particularly by Karl Göran Mäler are gratefully acknowledged. Useful comments were also provided during seminars organised by the International Trade Division and Technet at the World Bank.

geographically dispersed (Cantwell 1993 and 1994, Kumar 1995). *Second*, R&D is often carried out by networks of firms located in different countries, and even in developing countries (Contractor and Lorange, 1988, Vonortas, 1991, Pietrobelli, 1996). Some recent theoretical contributions support this evidence, by showing that there is a rationale for R&D co-operation between firms, even under rather unexpected conditions, i.e. when firms compete in the product market and when endowments of knowledge across partners are asymmetric (Jaquemin and d'Aspremont 1988, Bhattacharya, Glazer and Sappington, 1990 and 1992, Aghion and Tirole, 1994). *Third*, aggregate evidence shows that international trade generates R&D spillovers: the larger the share of imports from countries rich in R&D capital, the larger the developing countries' foreign R&D capital and the larger their rate of growth (Coe, Helpman and Hoffmaister, 1995). It is therefore important to understand how these spillovers may be generated at the micro level.

Thus, international flows of technology do not necessarily imply unilateral and hierarchical transfers of knowledge. Moreover, the geographical dispersion of R&D activities can take place in a non-internalised way, through co-operative arm length agreements between different firms.

This paper enquires into the possibility that technological flows between advanced and developing countries (LDCs) take the form of co-operative R&D arm-length agreements. It will do so at the theoretical level, by comparing the literature on hierarchical transfers of technology and the literature on R&D joint ventures. The empirical analysis is based on a data bank containing 632 inter-firm technological international agreements, where at least one of the partners is located in a developing country. These agreements are a subset of the Cati-Merit data bank[1].

This work takes a different perspective from other papers which have analysed international R&D activity in different respects. *First*, because it looks at technological agreements between a group of developing countries and a group of advanced countries. To the best of our knowledge, this is the first attempt to do so. Previous analyses were mostly focused on advanced countries, looking at the geographic dispersion of R&D investments of multinationals based in a single country (Kumar, 1995 gives a useful survey). *Second*, because practically all of the empirical literature has looked at the R&D activity carried out by subsidiaries, up to now, and not at arm-length agreements (a survey of the very little existing evidence on technological agreements is provided by Pietrobelli, 1996). *Third*, and consequently, because previous works mainly focused on the determinants of localisation of R&D, whereas, here, we also address institutional features of R&D cooperation at the firm level.

The next section examines the literature on international transfers of technology, where knowledge is transferred in a hierarchical way. The following section extends the analysis to the literature on R&D joint ventures. Section three

---

[1] For a description of the data base, see Duysters and Hagedorn, 1993. For a descriptive application to developing countries see Freeman and Hagedorn, 1994. See also the Appendix to this paper.

discusses other aspects and determinants of R&D agreements in developing countries. Section four reports the empirical results and section five concludes.

4.1 Knowledge and hierarchical transfer of technology

Multinationals are exporters of firm specific assets. As summarised in Markusen 1995, both the empirical evidence and the recent literature on multinationals and trade show that multinationals are more likely to emerge in industries with large investments in intangible assets, like advertising, R&D, skilled human capital and so on. All these assets can be related to a broad concept of knowledge, being it embodied in human capital, procedures, stocks of information.

Why do firm specific assets give rise to multinationals? Because of cost efficiency. Multiplant firms can exploit firm specific assets (which are centralised). More single plant firms, instead, would generate duplications. Moreover, firm specific assets, like knowledge, are relatively easy to transfer. Manuals or engineers can move around more cheaply than physical assets like plants. Finally, knowledge is a public good, hence a non rival input with very little or no duplication costs. In sum, firm specific assets create easily transferable ownership advantages.

This basic setting is used to explain both the substitutibility between multinationals and trade and the optimal institutions for technology transfer, i.e. arm length contracts or subsidiaries. Arm length contracts are generally cheaper ways of transferring technologies than subsidiaries. However, arm-length contracts generate a trade -off, as they may prevent the home firm from fully exploiting the rents deriving from proprietary knowledge. This issue is important for developing countries, as far as arm length contracts imply a larger involvement of locals in the production process, and therefore a faster absorption of foreign technologies.

To understand this point we will concentrate on three different models (Ethier, 1986, Wright, 1993 Ethier and Markusen, 1991) . These papers do not exhaust all the literature on the matter, but they illustrate the main angles of analysis.[2] We will also discuss a fourth model, Wang and Blomstrom, 1992, which, looks at the impact of spillovers to competitors in the host country and endogenises the optimal amount of knowledge which should be transferred. Table 1 illustrates the key knowledge specific assumptions of these models.

The common starting point is some R&D activity creating a firm specific asset, which can be a new product, a reduction in production costs or the improvement of product quality. This new knowledge is held by the parent company, who wants to exploit this asset on a foreign market[3]. R&D activity is endogenous with respect

[2] These papers mostly focus on the public good features of knowledge. The literature on licensing vs. subsidiary considers many other sources of imperfection. For an exhaustive summary see Markusen 1995.

[3] Ethier and Ethier-Markusen are general equilibrium models. We will not touch here upon the broader structures of these models and upon the impact of country characteristics.

Table 1. Models for hierarchical transfer of technology: main assumptions

Assumptions	Ethier 1986	Wright 1993	Ethier-Markusen 1991	Wang-Blomstrom 1992
1. Creation of new knowledge: Purpose	Reduce productions	Reduce production costs	Develop a new product	Improve product quality
2. Creation of new knowledge: endogeneity	Endogenous w.r.t. the transferring contract	Endogenous w.r.t the transferring contract	Endogenous, but exogenous w.r.t. the transferring contract	Exogenous
3. Creation of new knowledge: how	R&D effort increases the probability of innovation	R&D effort increases the probability of innovation	Two periods R&D race. All participants same chances to succeed	Not specified
4. Length of innovation	indefinite	indefinite	2 periods	indefinite
5. Flows of Innovation	one shot	one shot	Continuous	the home firm has a stock of superior knowledge
6. Cost of transferring knowledge	none	none	Zero, but there are other fixed costs related to the transfer of the technology	Fixed cost decreasing with vintage of the technology
7. Intellectual property rights	none	none	none	none
8. Form of knowledge spillovers	None: the transfer of technology can be done in a way that prevents transfer or dispersion of proprietary knowledge	None: the transfer of technology can be done in a way that prevents transfer or dispersion of proprietary knowledge	Only to the transferee and in the second period. None to third firms. Subsidiaries can break up in the second period	The subsidiary is captive. There are spillovers to third firms, partly costless and partly through costly learning effort.
9. Other	No localisation decision. Asymmetric information when the technology is licensed: the licensee cannot observe a priori the type of technology.	Asymmetric information when the technology is licensed: the licensee cannot observe a priori the type of technology.	Licensing is product specific, and subsidiaries are firm specific (i.e. they are entitled to the future products created by the home firm)	

to the transferring contracts in Ethier and Wright and exogenous in the other two models.

Ethier and Wright are rather similar in their structure and major assumptions (although the former is a general equilibrium model). The outcome of research is uncertain: R&D effort increases the probability of developing a low cost technology. Uncertainty creates asymmetric information, and consequently moral hazard in the technology transfer: the transferee cannot observe R&D effort and the transferor does not want to reveal its research effort, as it would disclose proprietary knowledge.

In this context it is possible to define incentive-compatible optimal arm length contracts only if certain conditions are met: when at least one of the two firms is risk neutral (because this firm will be indifferent between a state invariant payment and a state contingent payment with the same expected value), when product quality (which also cannot be observed ex ante) is state invariant (Ethier) and when the licensing contract is derived from a signalling game, where the signal is represented by a market share restriction (Wright). In other words, arm length contracts are feasible only when the number of contingencies is modest. Otherwise, an internalised transaction (i.e. a subsidiary) will emerge.

These two models allow the transferor to preserve its proprietary knowledge even when the technology transfer is not internalised (as far as the arm length contract is feasible). In contrast, Ethier and Markusen consider spillovers inevitable and assume away asymmetric information: knowledge is transferred at no cost (although there are other fixed costs for setting up a new production facility) and the transferee learns how to use the technology as soon as it adopts it. In the following period, the transferee could set up a new production unit and compete with the home firm, which would be forced to also set up a new production facility.

The problem, here, is to design an optimal two period licensing contract so that neither the licenser nor the licensee have an incentive to defect at the end of the first period. All rents can be extracted from the licensee, as many firms compete to obtain the license. The cost of setting up a new production unit is the pivotal variable. Given the public good characteristics of knowledge, this cost is lower the higher the share of knowledge capital on other costly assets. The more important the knowledge capital, though, the lower the cost of setting up a new production unit and the higher the licensee's incentive to defect in the second stage. Knowledge intensive multinationals will therefore prefer to set up a subsidiary.

This result holds if we assume that subsidiaries can fully internalise knowledge. Ethier and Markusen also consider the case where dispersion of knowledge takes place in a subsidiary. Indeed, the subsidiary's employees can defect in the second period and set up an independent production unit. The parent company will have to share part of the rent with the subsidiary's employees. The employee's revenues also consist of the entitlements on future innovations developed by the home firm. In this case, the subsidiary is an equilibrium outcome only for intermediate levels of knowledge capital. When knowledge capital rises (and thus the cost for the employees of setting up a new production unit declines), the home firm must increase the subsidiary's share of the rent. Beyond a given threshold, the multinational will no longer be willing to share its rent and a two period license will emerge instead of the subsidiary. In other words, the multinational is able to internalise knowledge spillovers only by giving away larger and larger shares of the rents generated by the present and future creation of innovative knowledge. Thus, up to a threshold level, the larger the potential spillovers, the larger the amount of knowledge transferred to the subsidiary.

A similar result is reached when subsidiaries generate spillovers to third competitors in the host country. This is the case analysed by the Wang-Blomstrom model. This model does not consider the licensing/subsidiary alternative, although it still analyses hierarchical transfers of technology. Subsidiaries are not efficient tools to internalise knowledge, even if they are assumed to be fully captive (employees cannot set up an independent production unit). The parent company has access to superior technological knowledge (R&D is exogenous here).This knowledge is transferred to a subsidiary in the host country (no defection can take place in the second stage). This model introduces a set of very interesting assumptions. *First*, transferring knowledge is costly and the more so the newer the technology. *Second*, spillovers occur in the form of externalities towards competitors in the host countries. Such spillovers are partly costless and partly generated by the learning effort of the competitive firm. *Third*, the parent company decides how much to invest in the transferring process, and implicitly how much technology should be transferred *Fourth*, the amount of technology transferred is an endogenous equilibrium resulting from the strategic interaction between the foreign subsidiary and the host country competitors.

The interesting implication of this model is that, once spillovers are introduced (internalisation is not sufficient to avoid dispersion of proprietary knowledge), the technological effort of domestic firms influence the equilibrium transfer of technology. As the shadow value of increasing investments in transfer activity increases with the decline of the technology gap (because the rent of the multinational increases with its technological lead and the marginal productivity of the investment in transferring activity is constant), the multinational will transfer more technology, the larger the learning effort of the local firm and the larger costless spillovers.

Hence, spillovers to competitors and the absence of intellectual property rights prevent firms from internalising their proprietary knowledge by setting up a subsidiary instead of an arm length contract. Moreover, spillovers generate a kind of technology race: the more efficient is the learning competitor in the host country (i.e. the would be licensee in the licensing vs. subsidiary models), the more technology the multinational is forced to transfer (helas! nothing more to hide). This result is somehow similar to Ethier and Markusen's, where internalisation is possible, but more and more costly (in terms of knowledge rents given away) the larger potential spillovers

Now, if the transfer of knowledge cannot be internalised or it can be done at a cost; if, in equilibrium, the amount of knowledge transferred is larger the more dynamic the local firm and the more appropriable is knowledge, could we reach a stage where the strategic interaction may induce firms to cooperate in the creation of new knowledge? Could we, therefore, move from hierarchical transfers to co-operation?

The literature on R&D co-operative agreements between firms, can provide some interesting suggestion to explore the likelihood of these agreements emerging in developing countries.

4.2 From hierarchy to equal footing: relevance of models on R&D joint ventures

The literature has investigated rather thoroughly market incentives to undertake research and development[4]. More recently it has inquired into the value of co-ordinating research through inter-firm agreements. There are many reasons for doing so: avoiding effort duplication, pooling risks, and, mainly internalising knowledge spillovers.

This joint research effort can be carried out on the basis of arm length contracts by firms that are and may remain competitors on the product market. Joint R&D is characterised by the sharing of privately held knowledge and by the performance of a joint or co-ordinated research effort. The profits generated by the innovation are shared between the partners, according to some sharing rule. Firms may compete or cooperate in the product market. In the first case, the R&D joint venture is limited to the research stage, in the second case it is extended to the product market, giving rise to monopoly profits.

The pioneering work by d'Aspremont and Jaquemin, 1988 provides the basic setting to understand this issue. This model considers two risk neutral competitors, which produce the same product and carry out research and development to reduce production costs. Each firm's R&D generates positive spillovers to the other firm. Three different scenarios are compared: *first*, competition in R&D and in the product market (Cournot competition); *second*, co-operation in R&D (joint decision on the level of R&D expenditures) and competition in the product market; *third*, co-operation in both research and product market (the monopoly case). If spillovers are sufficiently large, co-operation in R&D increases the total research effort and profits. The monopoly will increase even further R&D effort and profits, as the rents generated by the innovation are fully captured by the two firms. Being the duopolist identical, the sharing rule is symmetrical.

The central result of the d'Aspremont and Jaquemin, i.e. that large spillovers induce R&D co-operation, has been confirmed in models which have generalised and extended their basic framework[5]. This result also supports our intuition, but it is not sufficient to justify R&D arm length agreements in developing countries.

---

[4] See for example Dasgupta and Stiglitz 1980, Harris and Vickers 1985

[5] See, for example (this list is not exhaustive) Kamien, Muller and Zang, 1992, Vonortas 1994, Motta 1992, Rosenkrantz, 1995 and Chanduri, 1995, Suzumura, 1992.

D'Aspremont and Jaquemin assume that the duopolists are identical. In developing countries, instead, it is reasonable to assume that the local firm has an inferior endowment of knowledge and is less efficient in carrying out research effort. In other words, potential positive spillovers are larger form the advanced country firm to the developing country firm than vice versa. This is a problem, because the incentive to carry out joint R&D derives form bilateral spillovers. If spillovers in one direction are weaker, the efficient firm has a lower incentive to cooperate.

A related issue, is the asymmetric information that may arise within R&D joint ventures. When firms are not identical, and the outcome of R&D is stochastic, there may be problems of adverse selection, if the initial endowments of knowledge are not observable, and of moral hazard, if research effort is not observable. Thus, can R&D co-operation under arm length contracts still hold when technological endowments and information are asymmetric?

An analysis of R&D joint ventures under asymmetric technological endowments is carried out in Chaudhuri, 1995. Scale economies in R&D provide the rationale for co-operating: the probability of success raises with effort, i.e. it is higher if firms join forces instead of acting individually. Firms differ in their efficiency in delivering R&D effort. Whereas the (asymmetric) technological endowment of the two firms is common knowledge, effort cannot be observed. There is a moral hazard problem, as the technologically poor firm has an incentive to free ride. Chaudhuri compares profits under co-operative R&D and under competitive R&D. Under co-operative R&D the model derives a sharing rule that fulfils a fairness condition, i.e. the more efficient firm obtains a larger share of payoffs. He shows that if the two firms are equal in their technological parameters, and returns are sufficiently large, it is individually rational to opt for co-operative R&D. As far as the technological gap between the two firms increases, though, co-operating is no longer rational for the efficient firm, as the inefficient one has too large an incentive to free ride. R&D effort, though, is larger under complete merger than R&D competition and co-operation: indeed, there is no longer a free riding problem or a need to devise an optimal sharing rule under asymmetric information. In other words, arm-length R&D contracts are consistent with asymmetries in technological endowments, if the latter are not excessive.

The model by Bhattacharya, Glazer and Sappington, 1990 and its subsequent extensions[6] introduce a further asymmetry in the joint ventures. R&D activity is made of two stages. In the first one, the partners share a given amount of previously held knowledge. In the second one, they carry out R&D effort. Firms do not cooperate at this stage, as R&D efforts are privately decided (although there is a common licensing scheme). Similarly to the Chauduri 1995 model, initial endowments of knowledge affect the efficiency of the R&D effort. In this case, though, the initial endowments are not common knowledge. Partners may decide to disclose less knowledge than what they actually hold. R&D effort is also

---

[6] Bhattacharya, Glazer and Sappington, 1992 and d'Aspremont, Bhattacharya and Gerard-Varet

not observable. We have, therefore, two types of technological asymmetries (endowments of technological knowledge and efficiency of R&D effort) associated to two types of informational asymmetries (adverse selection in the initial sharing of knowledge and moral hazard in the supply of R&D effort). Is it possible to design an incentive scheme which induces full knowledge revelation without jeopardising individual R&D effort?

Bhattacharya et al, 1990 and 1992 assume that knowledge structures satisfy a Blackwell ordering, across the privately acquired knowledge levels of the participants. This means that, if all knowledge is disclosed, the most knowledgeable agent's knowledge is the only useful input for efficient development effort. In other words, knowledge sharing is always asymmetrical. Note that firms are not ranked according to (and payments and rewards do not depend upon) any cardinal measure of the levels of knowledge disclosure. There is simply a crude distinction between the more knowledgeable firm and all the others. Why should the leading partner be willing to disclose its knowledge to the lagging partners, given that he receives no useful knowledge in exchange? This will happen if his rewards are sufficiently high, which, in turns, implies two further requirements. *First*, that the lagging firms pay a sufficiently high fee in case of success. *Second*, that the probability of success is higher if other firms use the leading firm knowledge in their R&D effort, i.e. there are scale economies in R&D effort.

The problem, though, is that a scheme that leads to full knowledge revelation may deliver non-optimal amounts of R&D effort. Bhattacharya et al. show that (for risk neutral firms) it is possible to design incentive schemes which deliver first best knowledge disclosure and R&D effort. According to this scheme, the leading firm is always allowed to keep the full benefits form the innovation, if it is the winner of the R&D race. If lagging firms win, they will be allowed to use the innovation, following the payment of a fee to the leading firm.

This result shows that with asymmetric endowments of knowledge and when co-operation is restricted to knowledge sharing (and not extended to R&D effort), it is possible to implement efficient arm-length R&D contracts. Two qualifications are however necessary. The *first* one is that when knowledge held by the leading firm is particularly rare and valuable, conditions under which full disclosure takes place are more restrictive than otherwise. In fact, if the level of knowledge of the lagging firms is very low, their probability of success in the R&D contest are extremely low. Hence the leading firm will have no rewards from disclosing its knowledge. It would do better to keep it private and carry out R&D by itself. The *second* one, is that if the Blackwell ordering assumption is dropped (d'Aspremont, Bhattacharya and Gerard-Varet, 1994), and all firms' knowledge have a positive value, independently from rankings, first best outcome can be achieved under less restrictive hypothesis about individual rationality. In Bhattacharya et al. the licensing scheme delivers interim individually rational outcomes, i.e. positive expected profits conditional on the privately acquired knowledge. D'Aspremont et al. weaken individual rationality to an ex-ante requirement, i.e. positive expected profits before any acquisition of knowledge.

The analysis of the micro foundations of R&D agreements confirms our intuition that knowledge spillovers provide a fundamental incentive for firms to cooperate in R&D. We may therefore expect these agreement to emerge in industries where intangible assets prevail on other physical assets, i.e. where the public good characteristics of knowledge are dominating. This result still holds under asymmetric technological endowments, which are likely to characterise transactions between advanced and developing countries. But asymmetries reduce the incentive for the advanced partner to participate in the agreement and cause problems of information. The larger the asymmetries, the more likely the disadvantaged partner to free ride. Thus, we will not observe R&D agreements when the technological gap is very large.

4.3 Other determinants of R&D agreements in developing countries

Before moving to the empirical analysis we want to briefly touch upon a set of other issues that may help to single out other determinants of R&D agreements. First, we discuss whether R&D agreements are more likely to emerge in relation to vertical rather than horizontal investments. Second, we consider whether joint ventures (i.e. an equity based relationship between the firms) may overcome some of the shortcomings related to arm-length R&D contracts. Third, we consider factors that may explain the location of R&D agreements.

4.3.1 Vertical versus horizontal investments

This question is important, as incentives may change under these two different settings. *Horizontal* investments imply the reproduction abroad of production processes already carried out at home. They may concern the whole production process or only a few stages. *Vertical* investments refer to the setting up abroad of production processes or stages of production which are not carried out at home. This classification is straightforward when we consider the geographic dispersion of production stages. It is looser, when we consider the development of new products, targeted to the local market, that may partially reproduce production processes carried out at home.

All the literature analysed in section 1, dealing with hierarchical transfers of technology, only examines horizontal investments. The literature on R&D joint ventures, generally concerns the development of a new product or the setting up of a new, more efficient production process. In this respect, it can broadly be identified with vertical investments, although production processes are never decomposed in more than one stage. Yet, R&D JVs may focus on differentiating products to make them suitable for local demand. In this case, production processes are not altered and such investments could be classified as horizontal.

The available evidence on overseas R&D by subsidiaries of MNEs (evidence on R&D arm-length contract is not existent) shows that this is mostly carried out to adapt products to local market conditions, i.e. in relation to horizontal investments (Kumar, 1995, Papanastassiou and Pearce, 1994, Hirschey and Caves, 1981). Yet, there is also evidence that multinational firms increasingly tend to disperse geographically their production structure. Vertical investments are fostered by trade liberalisation, increasing labour costs in the industrialised world and raising global competition. At the same time, dispersion of R&D activity appears to be increasingly related to the need to exploit specific and localised technological knowledge or networks of innovation (Cantwell and Kotecha, 1993). These two trends may well be partially related. If a country tends to specialise in a given stage of production, we may at the same time expect this country to develop related technological knowledge and R&D facilities.

Unfortunately, the role of vertical investments and their relationship to the geographic dispersion of R&D activity is a very under-developed area of research. To the best of our knowledge, the only paper that investigates location issues for a multinational firm with multiple production stages is the pioneering work by Casson, 1985. Moreover, internalisation issues are not explored at all in this context.

Therefore, at this stage, we do not have the theoretical foundations to understand whether R&D arm -length contracts are more likely to emerge in relation with vertical rather than horizontal investments. Intuitively, vertical investments are not likely to increase the incentive to carry out R&D vis a vis horizontal ones. Production of components (or assembly of the final product) concerns the whole production process of the home firm (unless local firms also become competitors in the international market). Instead, horizontal investments normally concern only one market, thus only part of the home firm's revenues. In this context, the home firm has a higher incentive to cooperate in R&D, by disclosing as much knowledge and making as much effort as possible. At the same time, though, the value of the local knowledge created is higher (again, because it concerns the whole production process and not one single market). Thus, the costs of the spillovers are larger. The trade off between these two effects, and its impact on the nature of the contract, will depend on the fungibility for other competitors of the new knowledge created and on the degree of captiveness of the customer supplier-relationship. In principle, these are the same factors that influence the appropriability of knowledge in arm-length horizontal R&D contracts.

4.3.2 Does equity matter?

The models considered in sections 1 and 2 look at the alternative between a subsidiary and an arm length contract. None take into account the intermediate option, i.e. the setting up of a joint venture. Even models of R&D joint ventures never relate sharing rules to an equity structure.

In principle, joint ventures could be seen as a mean of internalising the relationship between two partners. We may expect equity sharing to emerge for longer term relationships and characterise stronger binds between the partners than arm-length deals. A long term relationship is likely to involve many facets and many possible contingencies, which cannot be foreseen when the joint venture is initially established. Equity agreements may define a straightforward and simple (although not necessarily optimal) sharing rule.

However, the partners in the joint venture are and remain independent units with partly common but partly conflicting objectives. This is clearly so when the partners supply an input to the joint venture (knowledge, raw materials, labour etc.). In this case, the maximisation of each partner's profits does not correspond to the maximisation of the partner's share of the joint venture's profits (given the equity shares). It can be shown that when it is possible to draft first or second best contracts between the partners, a sharing rule is defined endogenously and equity is totally irrelevant (Svejnar and Smith 1984, Barba Navaretti 1995, Aghion and Tirole, 1994). In other words, the sharing rule defines the returns to the partners, net of their initial investment and their equity share in the joint venture. Consequently, we expect equity arrangements to emerge when contingencies are such that first or second best contracts cannot be implemented.

4.3.3 Beyond internalisation: what affects localisation decision?

Up to here, we have mostly dealt with issues of internalisation, i.e. whether R&D arm length contracts in developing countries are efficient. We have not touched upon the choice of the country where to set up R&D facilities. Once we have established that R&D contracts are feasible when endowments of knowledge are asymmetric, which countries should emerge as suitable hosts for these types of contracts?[7]

There is a key difficulty in pursuing this argument, though. Particularly in developing countries, overseas R&D activity generally supports foreign investments initially decided for other reasons. It is therefore hard to single out the determinants of R&D from the determinants of the initial investment, particularly without the support of a theoretical model linking investments and trade decisions. Given the lack of theoretical models of this type, and given the scope of this paper, we will therefore not go beyond straightforward intuitive questions.

A *first* important question is the degree of asymmetry between developed and advanced countries. As we have discussed above, arm length R&D contracts are not likely to emerge when the technological asymmetry between firms is exceedingly large. This result can intuitively be extended at a country level. If the level of local technological knowledge (embodied in human capital, infrastructure, dynamism of local firms) is exceedingly low, the cost of carrying out R&D locally may be too high.

A *second* issue is whether specialised rather than diffused technological knowledge is more likely to attract R&D. Particularly in the case of vertical investments, we expect countries to specialise in the production of given components or products and consequently develop specialised knowledge. Is this

[7] Few papers have analysed the decision of MNEs based in the same home country about where to locate their R&D activities. See Kumar, 1995 for the US, Cantwel and Kotecha 1993 for France Zejan, 1990 and Hakanson and Nobel, 1993 for Sweden.

enough to attract R&D? Or should technological knowledge be diffused across more industrial sectors?

A *third* question relates to the local competitive environment. If local firms are not sufficiently competitive and able to benefit from knowledge spillovers, incentives to cooperate in R&D are likely to be low.

Fourth, protection of intellectual property rights is also an important determinants. The effectiveness of intellectual property rights protection affects the extent of knowledge spillovers. MNEs are probably not very keen to locate their R&D activity in countries with weak protection systems

Finally, there are issues that are more intertwined to the underlying motivation of the investment. For example, R&D linked to horizontal and generally market oriented investments, is likely to emerge where the size of the market justifies investments aimed at adapting products to local conditions. At the same time, cost saving factors like the availability of cheap specialised manpower and technical services are more important for R&D linked to vertical investments.

4.4 The empirical analysis

The objective of the empirical analysis is to examine under what conditions firms move from mere technology transfer or sharing to the co-operative development of new technologies. This will be done by looking at a sample of arm-length agreements and joint ventures of both R&D and non R&D types. Our counterfactuals are non-R&D agreements. As the data base does not include subsidiaries of multinationals, we cannot analyse whether and when the geographic dispersion of R&D by multinationals takes place via internalised transactions or arm-length agreements[8].

The sample includes 632 agreements in 29 developing countries established before 1989. These agreements can be classified according to whether they imply R&D activity between the partners or whether they don't. Tables 2 and 3 report the distribution of the agreements by country and industry respectively. The sample is the developing countries section (at least one of the partners is located in a developing country) of the Merit Cati data base which contains nearly 10,000 co-operative agreements. The data base provides information about the type and the scope of the agreement, the industrial sector, the year of establishment and closing down and some information about the firms undersigning it. A detailed description of the data base can be found in the appendix.

Table 4 shows the distribution of the observations according to the type of agreement. Each type of agreement has special characteristics, which may affect the incentive structure. At this level of analysis it is possible to group them according to three major criteria.

The *first* one is whether agreements imply R&D activities, or whether they don't (the appendix spells out in details methods of aggregation). The *second* criteria is whether the relationship between the partners is hierarchical. In R&D contracts we

[8] Under arm-length agreements we also include joint ventures and minority holdings.

have a hierarchical relationship when it is possible to clearly identify a customer and a supplier. A typical example is a developing country's automobile company that asks a European design company to design a new model. Instead, we have a non-hierarchical relationship, when both firms contribute to the development of the R&D programme on an equal footing base and when it is not a customer-supplier relationship. In non-R&D contracts, the relationship is hierarchical when there is a technology transfer from one company to another; licensing contracts are

Table 2.

Country	R&D	NO R&D	Total	Share of R&D on NO R&D
Algeria	1	3	4	25,00%
Argentina	4	3	7	57,14%
Brazil	4	30	34	11,76%
China	12	70	82	14,63%
Ecuador	0	2	2	0,00%
Egypt, Arab Republic of	0	6	6	0,00%
Gabon	0	3	3	0,00%
Hong Kong	2	7	9	22,22%
Hungary	2	17	19	10,53%
India	7	47	54	12,96%
Indonesia	5	12	17	29,41%
Iran, Islamic Republic of	0	2	2	0,00%
Israel	13	16	29	44,83%
Jamaica	0	1	1	0,00%
Korea, Republic of	23	157	180	12,78%
Malaysia	1	7	8	12,50%
Mexico	1	25	26	3,85%
Morocco	0	3	3	0,00%
Pakistan	0	1	1	0,00%
Philippines	1	3	4	25,00%
Poland	0	2	2	0,00%
Singapore	2	8	10	20,00%
South Africa	3	9	12	25,00%
Taiwan	8	42	50	16,00%
Thailand	4	8	12	33,33%
Turkey	4	17	21	19,05%
Venezuela	0	11	11	0,00%
Yugoslavia, Federal Republic of	4	18	22	18,18%
Zaire	0	1	1	0,00%
Total	**101**	**531**	**632**	**15,98%**

typically part of this group. It is non hierarchical when there is technology sharing between the partners. The *third* criteria is whether the agreement is characterised by equity sharing. In this group we have joint ventures and minority holdings. Equity agreements are divided between R&D and non R&D taking into account the scope of the joint venture.

The first step of the empirical analysis is to estimate a dichotomous choice model to explain the determinants of R&D agreements. The dependent variable is

Table 3.

SECTOR	R&D	NOR&D	TOT	RATIO
CORE	***48***	***189***	***235***	**20,43%**
BIOTECHNOLOGY	**10**	**12**	***22***	**45,45%**
Pharmaceutical	0	6	*6*	0,00%
Agricoltural	4	5	*9*	44,44%
Chemical	3	0	*3*	100,00%
Nutrition	1	0	*1*	100,00%
Environmental	1	0	*1*	100,00%
Basic Research	1	1	2	50,00%
NEW MATERIALS	**9**	**35**	***42***	**21,43%**
Electronic Materials	1	5	*6*	16,67%
Ceramics	3	7	*8*	37,50%
Fiber/plastics	2	10	*12*	16,67%
Powder Metallurgy	3	13	*16*	18,75%
INFORMATION TECHNOLOGY	**29**	**142**	***171***	**16,96%**
Computers	4	14	*18*	22,22%
Industrial Automation	4	43	*47*	8,51%
Microelectronics	5	48	*53*	9,43%
Telecommunications	13	26	*39*	33,33%
Software	1	11	*12*	8,33%
other i.t.	2	0	*2*	100,00%
NON CORE	***53***	***344***	***397***	**13,35%**
INSTRUMENTS	**1**	**4**	***5***	**20,00%**
HEAVY ELECTRICAL EQP.	**10**	**18**	***28***	**35,71%**
AVIATION/ DEFENSE	**11**	**10**	***21***	**52,38%**
Space	0	1	*1*	0,00%
Fighters	8	4	*12*	66,67%
Civil Aircrafts	3	5	*8*	37,50%
AUTOMOTIVE	**6**	**132**	***138***	**4,35%**
CHEMICALS	**15**	**80**	***95***	**15,79%**
FOOD & BEVERAGES	**2**	**7**	***9***	**22,22%**
CONSUMER ELECTRONICS	**1**	**23**	***24***	**4,17%**
ENGINEERING	**4**	**68**	***72***	**5,56%**
OTHER	**3**	**2**	***5***	**60,00%**
TOTAL	**101**	**533**	***632***	**15,98%**

dichotomous, in that it takes value 1 if agreement i is of the R&D type and 0 otherwise (R&D). We will use the *probit* technique, that allows us to test the probability that the dependent variable is explained by the independent ones. The model tested can be spelled out as follows:

(1) $$P(R\&D_i=1)= \alpha + \beta_i X_i + \mu_{ik} V_{ik} + \nu_{in} Y_{in} + \gamma_{ij} Z_{ij} + \varepsilon_i$$

where α is the constant, X_i is a vector of characteristics of agreement i, V_{ik} is a vector of characteristics of the firms k involved in agreement i, Y_{in} is a vector of characteristics of industry n where the products to which agreement i refers is classified, Z_{ij} is a vector of characteristics of country j where agreement i is located and ε_i is the error term.

Table 5 lists all the variables used (not all of them together) in the probit. The first group of variables represent *characteristics of the agreement*, (X_i) and in particular whether the relationship between the partners is hierarchical (HIERARCH) and whether the agreement is equity or non equity (EQUITY). The theoretical literature surveyed in sections 2 tells us that R&D agreements are more likely to emerge because of strategic interactions between firms. We therefore expect hierarchical contracts to dominate. Equity is likely to emerge when contracts are very complex and contingencies difficult to foresee.

Then, we test whether R&D is more likely to be related to product development/improvement or efficiency increases in the production process (HOVER).

Firm specific variables (V_{ik}) try to measure firms' dynamism in international contractual activities. We do expect developing countries firms to enter into an R&D agreement, after having interacted with foreign firms at a more basic level of knowledge (MOREAGRE). Also in learning there is a process of learning by doing, i.e. firms learn how to learn. Hence, they must become efficient in acquiring other firms' knowledge, before being able to develop new one. We also expect that the same firms that underwrite an R&D agreement are already used to operate together (MOREMORE). R&D agreements are usually long term and risky and a good relationship of trust between the partners probably makes life easier. Both variables reflect the assumption, clearly supported by the theory, that the extent and the scope of the technology transfer gradually increases with time.

Industry specific variables (Y_{in}) measure R&D intensity in the industry. We know from the theoretical literature, that R&D agreements are more likely to emerge the larger potential knowledge spillovers, therefore, the larger the share of knowledge capital (which is a public good) on total capital assets of the firm. R&D intensive industries are assumed to be the most knowledge intensive. The problem, though, is that we cannot exclude that R&D industries could also be extremely capital intensive. A variable which measures the average sector share of R&D expenditure on sales (REDINT), is a very noisy index of the public good nature of the firms' capital. Pavitt, 1984 provides a useful classification of industries into four groups: traditional, specialised suppliers, scale intensive and science based. The science based group isolates industries where knowledge capital is dominating (PAVITT).

Table 4.

	JOINT EXCHANGE OF TECHNOLOGY (equal footing)					HIERARCHICAL TRANSFER							ROW TOT.
	Equity		*Joint R&D or Sharing - no Equity*		*tot. joint exchange*	*Costumer - Supplier (joint R&D or production)*		*One-way technology transfer*		*Equity*		*tot un. trans*	
R&D	Joint Venture	36	Joint Research Pact	3		Research Contract	10	JDA +Licensing	4	Joint Venture + Licensing	10		
			Joint Development Agreement (JDA)	23									
			Bidding Consortium	10									
			JDA +Technology Sharing	5									
COLUMN TOT		36		41	*77*		10		4		10	*24*	*101*
SHARING / TRANSFER	Joint Venture	85	Technology Sharing	9		Co-Makership Contract	2	Second Source Agreement	2	Joint Venture + Licensing	72		
	Cross Holding	1	Mutual Second Source Agreement	0		Customer Supplier Partnership	70	Licensing	137	Minority Holding	90		
								Second Source Agreement +Licensing	8				
								Technology Sharing +Licensing	54				
								Customer Supplier Partnership +Licensing	1				
COLUMN TOT		86		9	*95*		72		202		162	*436*	*531*
SAMPLE													*632*

Table 5.a Variables used in the Probit Analysis

Variable	*Source*	*Description*
R&D (Dependent Variable)		Binary variable that assumes value 1 if the agreement is aimed at R&D activities, and 0 otherwise
A) Charachteristics of the agreement (Xi)		
hierarch	CATI	Binary variable that assumes value 1 if the contractual relationship among the partners in the agreement is one of equal footing, and 0 if there is an unilateral transfer of inputs
equity	CATI	Binary variable that assumes value 1 if the firms involved in the agreement join in a equity institution (joint venture, research corporation, cross and minority holding) and 0 otherwise
hover		Binary variable that assumes value 1 if the agreement is aimed at developing products for the final market and 0 otherwise
regio	CATI	Binary variable that assumes value 1 if the agreement is set forth by at least two firms based in the same region of the world, and 0 otherwise
subenv	CATI	Subject Environment. Discrete variable that assumes value 1 if the agreement relates to one product or one piece of technology, 2 if it relates to a few , 3 if it relates to a very broad range of products or technologies
opcon	Merit	Operational Contest. Discrete variable that assumes value 1 if the operational contest of the agreement is national, 2 if it relates to a few states, 3 to a continent, 4 if it is worldwide
B) Firms' Dynamics (Vik)		
moreagre	CATI	Binary variable that assumes value 1 if at least one of the firms involved in the agreement is engaged in other agreements , and 0 otherwise
mormore	CATI	Binary variable that assumes value 1 if the same firms involved in the agreement are engaged in more than one agreement , and 0 otherwise
C) Charachteristics of the industry (Yin)		
pavitt	CATI	Pavitt's index of industry 's technological charachteristics: 1 = traditional; 2 =specialised supplier; 3 = scale intensive; 4 = science intensive
redint	OCSE	Index of R&D intensity in the sector

Table 5.b Variables used in the Probit Analysis

Variable	***Source***	***Description***
D) Characteristics of the country (Zij)		
D1) Technological capabilities of the country		
gdpop	Barro-Lee	Real GDP per capita (1985 international prices), 1985
setech	United Nations	Number of scientist, engineers and technicians involved in R&D activities on total population, 1985
redgdp	United Nations	Total expenses in R&D on GDP
patent	United Nations	Patent applications on total population 1988-1990
indes	United Nations	Industrial designs applications on total population 1988-1990
trade	United Nations	Trade marks applications on total population 1988-1990
manu	World Bank	Share of Manufacturing on GDP, 1985
high	Barro-Lee	Total gross enrollment ratio for higher education.1985
persec		Percentage of sectors with agreements involving R&D activities on the number of sectors in which agreements take place in the country.
D2) Factor endowments of the country		
Kapital	World Bank	Cumulated gross domestic investment per cap. (1987 US$) 1970-1985
invgdp	Barro-Lee	Ratio of real domestic investment (private plus public) to real GDP. From Summer Heston
wages	I.L.O.	Wages in Manufacturing per hour, US dollars, 1985
diffwage	I.L.O.	Ratio of wages per hour in the country of the Pvs firm to the wages per hour in the country of partner(s) from the developed world. When there are more than a developed world partner, an average of these ratios is used.
lcost	I.L.O.	Cost of Labor in Manufacturing per hour, US dollars 1985
D3) Market Demand		
pop85	World BanK, U.N	Total population, 1985
owqi	Barro-Lee	Own-import weighted tariff rates on intermediate inputs and capital goods
owti	Barro-Lee	Own-import weighted non-tariff frequency on intermediate inputs and capital goods.
ex	Barro-Lee	Ratio of export to GDP (in current international prices)
im	Barro-Lee	Ratio of import to GDP (in current international prices)
growth	Barro-Lee	Average growth rate of GDP,1980 - 1984
lgdp	World Bank	Logarithm of real GDP, US dollars, 1985

We then move to *country specific* variables (Z_{ij}). First we look at indicators that proxy the technological capabilities or, more generally, the level of development, of the country. We know, from the theoretical survey, that R&D agreements are less likely to emerge when technological asymmetries between firms are large. Unfortunately, our data base is extremely poor in information about the firms undersigning the agreements. It is therefore impossible to devise any indicator about technological capabilities at a firm level. Thus, we must confine ourselves to the use of country level variables. These variables are also helpful to understand location choices. Large technological asymmetries between the home and the host countries are probably likely to increase the cost of setting up R&D facilities in developing countries.

Finally, we have standard country variables to explain the location choice. These variables should capture the size and the dynamics of the market, explaining market oriented investments and factor endowments, explaining cost saving and vertical investments.

The results of the probit are reported in table 6. We just show the most significant outcomes. Country data were not available for all the countries considered. Some of the initial observations had therefore to be dropped.

Equation 1 includes the largest sample for which we were able to draw satisfactory results. The characteristics of the agreement and the nature of the relationship between the partners are very significant. Indeed, R&D contracts are more likely to emerge when the relationship between the partners is non-hierarchical (HIERARCH) and when it is characterised by equity sharing (EQUITY). The non-hierarchical nature of R&D contracts, supports the basic co-operative structure emphasised by models of R&D joint ventures. At the same time, it shows that local firms must be able to actively interact with the foreign firm in developing the innovation. The positive sign of the equity variable could show that contingencies related to R&D are often too many and complex to be regulated by straightforward arm-length contracts. Hence, first or second best arm length R&D contracts cannot always be implemented, and thus equity sharing is an effective sharing rule. The presumed effectiveness of equity should not be taken for granted, though. On the basis of our evidence we are unable to tell whether partners in the joint venture subscribe other side agreements which modify the sharing rule defined by equity stakes and which perhaps reflect first or second best criteria.

The HOVER variable is also very significant, and with a negative sign. As it takes value 0 for non-market oriented activities, we may infer that R&D contracts are more likely to be implemented in connection with research aimed at incresing the efficiency of the production process.

The positive and significant coefficient of the Pavitt variable shows that R&D contracts are likely to emerge in science based industries, i.e. in industries where knowledge capital is a dominating firm's asset. The PAVITT variable performed better than REDINT (not shown in the table), which is unable to discriminate industries with both large tangible and intangible assets from industries which have only intangible assets like R&D and knowledge. This result, therefore,

Table 6. Probit Estimates

	1	Large Sample				2	NICs Sample			
Dependent Variable: R&D		Coef.	Std. Err.	z	P>|z|		Coef.	Std. Err.	z	P>|z|
cons		-3,152158	0,5231684	-6,025	0		-2,935445	0,6739141	-4,356	0
equity		0,3261943	0,1961113	1,663	0,096		0,7264103	0,2908486	2,498	0,013
hierarch		1,636574	0,1862822	8,785	0		1,443977	0,2542912	5,678	0
hover		-0,9453465	0,2146415	-4,404	0		-1,179904	0,2722075	-4,335	0
pavitt		0,372248	0,1101581	3,379	0,001		0,4508088	0,1702233	2,648	0,008
persec		1,58837	0,5352644	2,967	0,003		-	-	-	-
redgdp		32,77965	11,22378	2,921	0,003		29,42076	12,21903	2,408	0,016
subenv		0,1702056	0,150653	1,13	0,259		-	-	-	-
wage		-0,5091896	0,1528295	-3,332	0,001		-	-	-	-
Log Likelihood =	-137,23426					-70,833515				
Pseudo R2	0,3825					0,4244				
Number of obs =	514					269				

N. B.

Countries in probit 1:	Argentina	Hong Kong	Jamaica	Philippines	Thailand	Countries in probit 2:	Hong Kong	S. Korea
	Brazil	Hungary	S. Korea	Poland	Turkey		Israel	Taiwan
	China	India	Mexico	Singapore	Venezuela		Singapore	
	Egypt	Israel	Pakistan	S. Africa	Yugoslavia			

Table 7. Distribution of moreagre and moremore

R&D	0	1	Total
moreagre			
0	221 41,62%	38 37,62%	259
1	310 58,38%	63 62,38%	373
Total	531 100,00%	101 100,00%	632
R&D	**0**	**1**	**Total**
moremore			
0	451 84,93%	91 90,10%	542
1	80 15,07%	10 9,90%	90
Total	531 100,00%	101 100,00%	632

stresses the role of knowledge capital and other intangible assets. R&D contracts emerge in industries where potential knowledge spillovers are large.

Two other significant variables measure the level of diffusion of R&D activity in the host country. REDGNP is the share of R&D expenditure on GNP in the country. PERSEC measures the share of sectors with R&D agreements on total sectors with agreements. In other words, it measures whether only few specialised sectors are active in R&D or whether R&D agreements are the domain of a larger share of the manufacturing sector in the country. R&D agreements are therefore more likely to emerge the larger and the more widely diffused R&D activity in the host country.

WAGE is the only other country variable reflecting overall locational decisions of multinationals that turns out to significantly affect the R&D choice. The negative coefficient is not easy to explain. As we are considering a sample of developing countries, and as far as wages do reflect the education level of the labour force, we would expect R&D to be located in countries with higher average wages (i.e. WAGE to enter with a positive sign).

Wages are a very imperfect measure of the labour cost. However, the variable LCOST which also includes non-wage components of the labour cost (which may vary considerably from country to country) and which is available for a smaller set of countries turns out to be non significant (probits including this variable are not shown here). If we accept that wages are not a satisfactory proxy of the cost of labour, the poor performance of the LCOST variable, implies that differences in labour costs between developing countries do not affect significantly the probability of R&D contracts.

Our theoretical assumptions and probit 1 support the hypothesis that R&D agreements are more likely to emerge when asymmetries are low. These type of agreements would therefore be more frequent in Newly Industrialising Countries (NICs) than in other country, where they are likely to be subscribed only in exceptional cases. If this is so, our theoretical predictors are likely to perform better in NICs. Indeed, probit 2 which includes only observations for five NICs, Hong Kong, Israel, South Korea, Taiwan and Singapore is the best performing one.

Surprisingly, the two variables measuring firm dynamics (MOREAGRE and MOREMORE) are never significant, in any of the probit tested. Table 7 shows the distribution of R&D and non R&D contracts with respect to these two variables. The majority of R&D agreements have indeed been undersigned by developing countries firms that have other agreements. However, this result seems to pertain to a more general pattern, whereby the majority of firms which enter *any type* of international agreement, also take part in other agreements. This is why the MOREAGRE variable is not significant in the probit. In contrast to our expectations, the majority of R&D contracts are undertaken by firms which have no other agreements between themselves (MOREMORE is 0 for the majority of R&D contracts). This result may perhaps reflect a problem of the database. As the data base only includes agreements reported in the press, the probability that all the agreements undertaken by two or more firms are included is relatively low.

Finally, the very little significance of country specific variables calls for further enquiry. Country variables have very little variability in probit estimates. All agreements within a given country have the same value of the country variables. Thus, the impact of these factors is likely to be under-estimated. Unfortunately, the number of countries is not sufficient to carry out a cross section at a country level. In table 9, we report correlation ratios between two country specific measure of R&D agreements and country variables. REASEARCH is the number of R&D agreements in a country and REDRA is the share of R&D agreements on total agreements observed in a country. Table 8 lists variables used only in these correlations. These variables measure some average characteristics of the agreements observed in each country.

The most significant variables are again those proxying the endowment of knowledge of the host country. Labour cost is significantly and negatively correlated with the share of R&D contracts. The size of the population and the rate of growth, which proxy the size of the market, are significantly correlated with the number of R&D contracts. R&D contracts are also correlated with the number of firms doing R&D (hence R&D agreements are not concentrated in the hands of few particularly dynamic firms), with the average number of agreements per firms, and with the number of industries which have R&D contracts. This results confirm the result of the probit, that R&D agreements are more likely to emerge in countries where research is more widely spread in the economic environment.

Table 8. Variables used only in Country Correlations

Variable	Source	Description
research	CATI	Number of R&D agreements in the country
redra	CATI	Percentage of R&D agreements on all the agreements involving at least a firm in the country
firmred	CATI	Percentage of firms involved in R&D on the firms in the country involved at least in one agreement
meanagr	CATI	Average number of agreements per firm in the country
foot	CATI	Percentage of non-hierarchical agreements on all the agreements involving at least a firm in the country
pvsimra	CATI	Percentage of agreements in which the PVS partner provides productive inputs on all the agreements involving at least a firm in the country

NB. The country data used in the agreements database refer to the country of the PVS firm involved in the agreement.

Table 9. Correlations of country variables with two R&D activity indicators

	redra		research		No. Obs
research	0,3681	**	1		29
pavitt	0,3239	*	0,2109		29
meanagr	-0,0216		0,6642	***	29
firmred	0,949	***	0,4894	***	29
foot	-0,2651		-0,1885		29
persec	0,8478	***	0,6877		29
kapital	0,3838		0,086		19
invgdp	0,0208		0,1022		29
lcost	-0,7107	**	-0,3068		9
wages	-0,0667		0,1272		20
growth	0,0126		0,518	***	29
high	0,4709	***	0,2757		29
im	0,0913		-0,0303		29
ex	0,0699		-0,0499		29
owqi	-0,1799		0,0162		24
owti	-0,0398		0,0893		24
manu	0,3458	*	0,5075	***	27
trade	0,1832		0,134		25
indes	0,1744		0,7832	***	20
patent	0,2045		0,2914		27
setech	0,3278		0,4153	**	24
redgdp	0,1108		0,0828		24
gdpop	0,2433		0,1295		29
pop85	-0,0055		0,3453	*	29

Note * 90% significance ** 95% significance *** 99% significance

4.5 Conclusions and future research

International dispersion of R&D activity also concerns developing countries. Most of it is concentrated in the NICs, but all the same our data base reports important and considerable exceptions.

The rationale for this result can in principle be linked to straightforward motives of the localisation of international investments: adapting products to developing countries markets, or improving the off-shore production of components. There is more in the story, though. Both theory and our empirical evidence show that R&D can be carried out via arm-length agreements, even between partners with asymmetric endowments of knowledge. Thus, developing countries firms are actively involved in international R&D, which is not the sole domain of subsidiaries of multinational.

This result takes us beyond the standard framework applied to the relationship between knowledge and technology transfer. Multinationals are generally assumed to have an effective internalisation option, i.e. if they want to avoid technological spillovers they set up subsidiaries instead of (less costly) arm-length agreements. Hence, to prevent knowledge spillovers, multinationals carry out R&D in developing countries with subsidiaries. Consequently, we may expect internalised transactions to emerge in knowledge intensive industries, where spillovers are larger.

In fact, subsidiaries are not water tight. Spillovers may occur towards host country competitors, or the subsidiaries employees could set up a new company. The inability to avoid spillovers and competitive strategic interaction force multinationals to cooperate in the creation of new knowledge. We revert the standard result. The larger the spillovers, the more competitive local firms and the more likely arm length transactions to take place.

The empirical analysis fully supports this result. R&D agreements are mostly non-hierarchical (firms cooperate on an equal footing base), they are carried out in research intensive industries, where knowledge capital is dominating, they emerge when asymmetries in technological endowments between the host and the home country are not too large and when technological activities are widely dispersed in the production structure of the host country.

Theoretical support to our conclusions derives form a straightforward survey of the existing literature. The fine link between competition and co-operation in the creation and dispersion of knowledge in this context needs to be analysed more thoroughly and formally than what has been done here.

We have also shown that R&D agreements are generally linked to vertical investments. We have practically no theoretical result on the relationship between knowledge and vertical investments; this is another obvious area of future research. Strategic decision on the localisation of R&D activities are strictly intertwined to more general localisation decision. To understand this issue more clearly, it would be necessary to analyse R&D agreements in a wider settings, like general equilibrium models, that also take trade into account.

Finally, empirical results would greatly benefit if it were possible to couple data on the agreements with data about the firms undertaking them.

4.6 Appendix: background to the empirical analysis

4.6.1 The Merit / Cati Data Base

Our sample has been extracted from Merit/Cati data bank on technological agreements. This data bank concerns nearly 10.000 agreements among firms located in more than 60 countries established before 1989. The main source of the data is the press, mostly newspapers and business journals. Information about the existence of agreements also has been gathered from companies' annual reports and business yearbooks[9]. The use of these sources inevitably brought along some bias in the data set, especially for what concerns the industry, the size and the country of the firms involved. Agreements regarding "core" technologies[10], and large firms based in Anglo-Saxon countries tend to be over-represented. Moreover, news about the dissolution of non-equity agreements fail to appear systematically in the press.

The data bank considers technological co-operative agreements. A technology transfer or a common innovative activity must be the aim of at least a relevant part of the agreement. The definition of co-operative agreement utilised in the data bank is "common interests between independent (industrial) partners which are not connected through (majority) ownership" (Duysters and Hagedoorn, 1994). This broad definition includes a wide range of contractual forms[11], from customer-supplier relationships and licensing (in which the common interest stems only from the royalties the licensing firm gains from the licensee's profits) to more closer form of co-operation among totally independent partners like bidding consortia, joint development agreements, joint research pacts, or agreements involving equity sharing, like joint ventures or minority holding.

Unfortunately, the incidence of missing values in some variables is heavy, especially for data about the firms. Some information was provided in the form of short comments, and thus not directly codifiable and ready for statistical use. The

[9] The Financial Time Industrial Companies Yearbooks and Dun & Bradstreet's "Who Owns Whom"

[10] In the CATI data bank, a first classification of technologies divides them between Core and non Core ones. It is assumed that the more advanced Manufacturing industries use mainly Core Technologies. In this group we find industries using information technology like computers and software, microelectronics, industrial automation, and telecommunications; new materials industries like polymers, ceramics, composites and new metal blends, the pharmaceutical industry, the fine chemical industry, and the more advanced agricultural sectors, in which biotechnology is widely used, like horticulture and cattle breeding. Finally, environment defensive systems employing biotechnologies are also counted in this group. It is assumed that other industries in the manufacturing sector do not use Core technologies. Thus, strangely enough, high-tech industries like the aereospatial and aereonautic ones are excluded from the Core technology group.

[11] For a complete list of the contractual forms used in the CATI data bank, see Duysters and Hagedoorn, 1993, pp. 12-13.

full data set includes the number and the name of the companies undersigning the agreement, the involvement of banks, research institutes, universities and venture groups, the year of establishment, planned and actual year of break up, the capital invested, the sector in which the technology is applied, the contractual form of co-operation, the number of product or pieces of technology affected by the agreement, the broadness of the market interested by the alliance (ranging from a single country to the whole world), the distance of the technology involved from the target market, the motives underlying the alliance, the direction of technology or capital flows. As regards the companies involved, there are (often incomplete) data about their home country, year of establishment and dissolution, and number of employees. Some variables are based on classifications defined by the authors and are not straightforward. For instance, the three digit sector code used is ad hoc and is often not consistent with international industrial and trade classifications.[12] There is also a very detailed taxonomy of the motives underlying the choice of the contractual form.[13] Three different sets of motives were singled out, according to whether the agreement was a research one, a non R&D joint venture, or a direct investment. A number of forms of co-operation were directly related to a single motive.

4.6.2 Constructions of the variables from the data base

We generated our dependent variable and a few explicative variables by aggregating the agreements according to the following criteria. In order to divide the agreements between *R&D* and Non-R&D we had to move in two steps. Some of the agreements are obviously of an R&D type; these include, joint research pacts, joint development agreements and R&D contracts. In contrast, bidding consortia and most of all, joint ventures are in a grey zone. To classify the agreements we looked at the underlying motives, among which we could easily identify those related to R&D activities.

For the variable *Hierarch*, we once more relied on the definitions of the contractual forms. We derived a partition of the agreements into a group of equal footing relationships, in which all the parties were subject to symmetrical mutual contributions, and a group of unilateral transfers in which a hierarchical relationship between customers and suppliers was recognisable. The variable *Equity*, includes joint ventures and minority holding agreements. For the variable *Hover*, we referred to the variable "distance to the market" of the CATI data bank.

---

[12] See Duysters and Hagedoorn, 1993, p. 10.

[13] See Duysters and Hagedoorn, 1993, p. 17.

References

Aghion, Phillippe and Jean Tirole, 1994, 'Opening the Black Box of Innovation', *European Economic Review,* 38, pp. 701-710, reprinted in this volume.

d' Aspremont, Claude, and Alexis Jaquemin, 1988, 'Co-operative and Noncooperative R&D in Duopoly with Spillovers'. *The American Economic Review*, 78, No. 5, pp. 1133-1137.

d' Aspremont, Claude, S. Bhattacharya and L. Gerard Varet, 1993, 'Knowledge as a Public Good: Efficient Sharing and Incentives for Development Effort'. 1993, Universita' Bocconi, IGIER Working Papers, n. 58.

Barba Navaretti, Giorgio, 1995, 'Joint Ventures: Conflict or Cooperation? Distribution of Profits under Asymmetric Information', *Il Giornale degli Economisti*, April-June 1995.

Bhattacharya S., Jacob Glazer, and David E. M. Sappington, 1990, 'Sharing Productive Knowledge in Internally Financed R&D Contests'. *Journal of Industrial Economics,* 39, No. 2, pp. 187-208.

Bhattacharya S., Jacob Glazer, and David E. M. Sappington, 1992, 'Licensing and the Sharing of Knowledge in Research Joint Ventures'. *Journal of Economic Theory*, 56, pp. 43-69.

Brainard, S. Lael, 1993,' A Simple Theory of Multinational Corporations and Trade with a Trade-off between Proximity and Concentration'. NBER Working Paper No. 4269.

Cantwell, John, 1993, 'Multinational Corporations and Innovatory Activities: Towards a New Evolutionary Approach'. University of Reading, Department of Economics, Discussion Papers in International Investment and Business Studies, N.172 Series B vol.V.

Cantwell, John, 1994, 'The Globalisation of Technology: What Remains of the Product Cycle Model?', University of Reading, Department of Economics, Discussion Papers in International Investment and Business Studies, N.185 Series B vol VII.

Cantwell, John and Usha Kotecha, 1993, 'The Internationalisation of Technological Activity: the French Evidence in a Comparative Setting'. University of Reading, Department of Economics, Discussion Papers in International Investment and Business Studies, N.179 Series B vol. VI.

Casson, Mark, 1985, 'Multinationals and Intermediate Product Trade', in P.J. Buckley and M. Casson, *Economic Theory of the Multinational Enterprise: Selected papers'*, London, Macmillan.

Chauduri, Prabal R., 1995, 'Technological Asymmetry and Joint Product Development'. *International Journal of Industrial Organisation,* 13, pp. 23-29.

Coe, D.T., E. Helpman, A. W. Hoffmaister, 1995, 'North South R&D Spillovers', Centre for Economic Policy Research (CEPR) Discussion Paper Series, N.1133

Contractor, F.J. and P.Lorange, 1988, *'Cooperative Strategies in International Business'*, Lexington Books

Dasgupta, P and J. Stiglitz, 1980, 'Uncertainty Industrial Structure and Speed of R&D', *Bell Journal of Economics*, 11, pp. 1-28.

Dunning, John, 1981, *'International Production and the Multinational Enterprise'*, London, George Allen and Unwin

Duysters, Geert and John Hagedorn, 'The Co-operative Agreements and Technology Indicators (CATI) Information System', Merit, Maastricht, mimeo.

Ethier, Wilfred J., 1986, 'The Multinational Firm'. *Quarterly Journal of Economics*, 101, No. 4, pp. 805-833.

Ethier, Wilfred J. and James R. Markusen, 1991, 'Multinational Firms, Technology Diffusion and Trade'. NBER Working Paper No. 3825

Freeman, Chris and John Hagedorn, 1994, 'Catching up or Falling Behind: Patterns in International Interfirm Technology Partnering', *World Development*, vol. 22, No. 5, 771-780.
Hakanson, L. and R. Nobel, 1993, 'Determinants of Foreign R&D in Swedish Multinationals', *Research Policy*, 24,22, pp. 397-411.
Harris, C. and J. Vickers, 1985, 'Perfect Equilibrium in a Model of a Race'., *Review of Economic Studies*, 52, pp. 193-209.
Hirschey R, and R. C. 1981 Caves, 'Research and Transfer of Technology by Multinational Enterprises'. *Oxford Bulletin of Economics and Statistics*, 43(2), pp. 115-130.
Horstmann, Ignatius J. and James R. Markusen, 1987, 'Strategic Investments and the Development of Multinationals'. *International Economic Review*, 28, No. 1, pp. 109-121.
Kamien, Morton I., Ethan Muller and Israel Zang, 1992 'Research Joint Ventures and R&D Cartels'. *The American Economic Review*, 82, pp. 1293-1306.
Kumar, Nagesh, 1995, 'Intellectual Property Protection, Foreign Direct Investments and Location of Overseas R&D Activities by Multinational Enterprises'. United Nation University, Institute for New Technologies, Maastricht, mimeo.
Markusen, James R., 1995 'The Boundaries of Multinational Enterprises and the Theory of International Trade', *Journal of Economic Perspective* Vol. 9, No. 2, pp. 169-190
Motta, Massimo, 1992, 'Co-operative R&D and Vertical Product differentiation'. *International Journal of Industrial Organisation* , 10, pp. 643-661.
Papanastassiou, M. and R. Pearce, 1994, 'The Creation and Application of Technology by MNEs' Subsidiaries in Europe and their Role in a Global-Innovation Strategy', University of Reading, Department of Economics, Discussion Papers in International Investment and Business Studies, N.184, Series B, vol VI.
Pavitt, Keith, 1984, 'Sectoral Patterns of Technical Change: Towards a Taxonomy and a Theory', *Research Policy*, 13, 343-373.
Pietrobelli, Carlo, 1996, 'Emerging Forms of Technological Cooperation: the Case for Tehnology Partnership', Unctad, Division for Science and Technology,
Rosenkranz, Stephanie, 1995, 'Innovation and Co-operation under Vertical Product Differentiation'. *International Journal of Industrial Organisation*, 13, pp. 1-22.
Suzumura, Kotaro, 1992, 'Co-operative and non Co-operative R&D in Oligopoly with Spillovers' *The American Economic Review*, 82, No.5, pp. 1307-1320.
Svejnar, J. and S. Smith, 1984, 'The Economics of Joint Ventures in Less Developed Countries', *Quarterly Journal of Economics*, 99, No.1, pp. 149-167.
Vonortas, N.S., 1991, *'Research Cooperation in R&D INtensive Industries'* Academic Publishing Group, Aldershot
Wang, Jian-Ye and Magnum Blomström, 1992 'Foreign Investment and Technology transfer: a simple Model', *European Economic Review* 36, pp. 137-155.
Wright, Donald J., 1993, 'International Technology Transfer with an Information Asymmetry and Endogenous Research and Development'. *Journal International Economics*, 35, pp. 47-67.
Zejan, M., 1990 'R&D Activities in Affiliates of Swedish Multinational Enterprises', *Scandinavian Journal of Economics*, 92(3), pp. 487-500.

5. Research and Productivity

Boyan Jovanovic, Yaw Nyarko*

5.1 Introduction

We propose an informational theory of how research raises productivity. We model research as a signal on an unknown parameter of a technology. Such a signal teaches the firm how to operate the technology it uses now, or one that it intends to use in the future. The paper deals with two topics. The first is the construction of the firm's knowledge stock, and the second is the interpretation of existing estimates of the rate of return to basic and applied research.

Research gives rise to the invention of new processes and products, and is generally considered to be essential in generating sustained growth at the world level. But beyond this recognition, many questions remain to be resolved at the theoretical level. What role does research play in assimilating the results of the research done by others? How does the research process interact with the process by which the fruits of old research become obsolete? When research effort is directed towards the development of a particular technique or product, how far can the results of that research help with the development of other techniques and other products for which it was not originally intended? How does research interact with other forms of learning such as schooling, training, and learning by doing? In particular, all these diverse sources of information somehow combine to form a stock of knowledge?

Our approach treats all form of learning (including research) as the acquisition of information that is valuable to solving one or more of a sequence of problems. Information theory then tells us unambiguously how different information is combined, and what its effects on productivity will be. This then is a framework in which the questions we posed above will have a clear answer.

In the paper we answer only some of those questions, but our model will suggest ways of tackling the rest. Our specific interest is the interplay of the decision of choosing a research direction and the decision of choosing what technology to produce with.

---

* New York University.

We thank the C.V. Starr Center for Applied Economics for technical and financial help. Zvi Griliches and Bronwyn Hall for comments, and Chung Tse for helping with the research.

5.1.1 Lagged R&D weights in the firm's knowledge stock

When constructing a firm's research capital (K_t) from its past R&D spending (R_{t-j}^*), the usual framework (e.g. Griliches 1986) is:

(1) $$K_t = \sum_{j \geq 0} \mu_j R_{t-j}^* ,$$

and the firm's output (sales, or value added) is

(2) $$Q_t = [\lambda^t K_t^{\alpha}] C_t^{\beta} L_t^{\gamma} ,$$

where C and L stand for physical capital and labor. Griliches interprets $[\lambda^t K_t^{\alpha}]$ as follows: λ^t is "external" technical change, and K_t is knowledge generated internally.

We derive explicit expressions for λ and K in terms of the parameters of the model. They both depend on how fast the firm upgrades its technology: λ rises with the speed of upgrading because the adoption of better blueprints is faster. But upgrading destroys knowledge specific to abandoned technologies, accelerates the decay in the μ_j's, and lowers K.[1] To a first approximation, the μ_j's decay geometrically, as is indeed typically assumed in practice.

5.1.2 The rate of return to basic research

We distinguish applied from basic research and model an industry equilibrium in which firms that follow different research strategies coexist in the long run.[2] In the model the private rate of return to basic research can seem larger than that to applied research (just as Mansfield (1980), Link (1981) and Griliches (1986) found) even though it really isn't. This is because basic research signals a fast-track upgrading strategy.

In interpreting existing estimates of the rate of return to research, we focus on Griliches's (1986) method which allows for the weight on the firm's applied research spending, R_t^a, to differ from the weight on its basic research spending, R_t^b. He defines a composite research index, R_t^*, as follows:

(3) $$R^*_t \equiv R_t^a + (1 + \pi) R_t^b = R_t^*(1 + \pi s_t),$$

where

(4) $$s_t \equiv \frac{R_t^b}{R_t^a + R_t^b}$$

[1] This point has been made before [Fellner (1970, appendix C), and Griliches (1973, fn.9)], but seems to have escaped modelling.

[2] Earlier work on the direction of research focused on risky versus safe research, possibly under oligopoly - e.g. Bhattacharya and Mookherjee (1986).

is the share of spending on basic research in total spending on research. He interprets π as the premium on basic research, and his estimates of π (reported in his table 2) range from 0.27 to 0.52. That is, a dollar of basic research buys as much productivity as \$ 1.27 - \$ 1.52 of applied research buys, which suggests the firm might make money by shifting some of its research dollars to basic research.

5.2 The Theoretical Framework

5.2.1 The general framework.[3]

Let the output of the firm Q, be produced with capital, C, labor, L, and materials, M. Let q denote an efficiency parameter, so that the production function is:

(5) $$Q = F(C, L, M; q) .$$

How q enters F(.) will depend on the factor of production in which the learning is "embodied".[4] The model of the growth of q that we shall now present is, we hope, applicable to various types of learning.

Assume that q depends (a) on the technological *grade* n that the firm chooses to use, and (b) on how closely a grade-specific decision, z_n, matches an ideal, y_n :

(6) $$q = G(y_n - z_n ; n) .$$

Suppose that q is largest when $z_n = y_n$. We may think of $(y_n - z_n)$ as a *mistake* that reduces q below its maximal level of G(0, n). For example n might stand for the grade of computer equipment to use, while z_n might stand for the speed of the production line in a factory, the organization of the workspace, the types of raw materials purchased, the type of labor hired and its assignment to the tasks, the choice of factor proportions, and so on. All of these aspects of operating decisions are collapsed into a scalar y_n, which fluctuates randomly over time around a mean, denoted by θ_n .

If the firm could learn θ_n by copying the production methods of another firm that uses grade n, or if it could copy that other firm's z_n , the firm would not need to do any research. Indeed, there are features of technologies that can be reverse engineered, and are even explained in manuals, and that do not have to be learned through costly research effort or by trial and error. But when the technology is new and is not used by other firms, the firm in question will have to discover it on its own. And even when the technology is in widespread use, (as, for example, hybrid

[3] We have used this framework elsewhere to study optimal choice of technologies (J-N 1994), learning curves on a given technology (J-N 1995a) and the training of inexperienced workers by experienced workers (J-N 1995b).

[4] Bakh and Gort (1993, especially Table 4) ask how q enters the production function of the *plant*. If productivity growth was due to better management, we might write Q = qF(C, L, M), as in (3). If it was due to better machines, one would write Q = F(qC, L, M).

corn was) other features of the optimal decision z_n are specific to the situation at hand, to the nature of the factors of production, the raw materials, the workspace, the specifications of the output, and so on. These, the firm will have to infer for itself.[5]

5.2.2 Productivity growth on a given technological grade

Suppose that

(7) $$q_n = G(y_n - z_n ; n) \equiv \gamma^n [1 - (y_n - z_n)^2] .$$

Here $n \geq 1$ is an integer. The maximal level of q_n, attainable under ideal conditions, is γ^n. Each period τ leads to a new value of y_n :

(8) $$y_{n,\tau} = \theta_n + w_{n,\tau} \qquad \tau = 0, 1, 2, \ldots$$

where the w_n, are independent, normally distributed random variables with mean zero and variance σ_w^2. Noise, w, is essential if learning and productivity growth on a *given* grade n is to be *gradual.*

Let t be the cumulative number of periods, = 0, 1, ... , t-1. The firm must choose $z_{n,t}$ *before* seeing $y_{n,t}$. Let E_t (.) denote conditional expectation. A risk-neutral firm will maximize $E_t (q_t)$ and choose the following decision:

(9) $$z_{n,t} = E_t(y_{n,t}) = E_t(\theta_n) .$$

This decision is dynamically optimal as well, because the information that the firm gets does not depend on the z chosen. So, we get the reduced form

(10) $$q_{n,t} = \gamma^n (1 - [\theta_n - E_t(\theta_n) + w_{n,t}]^2) .$$

Now let $Var_t(\theta_n) \equiv E_t [\theta_n - E_t(\theta_n)]^2$ be the posterior variance over θ_n. Expected q at t then is

(11) $$E_t(q_{n,t}) = \gamma^n (1 - Var_t(\theta_n) - \sigma_w^2) .$$

It is increasing in t, but *bounded:* As n gets large, $E_t(q) \to \gamma^n[1 - \sigma_w^2]$. To get long-run productivity growth, the firm must therefore keep upgrading its technology, and this is the process we specify next.

5.2.3 The transfer of knowledge across grades of technology

There is no direct cost of switching to a different grade of technology, and no adjustment costs in z. Successive grades are linked as follows:

(12) $$\theta_{n+1} = (\sqrt{\alpha})\, \theta_n + \epsilon_{n+1}$$

---

[5] Evenson and Kislev (1975) and Mansfield *et al* (1981) find that imitation is costly, although cheaper than inventing something oneself.

for all n, where ϵ_n is an i.i.d. normal variate with mean zero and variance σ_ϵ^2. If the firm has not yet tried grades n+1, n+2, ..., eq. (12) implies that

$$(13)\qquad Var_t(\theta_{n+1}) = \alpha Var_t(\theta_n) + \sigma_\epsilon^2,$$

If $\alpha = 1$ and $\sigma_\epsilon^2 = 0$, then $\theta_n = \theta_k$ for all k, which corresponds to the case where knowledge is *general* and freely transferable across grades of the technology. But if $\alpha = 0$, knowledge is *grade-specific*.

We assume that at date t = 0, the firm finds itself on grade n = 1, and that it has never used a higher grade. We assume that the prior over θ_1 at date 1 is normal. Eq (8), and the normality of w_{nt} imply that the posterior belief at each date over the parameter of any grade, θ_n, will also be normal. If the prior variance over θ_n is x, then spending period t on grade n and hence observing $y_{n,t}$ gives, via Bayes rule, a posterior variance of $h_1(x)$, where

$$(14)\qquad h_1(x) \equiv \sigma_w^2 x/(\sigma_w^2 + x).$$

On the other hand, if at date t the firm chooses instead to produce using the next-highest grade, grade n + 1, its *prior* variance over θ_{n+1} would, in view of (13), be given by $h_2(x)$, where

$$(15)\qquad h_2(x) \equiv \alpha\, x + \sigma_\epsilon^2,$$

and at the end of the period, following an observation of y_{n+1}, the posterior variance over θ_{n+1} would then be h(x), where

$$(16)\qquad h(x) \equiv h_1(h_2(x)).$$

5.2.4 Research

We assume that the firm can direct its scientists to research *any* grade $k \geq n$. If it directs them to k = n, we shall call such research *applied*, whereas if it directs them to k > n, we shall call such research *basic* (more on this shortly). If, at t, research takes place on grade k, the firm observes a signal

$$(17)\qquad y_{k,t} = \theta_k + \eta_{k,t},$$

where $\eta_{k,t}$ is normally distributed with mean zero and variance σ_r^2. Three points should be noted about how we are modelling research:

(i) Information from research is the same as the information obtained under "learning by doing" except that the precision of the information from research, $1/\sigma_r^2$, may differ from the precision obtained from learning by doing, $1/\sigma_w^2$.

(ii) Research dollars spent this period yield information that one can use next period. Given the lags involved in the maturation of research projects (see Pakes and Schankerman 1984), this means that the length of the period should be fairly long - at least a year.

(iii) Unless $\alpha= 0$, this signal will provide information on the parameter θ_n of all grades of technology, and not only on grade k. This includes grades n that are strictly *less than* k.

5.3 Basic and Applied Research, and the Choice of an Upgrading Policy

We define *applied* research to be spending aimed at learning more about the technological process a firm is already using, or about a good that it is already producing. *Basic* research, on the other hand, is directed towards processes not yet in use, or goods not yet produced. Our definition of basic research agrees with that of the National Science Foundation (1959, p. 124) -- "...original investigation for the advancement of scientific knowledge....which do(es) not have immediate commercial objectives". It is also consistent with some findings in Mansfield's (1981) study of the effects of basic research among chemical and petroleum firms. In eq.(2) Mansfield reports the finding that holding a firm's R&D expenditures constant, its innovative output is positively and significantly related to the share of its R&D spending going to basic research.

After (a) making some basic assumptions, this section will (b) establish a direct relation between the firm's research policy and its upgrading policy, and (c) characterize the optimal upgrading policy.

5.3.1 A preliminary assumption about upgrading

We begin by assuming that the firm will upgrade at a constant rate (in section 4 we shall turn this assumption into a result). Fix a date t. Suppose that a firm currently operates grade N of technology - the highest it has ever operated. Assume that the firm upgrades its technology at the constant rate *u*, so that, since the initial grade has been normalized to be zero,

(18) $$N = u\,t.$$

The firm's long-run growth in q - call it g - is:

(19) $$g = \gamma^{u} - 1 \approx u\, ln\,(\gamma)\,.$$

If we knew γ, we could identify *u* from the firm's long-run rate of Productivity growth:

(20) $$u \approx \frac{g}{ln(\gamma)}$$

Suppose then that we identify basic research as associated with a "high *u*" upgrading strategy, and applied research with a "low *u*" upgrading strategy. More precisely research ahead of the firm's current technological position is basic, and research on refining the firm's current technology is applied.[6] This will allow us to explain the puzzling finding by Mansfield (1980), Link (1981), and Griliches (1986) that basic research has a higher private rate of return than applied research. In our model, given (20) and our definition of basic research, the productivity growth of firms doing basic research will for ever exceed that of firms doing applied research.

5.3.2 The relation between the research policy and the upgrading policy

To simplify the argument, suppose that the firm maximizes the expected return in the current production period, and that it directs its research with this sole aim in mind. In other words, assume that the firm's discount factor is zero (the result easily generalizes to the case in which the discount factor is positive, but small). Proposition A.1 in the appendix shows that under mild restrictions on the parameters, all of the firm's research effort would optimally be directed towards that technological grade which it is about to produce output with. Under these restrictions, the upgrading policy and the research policy are one and the same.[7]

The question now is: Why would two firms choose such different research strategies with the resulting permanent difference in their productivity growth? We address this next.

5.3.3 The optimal upgrading policy

Fix a date and let x be the *before research* posterior variance on the status quo grade N. Define $h_r(x) = \sigma_r^2 x/(\sigma_r^2+x)$, the learning by research function. We shall normalize the payoffs by dividing them by γ^N. Section 3.2 says that the firm will research the same grade it is just about to produce on. If the firm sticks to its status quo technology, its research will yield a posterior variance $h_r(x)$, and its expected return will be

$$l(x) \equiv 1 - \sigma_w^2 - h_r(x).$$

Assume that an upgrade can only be made by exactly one step. If it switches to the next grade and researches it before producing, the firm's posterior variance on *that* grade will be $h_r[h_2(x)]$, and its expected return will be

$$r(x) \equiv \gamma(1 - \sigma_w^2 - h_r[h_2(x)]).$$

[6] Evidence supporting this view is in Mansfield (1981), who finds that innovative firms put more weight in their research portfolios on basic research.

[7] Of course, this statement can not literally be true in that the typical research project takes a year or two to bear fruit in the form of sales (Pakes and Schankerman, 1984). The statement can make sense only if the period is long enough.

The functions l (x) and r(x) are downward-sloping and convex in x. In particular, fig. 2 assumes that there is a value $\hat{x}$ such that the following are true on [0, $\hat{x}$]: r(0)< l(0), l(x) curve is steeper than the r(x) curve and r ($\hat{x}$)> l ($\hat{x}$). (In the appendix we explain the conditions on the parameters which results in this.) There will therefore exist an x^* in (0, $\hat{x}$) such that r(x)< l (x) for x on [0,x^*) and r(x)> l (x) for x in (x^*,$\hat{x}$], as shown in figure 1.[8]

Suppose that a firm has posterior variance of x on the status quo grade. Suppose that the firm researches and then produces on the next grade. Its posterior variance on the new grade will therefore be, after observing the learning by doing signal, given by the function

$$h(x) \equiv h_1(h_r(h_2(x))).$$

The h function can be shown to be strictly concave with h(0) > 0 and h (∞) < ∞, as drawn in Figure 2. It therefore has a unique positive fixed point which we denote by x^{**}.

Implicit in figure 1 is our assumption that $x^* < x^{**}$. If the firm upgrades in each period the dynamics of the posterior variance will be given by iterates of the h function, which will converge monotonically to x^{**}. Fig. 2 shows the following:

Figure 1.

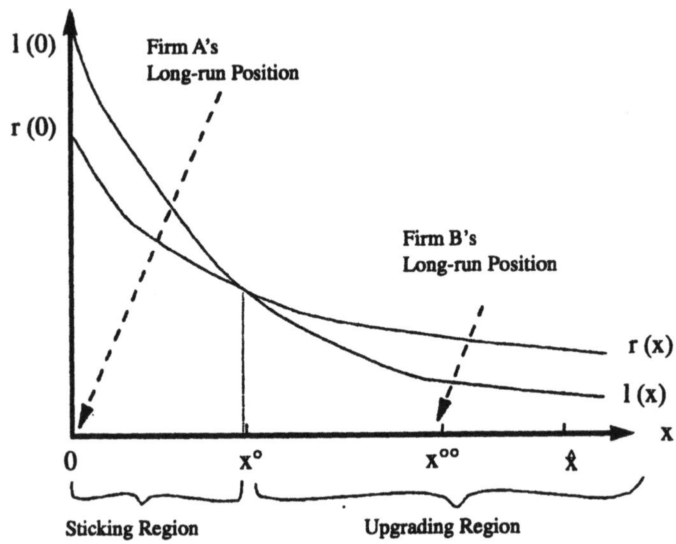

[8] Figure 2 shows that the set [0,$\hat{x}$] is stable and absorbing: From any initial value, the posterior variance will enter [0, $\hat{x}$] in finite time (regardless of the policy that is chosen) and once in that set, it remains there.

When x is low, and in particular when $x \leq x^*$, the firm chooses to stick in each and every period. This will cause the posterior variance to converge to zero. When x is large, (i.e., $x \geq x^*$, the firm will choose to upgrade in every period. Figure 2 shows that the posterior variance then converges to x^{**}.

We summarize the results of section 3.3. in the following proposition:

Proposition 3.3.1. Assume the payoffs are as in fig.2. Suppose two firms, I and II, are using the same grade n and have initial posterior variance, x'_0 and x''_0 respectively, with $x'_0 < x^* < x''_0$. Then firm I initially has a higher expected output, but it will never upgrade. Firm II, will keep upgrading, and growing forever.

These ideas are central to the model in the next section.

5.4 A Long-run, "Coexistence" Equilibrium

We now look at equilibrium. Since the productivity of the upgrading firm keeps growing relative to that of the other firm, two related questions arise: (A) wouldn't the laggard eventually be driven out? and (B) would we not therefore have only upgrading firms in an equilibrium, and therefore no heterogeneity in the long run in the type of research being done? We now model an equilibrium in which the answer to (A) is yes, but the answer to (B) is no.

Suppose that the economy grows as γ. The demand for the industry's product therefore grows as γ. We shall now show that there is a steady state equilibrium in which two kinds of firms coexist, in which industry price and the number of firms of each type is constant, and in which the industry output supplied also grows as γ.

Figure 2: The evolution of x for an upgrading firm

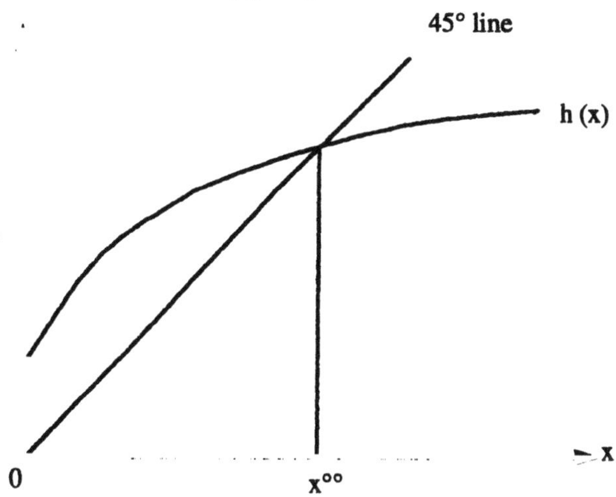

At date t, the aggregate quantity Q is related to the industry's price p by the demand curve $Q = \gamma^t D(p)$. At t = 0, suppose the status quo for all firms is n = 0. As far as x is concerned, firms are of 2 types: A, and B; the initial beliefs of type A's are x = 0, while those of type B's are $x = x^{**}$. Assume that the situation is as depicted in figure 1. Then type A's always stick to their old technology, hence their x = 0, always. Type B's always upgrade, hence their $x = x^{**}$, always. We shall look for a steady state in which firms are (essentially) of 2 types, A (the "transients") and B (the upgraders), having measures μ_A and μ_B. If the product price can be shown to be constant over time, industry demand will grow as γ^t, and the population of upgrading firms will maintain a fixed share of that demand, $\mu_B/(\mu_A+\mu_B)$.

Entry and exit of transients: Entry costs at date t are $\gamma^t c_0$, and operating costs are $\gamma^t c_1$. At date t, we assume that there is an infinite number of potential entrants who understand *perfectly* well the previous period technology used as the status quo of the industry leaders.[9] This grade is grade n = t - 1, and for these potential entrants, x = 0 on this grade. For simplicity, assume that the discount factor is zero. Since we are in a situation as depicted in figure 1, these potential entrants will use grade t - 1 if they decide to enter. For the expected returns to entry to be zero at date t, we need that

$$p\gamma^{t-1}(1 - \sigma_w^2) = \gamma^t (c_0 + c_1), \tag{21}$$

which means that

$$p = \frac{\gamma (c_0 + c_1)}{1 - \sigma_w^2} \tag{22}$$

A transient firm exits as soon as its operating costs, $\gamma^{t+T} c_1$, exceed $p\,\gamma^t (1 - \sigma_w^2)$ - its expected revenue on the technological grade it chose when it entered. That is, its lifetime in the industry is the smallest integer T satisfying:

$$p(1 - \sigma_w^2) \leq \gamma^T c_1. \tag{23}$$

In the equilibrium that we shall construct, the measure of upgraders is μ_B, and the measure of transients is μ_A. However, there are different types of transients, one for each age (0, 1, ... , T-1), and the measure of each different type is 1/T. Therefore, normalized by γ^t, steady state output demanded must equal output supplied:

$$D(p) = \mu_A \{T^{-1} \sum_{\tau=1}^{T} \gamma^\tau\} (1 - \sigma_w^2) + \mu_B (1 - \sigma_w^2 - x^{**}). \tag{24}$$

[9] This simplifies the analysis a lot. In general, what we need here is that the entrants have somehow managed to learn the previous period technology well enough to put them in the "stick" region of figure 1.

The term in braces multiplied by $(1 - \sigma_w^2)$ is the average productivity of transients. The measure μ_B is a parameter, given by the initial conditions - the initial number of type B firms.

An *equilibrium* is a vector (μ_A, μ_B, p, and T) of nonnegative scalars satisfying (22), (23) and (24). A further requirement that equilibrium must satisfy is that the type B firms do not want to exit:

$$(25) \qquad p(1 - \sigma_w^2 - x^{**}) > c_1.$$

There is a continuum of coexistence equilibria here, and it helps to think of them as being indexed by μ_B. We can imagine that at date zero, there is a measure of μ_B type-B firms in the industry. Thus μ_B can be thought of as an initial condition. Since T and p do not depend on either μ_A or μ_B, there is a linear, negative relation between μ_A and μ_B as we go across equilibria. This linear relation emerges from the condition (24) that defines μ_A implicitly in terms of μ_B. Note that the coexistence equilibrium involves a steady entry and exit of transients in each period.

But if the initial condition μ_B is large, equilibrium will not be of the coexistence type, and we will have $\mu_A = 0$ -there are so many upgraders that there is no room in the industry for any transients. In this case, p will no longer be given by (22), because type A's will now strictly prefer not to enter. Instead, it is the upgraders who will be indifferent between staying in and exiting, (25) will hold as an equality, and μ_B will have a unique solution:

$$(26) \qquad \frac{1}{(1 - \sigma_w^2 - x^{**})} \quad D\left(\frac{c_1}{(1 - \sigma_w^2 - x^{**})}\right)$$

So, there is a unique "non-coexistence" equilibrium in which there are no transients, and hence this steady state equilibrium involves no entry and exit.

This section has shown how that there *can* exist an equilibrium in which two types of firm can coexist. In any such equilibrium, there are firms that opt for basic research and who also opt for a fast-upgrading, high growth strategy, and at the same time, there are transient firms that choose to cash in on their grade-specific knowhow.

5.5 The Observed Relation Between Productivity, and Basic and Applied Research

5.5.1 Interpreting productivity growth regressions

In the above equilibrium, basic research is positively correlated with the speed of upgrading and this can cause a bias on the coefficient of basic research in an OLS productivity regression. In fact, combined with the argument of section 3.2, this definition of basic research implies

Proposition 5.1 Suppose that the conditions of Proposition A.1 (in the appendix) are met, and that the firm is interested in maximizing its current payoff. A firm that always does basic research will in the long run grow faster than a firm that always does applied research.

Let's see how this bears on the actual panel data regressions. Suppose a sample of firms is drawn from the long run equilibrium modelled in section 4. From **(19)** we find that since its $u_B = 1$, a type B firm's long-run growth rate is $ln(\gamma)$, while a type A firm's long-run productivity growth is zero, because $u_A = 0$. That is,

$$\textbf{(27)} \qquad g_{i,t} = [\,ln(\gamma)\,]\,u_{i,t}$$

where each firm i is of type either A or B, and $u_{i,t}$ is its basic research intensity at date t. In the steady state equilibrium that section 3.3. presents, $u_{i,t} = 0$ for all t if i is of type B, and $u_{i,t} = 1$ for all t if it is of type A. Equation **(27)** corresponds to the productivity growth regressions reported in table 3 of Griliches (1986), with the variable u corresponding to his variable "BR 72", and to what we call $s_{i,t}$ in **(4)**. This variable comes in with a coefficient between 0.034 and 0.056. Since the dependent variable is measured in percentage points of growth per year, this is a very large effect.

The argument can also be put in terms of the "omitted variables bias" language. Basic research is positively correlated (in our example perfectly correlated) with the speed of upgrading, and this is an omitted endogenous regressor. Hence we estimate a positive v_3. Indeed, Griliches (1986, p. 153) considers this to be one possible interpretation of his results when he says that firms have different opportunities to do basic research. In our paper, research opportunities are determined by initial conditions - a firm's initial knowledge effectively changes its opportunities to doing applied and basic research. So our model provides a context in which this interpretation is exactly right.

5.5.2 Other possible explanations

Other possible explanations for the basic research coefficient have been mentioned, and we list them here.

5.2.a Griliches (1986) mentions two other reasons why the returns to basic research would appear high: (A) Basic research is probably riskier than applied research, and so the higher coefficient could be a risk premium to basic research, and (B) the higher rate of return to basic research could represent a higher social but *not* a higher private return - this could occur if the basic research spenders learn from other basic research spenders.

5.2.b Nelson (1988) provides a "coexistence" model with firms that follow different research strategies - innovation and imitation. An imitator acquires an innovator's invention after a - period delay. It is therefore natural in Nelson's model to think of innovators as doing basic research, and of imitators as doing applied research. All firms grow at the same rate; the productivity differential between the two types of firm is in levels only. Therefore, while the Nelson model can help us understand the results of table 2 of Griliches (1986) which are about *levels* of productivity, it is not consistent with the results in Griliches's table 3 that show a strong association between the share of basic research and productivity *growth*. The latter finding is confirmed in table 3 of the study by Lichtenberg and Siegel (1991).

5.2.c There could be reverse causation: More productive firms may do more basic research as a "luxury item". Certainly, productive firms generate larger cash flows, and there is some evidence in Hall (1990) and elsewhere that cash flow seems to affect R&D budgets. As with Nelson's model, however, this argument can shed light on the results of table 2 of Griliches (1986), but not, it would seem, those in his table 3 that show a positive relation between the share of basic research in a firm's R&D budget, and its productivity growth.

5.5.3 Auxiliary implications

Our interpretation is supported by Mansfield's (1981) finding that innovative firms skew their research portfolios towards basic research. But there are other, yet untested implications: Technology is usually embodied in equipment. Hence upgrading should be accompanied by equipment investment. A test of proposition 5.1.a would then check if firms doing more basic research also engage in more equipment investment. Tangentially related to this is Lach and Schankerman's (1989) finding that in the time series at the firm level, *total* R&D raises *total* investment. And Miyazaki (1994, esp. table 7), using data on company research disaggregated by type, finds a lot of persistence, among close product market competitors, in the type of research carried out. This indirectly supports the implication that firms that start out doing applied research will always do applied research, and similarly for firms doing basic research.

The model of section 4 is consistent with the finding by Doms *et al* (1994) that firms using advanced technologies have higher survival rates. The exit hazard is zero, except at age T, where it is one. There is some evidence of a non-monotone hazard of plant exit in Table 5 of Agarwal and Gort (1993).

5.6 The Index of Research Capital

The μ's are hard to estimate, because a firm's R&D spending tends to be smooth over time (Hall 1995). Most researchers therefore assume geometrically declining μ's - typically at 15%. Some have inferred obsolescence from the decay in the number of citations that a patent gets over time, or the number of citations of academic articles.[10] And others have tried to infer depreciation and obsolescence by following the prices of used equipment of various vintages.[11] We now derive the μ_j in our model.

Fix a date t. Suppose that a firm is currently operating a grade N of technology. Suppose that as of date t, for m = 1, ... , N, there have been τ_m units of learning by doing signals (i.e., the firm operated grade m for τ_m periods), and there have been ρ_m units of research on grade m. What is the number, ρ_m', of learning by doing signals that this is equivalent to? Well, we require that for each initial posterior variance x,

(28) $$1/x + \rho_m'/\sigma_w^2 = 1/x + \rho_m/\sigma_r^2 .$$

Then $\rho_m' = [\sigma_w^2/\sigma_r^2]\,\rho_m$. Hence a research signal is equivalent to σ_w^2/σ_r^2 learning by doing signals. Seeing τ_m learning by doing signals and ρ_m research signals is the same as seeing R_m learning by doing signals, where

(29) $$R_m = \tau_m + [\sigma_w^2/\sigma_r^2]\,\rho_m .$$

We shall call R_m *effective* signals.

5.6.1 The relation between R_m and R_t^*

As defined in (29), R corresponds closely to R^* in (3), but for three differences. First, the units in which they are measured are different: R is in units of numbers of signals, while R^* is in units of real dollars. This is not a problem if the price of research relative to other goods is constant, or if we have the correct price index with which to deflate gross R&D investment at each date.

Second, R_t^* has a time subscript, whereas R_m has a grade subscript. Hence to relate the two, we need to know what grade was researched at what date. This in turn will be related to the speed with which the firm upgrades its technology.

Third, R_m contains signals τ_m that the firm did not purchase through its research outlays. How much information the firm acquires in this way again depends on the its upgrading policy: A firm that upgrades rapidly will have fewer such signals *per grade* because it will be spending less time on any one grade.

If we can somehow identify the speed with which the firm upgrades its technology, and if we can adjust for these measurement problems, our theory of how the firm's knowledge relates to its effective signals (R_m) will have implications for how the firm's research capital stock depends on its past R&D spending, (R_t^*). This is the task to which we now turn.

5.6.2 R&D signals on past grades only

In general, the firm will have signals on grades that it is not currently using. But we shall begin with the simpler case in which such signals are not available. Then

---

[10] Adams (1990), Caballero and Jaffe (1993), Trajtenberg, Henderson and Jaffe (1992).

[11] Hulten and Wykoff (1981), Oliner (1993) are interested in depreciation of equipment, but in new technologies are embodied in equipment, then their estimates say something about the depreciation of research dollars too, so long as one recognizes the time that the research dollar has to "wait" before its fruits are embodied in a machine.

we shall generalize the index to allow for the additional information about future grades.

Let N denote the grade on which production is about to take place. Suppose that at date t there have been R_m effective signals on grade m for m = 1, ..., N [with each R_m coming from any combination of learning by doing and research satisfying **(29)**]. Let us suppose that there has been no information on any grade in excess of N. Suppose that before any information was received, the initial prior variance over θ_1 at date 1 is x_1. The expected productivity on grade N will then be

(30) $$Eq_N = \gamma^N(1-\sigma_w^2-x'_N),$$

where[12]

(31) $$x'_N = h_1^{R_N}(h_2[h_1^{R_{N-1}}(\ldots(h_1^{R_2}(h_2[h_1^{R_1}(x_1)])\ldots).$$

This is the general expression for the stock of knowledge about θ_N when R&D signals are not available on any grade higher than the currently used grade N .

How do **(30)** and **(31)** relate to **(1)** and **(2)**? The correspondence would be quite close if we could define

(32) $$\lambda^t K_t \equiv E_t(q_N).$$

The presence of the trend λ^t means not only that there is upgrading, but that it occurs regularly. Earlier we assumed upgrading at a constant pace:

(33) $$N = u\,t.$$

Then **(32)** holds with the following definitions for λ and K:

(34) $$\lambda = \gamma^u,\ \text{and}\ K_t = 1 - \sigma_w^2 - x'_N.$$

Our model therefore gives a specific meaning to the production function in **(1)**. Does **(2)** also have an interpretation? The answer is yes, in terms of a linear approximation of K_t :

Theorem 6.a (The Linearized Research Capital Index K^*): Fix an R and an initial variance x_1.[13] Then under the definition of K_t in **(34)**, K_t can be linearly approximated by K_t^* , where

(35) $$K_t^* = A\sum_{m=1}^{N}\delta^{n-m}R_m + c = A\sigma_w^2\sum_{m=1}^{N}\delta^{n-m}(a\tau_m + b\rho_m) + c$$

is the Taylor series linear expansion of **(34)** around $R_m = R\ \forall m$ when x_1 is close to $\tilde{x}$ [the fixed point of the map $h_2(h_1^R(.))$]. On the right-hand side of **(35)**, the parameters a, b, c, and A are given by

---

[12] To handle the possibility that R is zero for some grades that were never used or researched, we adopt the convention that $h_1^0(x) = h_r^0(x) = x$.

(36) $a = 1/\sigma_w^2; \quad b = 1/\sigma_r^2; \quad c = 1 - \sigma_w^2 - \tilde{x} - AR\sum_{m=1}^{N}\delta^{n-m}, \quad A = [h_1{}^{\tau}(\tilde{x})]^2/\sigma_w^2,$

and the discount factor δ is strictly less than one and is given by

(37) $\delta \equiv \alpha h_1{}^{R}(\tilde{x})^2/\tilde{x}^2.$

Proof: In the appendix.

From the definition of h_1 in **(14)**, one can show that $h_1{}^R(.)$ is an increasing and concave function, and so the ratio $h_1{}^R(\tilde{x})/\tilde{x}$ is decreasing in $\tilde{x}$. Since $\tilde{x}$ depends on the upgrading policy, so will δ. We shall analyze some properties of K_t^* in section 7. Next, we show that the index retains it basic character even when the firm has signals over grades that it has not yet used.

5.6.3 Signals on past *and* future grades

Now suppose that in addition to signals on grades n used in the past there are also signals about grades that have not yet been used. Let x'_N denote, as before, the posterior variance on the current grade, grade N, given only signals on *past* grades. Eq. **(31)** and Theorem 6.a above are expressions for x_N'.
Suppose that ρ_{N+m} units of effective signals are obtained on grade N + m for m=1,2,... . To make headway we shall suppose that all the future signals are concentrated on one grade, grade N+k, and that there are ρ_{N+k} such signals.

Theorem 6.b (The Linearized Research Capital Index K^{**}): Let R, x_1 be as in Theorem 6.a. Then K_t can be linearly approximated by K_t^{**}, where

(38) $K_t^{**} = c_1 K_t^* + c_2\rho_{N+k} + c_3,$

where the approximation is around the points stated in Theorem 6.a and around ρ_{N+k} close to some $\tilde{\rho}$, and where $c_1 = \partial G/\partial x$, $c_2 = -\partial G/\partial \rho$ (both evaluated at $(x,\rho) = (\tilde{x},\tilde{\rho})$), and $c_3 = (1-c_1)(1-\sigma_w^2) - G(\tilde{x},\tilde{\rho}) + c_1\tilde{x} - c_2\tilde{\rho}$.

Proof: In the appendix.

So, the geometric discounting of past research signals is present in K_t^{**} just as it was in K_t^*.

5.7 Endogenous Obsolescence: Upgrading and the Research Capital Index

The firm's upgrading speed *u* will raise the depreciation rate by lowering δ. Before showing this, we first establish that changing *u* is equivalent to keeping *u* fixed and changing some of the other parameters.

[13] Implicit in the following formula is fact that 0.

5.7.1 An invariance property

Redefining the size of the step turns out to be equivalent to changing some other parameters:

Proposition 7.a: (re-scaling of the technological grid) Suppose firm A upgrades by one step each period, while firm B upgrades by *u* steps. Suppose they both start on grade n = 1, and with the same x_1. If

i) Firm B faces the parameters γ_B, α_B, and $\sigma_{B\epsilon}{}^2$, and effective signals $R_{B,um}$ (for m = 1, 2, ...)

ii) Firm A has the parameters $\gamma_A = \gamma_B{}^u$, $\alpha_A = \alpha_B{}^u$, $\alpha_{A\epsilon}{}^2 = (1-\alpha_B{}^u)/(1-\alpha_B)\,\sigma_{B\epsilon}{}^2$, and effective signals $R_{A,m}$ (for m = 1, 2, ...)

iii) $\sigma_w{}^2$ is the same for both firms,

then if $R_{A,m} = R_{B,um}$ (for m = 1, 2, ...) firm A's and firm B's productivities, as well as their indexes K^* and K^{**} will be the same at each date.

Proof: In the appendix.

5.7.2 Faster upgrading speeds up the depreciation of knowledge

To get the expressions in terms of *u* one merely replaces α with α^u everywhere and $\sigma_\epsilon{}^2$ with $(1-\alpha^u)/(1-\alpha)\,\sigma_\epsilon{}^2$.

Proposition 7.b: An increase in *u*: i) Lowers K [in **(34)**] by raising x'_N [in **(31)**] for any $x_1 < \sigma_\epsilon{}^2/(1-\alpha) \equiv \hat{x}$; ii) Leaves *a* and *b* unaffected [in **(36)**]; iii) Lowers δ [in **(37)**] - the relative weight of one period lagged versus current signals.

Proof: In the appendix

The faster depreciation of knowledge and the reduction in K is the result of a faster depreciation of technologically specific knowhow when the step-size *u* is larger. This is the downside of rapid technological upgrading - we never stick with anything long enough to understand it perfectly, and when we upgrade to a relatively advanced method, we find that we do not understand it too well. The *upside* of a larger *u* is that our log-run productivity growth is higher.

5.8 Conclusion

In this paper, we have postulated an explicit formulation of how information enters the production function, and we have used it to answer two questions. The first question concerns the distinction between basic and applied research, and how differences in firms' research portfolios might in be related to their productivities. We have shown how different research portfolios can be an optimal policy for similarly situated firms, and that firms that follow different policies can coexist in a long-run equilibrium. We have shown how basic research can seem to have a higher rate of return than applied research. The reason is that firms doing basic research are upgrading their technologies at a faster rate than firms doing applied research.

The second question we have looked at is the construction of the firms knowledge stock. We find that as a first approximation, one can assume (as is typically done in practice) that the lags decay geometrically. The decay rate - the rate of obsolescence - is not exogenous, however. It depends on how fast new technological grades are coming into use. Qualitatively, this point is well known, of course; we have simply derived an explicit formula for how obsolescence works.

5.9 Appendix

First, we prove the claim made in section 3.2, that a myopic firm will always research that grade which it is about to use to produce with. Surprisingly, under some conditions, the firm could extract maximal information about a technological grade by researching a *different* grade! Further assumptions are needed to rule out this paradoxical case. They are provided in the following Proposition:

Proposition A.1. (Monotonicity of research grades): Fix any date t. Suppose that the status quo grade (i.e., the highest grade that the firm has ever operated) is N. Fix any grade $n \geq N$. Let Var $[\theta_n | y_k]$ denote the posterior variance on grade n conditional on researching grade k. Then under condition (M) below, Var $[\theta_n | y_k]$ is non-increasing in k for all $k \leq n$ and is strictly increasing in k for all $k \geq n$, and hence attains a minimum at $k = n$. In particular, condition (M) guarantees that to minimize x_n the firm will choose to research that grade $k = n$, and not any other grade.

Proof: This straightforward but lengthy proof is available on request.

Figure 1 is drawn under condition (M) and shows how the productivity, Eq_n, on a fixed grade n varies as the research grade k varies. Condition (M) ensures that the myopic firm will optimally choose to research that grade on which it is about to produce.

Figure 1: the productivity, $E[q_n | y_k]$ as a function of the research grade k

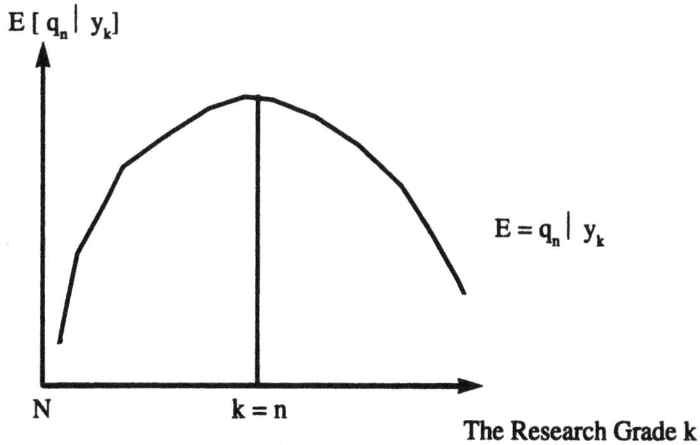

(M) *Either* (i) $\alpha < 1$, $\sigma_r^2 - \sigma_\epsilon^2/(\alpha-1) < 0$ and $x_0 < \hat{x} \equiv \sigma_\epsilon^2/(1-\alpha)$.
or (ii) $\alpha=1$.
or (iii) $\alpha>1$ and $\sigma_r^2 - \sigma_\epsilon^2/(\alpha-1) < 0$.

Proof of Theorem 6.a: Consider x_1 as fixed. Then from **(28)** we may consider x'_N a function, F say, of the sequence of effective signals: $x'_N = F(R_1,...R_N)$. Fix an R. Let $\tilde{x}$ be the fixed point of the map $h_2(h_1^R(.))$, so that $\tilde{x}= h_2(h_1^R(\tilde{x}))$. Hence when $\tilde{x}_1=\tilde{x}$ and $R_m = R\ \forall m$, $F(.)=\tilde{x}$; when $\tilde{x}_1$ is close to $\tilde{x}$, so too will F(.) be close to $\tilde{x}$. The Taylor series expansion of $\tilde{x}_N$ around R_m=R for m=1,...,N, is therefore:

$$\tilde{x}_N = \Sigma^N_{m=1}(\partial\tilde{x}_N / \partial R_m)R_m + C \text{ where}$$

(39) $$C = \tilde{x} - \Sigma^N_{m=1}(\partial\tilde{x}_N / \partial R_m)R$$

and where $\partial\tilde{x}_N/\partial R_m$ is evaluated at R_m=R for m=1,...,N.

Given any function f(x) of x let $f'(x) \equiv \partial f(x)/\partial x$; i.e., a prime denotes a derivative with respect to the x variable. Define

$$A \equiv - \partial h_1{}^{\tau}(\tilde{x})/\partial\tau \quad \text{evaluated at } \tau = R \text{ and}$$

$$\delta \equiv \alpha h_1{}^{R'}(\tilde{x}) .$$

Then it is easy to check that when evaluated at R_m=R for m=1,...,N and at $x_1 = \tilde{x}$,

$$\partial\tilde{x}_N/\partial R_m = -A\delta^{N-m}.$$

It therefore remains only to find expressions for A and δ. It is easy to verify that

$$A \equiv - \partial h_1{}^{\tau}(\tilde{x})/\partial\tau = [h_1{}^{\tau}(\tilde{x})]^2/\sigma^2 \text{ and } \delta \equiv \alpha h_1{}^{R'}(\tilde{x}) = \alpha[h_1{}^R(\tilde{x})/\tilde{x}]^2.$$

The first equality in Theorem 6.a. therefore follows from this and **(39)**. The second equality in Theorem 6.a. follows from the definition of R_m.

It therefore remains only to show that δ<1. First, since $h_1{}^R(\tilde{x}) < \tilde{x}$, δ<1 whenever α≤1. Next, since $\tilde{x}$ is a fixed point, $\tilde{x}= \alpha h_1{}^R(\tilde{x})+ \sigma_\epsilon{}^2$ so $h_1{}^R(\tilde{x}) = (1/\alpha)\ [\tilde{x}-\sigma_\epsilon{}^2]$. Hence $\delta = \alpha h_1{}^R(\tilde{x})^2/\ \tilde{x}^2 = (1/\alpha)\{[\tilde{x}- \sigma_\epsilon{}^2\ ]/\tilde{x}\}^2$. Since $\tilde{x}-\sigma_\epsilon{}^2 \geq h_2(0)- \sigma_\epsilon{}^2>0$, this latter expression is less than one whenever α< 1.

Proof of Theorem 6.b: If all the future signals are concentrated on one grade, grade N+k, we may write with abuse of notation $x_N'' = g(x_N', \rho_{N+k})$. A linear expansion of this around some $(\tilde{x},\tilde{\rho})$ results in:

$$x_N'' = c_1x_N' - c_2\rho_{N+k} + \bar{c}$$

where

$$c_1 = \partial g/\partial x, \qquad c_2 = - \partial g/\partial\rho \qquad \text{both evaluated at } (x,\rho)=(\tilde{x},\tilde{\rho}) \text{ and}$$
$$\bar{c} = g(\tilde{x},\tilde{\rho}) - c_1\tilde{x} + c_2\tilde{\rho} .$$

By definition the capital without future signals is $K_t^* = 1- \sigma_w{}^2-x_N'$. Hence $x_N' = 1-\sigma_w{}^2-K_t^*$. The research capital with *future* signals is therefore

$$K_t^{**} = 1 - \sigma_w^2 - x_N'' = 1 - \sigma_w^2 - \{c_1(1 - \sigma_w^2 - K_t^*) - c_2\rho_{N+k} + \bar{c}\}$$
$$= c_1K_t^* + c_2\rho_{N+k} + c_3$$

where $c_3 = (1-c_1)(1 - \sigma_w^2) - \bar{c}$. This proves Theorem 6.b.

Proof of Proposition 7.a: Since it depends only on σ_w^2 (which is the same for the two firms) the map h_1 does not depend on the grid size. But the $h_2(x)$ map does. Performing *u* iterations of h_2 from x, and calling the resulting map $h_2^u(x; B)$, gives, for the parameters faced by firm B,

$$h_2^u(x; B) = \alpha_B^u x + [(1-\alpha_B^u) / (1-\alpha_B)] \sigma_{B\epsilon}^2$$

This is how firm B's x changes when it upgrades by *u* steps. For firm A, on the other hand, $h_2(x; A) = \alpha_A x + \sigma_{A\epsilon}^2$. Given the relation between the two sets of parameters assumed in the proposition, we see that for each x, $h_2(x; A) = h_2^u(x, B)$. And since the map h_1 is the same for them both, their (x_t) sequences will be identical if they have the same set of effective signals.

Proof of Proposition 7.b:

(i) In **(31)**, h_1 does not depend on u. Also, $h_1(x) \leq x$, and $h_2(x) < \hat{x}$ whenever $x < \hat{x}$. Hence the argument in each of the h_2 functions on the right-hand-side of **(31)** is less than $\hat{x}$. Then since $h_2^u(x) = \alpha^u x + [(1- \alpha^u)/(1-\alpha)] \sigma_\epsilon^2$ is increasing in *u* whenever $x < \hat{x}$, x'_N must be increasing in *u*.

(ii) The parameters *a* and *b* are the relative weights on the learning by doing versus the research signals. These are independent of α and σ_ϵ^2, and hence independent of u.

(iii) The formula for δ is

(40) $$\delta \equiv \alpha^u h_1^R(\tilde{x}_u)^2/\tilde{x}_u^2 = \alpha^u[\sigma_w^2/(\sigma_w^2+R\tilde{x}_u)]^2 .$$

Now, $\tilde{x}_u$ is the fixed point of the function $h_2^u(h_1(x))$. This is a strictly increasing and strictly concave function of x which is strictly positive at x=0.

Further, $h_2^u(h_1(\hat{x})) < \hat{x}$. Hence $\tilde{x}_u$ lies in $(0,\hat{x})$. Also, $h_2^u(x)$ is increasing in *u* for each $x \in [0,\hat{x}]$. Hence $\tilde{x}_u$ is strictly increasing in *u*. (Indeed, one can prove that $\tilde{x}_u$ increases monotonically to $\hat{x}$.) Hence from **(40)** we conclude that when $\alpha<1$, δ is decreasing in *u*.

References:

Adams, James, "Fundamental Stocks of Knowledge and Productivity Growth," *Journal of Political Economy* 98 (1990): 673 - 702.

Agarwal, Rajshree, and Michael Gort, "The Evolution of Markets and Entry, Exit, and Survival of Firms," Discussion paper 9307, SUNY Buffalo, August 1993.

Bakh, Byong-Hyong and Michael Gort. "Decomposing Learning by Doing in New Plants". *Journal of Political Economy* 101, no. 4 (August 1993): 561 - 583.

Bhattacharya, Sudipto, and Dilip Mookherjee, "Portfolio Choice in Research and Development," *The Rand Journal of Economics* 17, no. 4 (Winter 1986): 594-605.

Caballero, Ricardo, and Adam Jaffe, "How High are the Giants' Shoulders: An Empirical Assessment of Knowledge Spillovers and Creative Destruction in a Model of Economic Growth," *NBER Macroeconomic Annual*, 1993.

Chari, V.V., and Hugo Hopenhayn, "Vintage Human Capital," *Journal of Political Economy* 99, no. 6 (December 1991): 1142 - 65.

Chow, Gregory, *Econometrics*, New York: McGraw-Hill, 1983.

Doms, Mark, Timothy Dunne, and Mark Roberts, "The Role of Technology Use in the Survival and Growth of Manufacturing Plants," U.S. Bureau of the Census, Center for Economic Studies, 1994.

Evenson, Robert, and Yoav Kislev, *Agricultural Research and Productivity*, New Haven: Yale University Press, 1975

Fellner, William, "Trends in the Activity Generating Technical Progress," *American Economic Review* 60, no. 1 (March 1970): 1 - 29.

Foster, Andrew, and Mark Rosensweig. 1994. "Learning by Doing and Learning from Others: Human Capital and Technical Change in Agriculture". Unpublished paper, University of Pennsylvania.

Griliches, Zvi, "Research Expenditures and Growth Accounting," in *Science and Technology in Economic Growth*, B.R.Williams, ed., MacMillan Press, 1973.

Griliches, Zvi, "Productivity, R&D, and Basic Research at the Firm Level in the 1970's," American Economic Review 76, no. 1 (March 1986): 141 - 54.

Hall, Bronwyn, "The Impact of Corporate Restructuring on Industrial R&D," *Brookings Papers on Economic Activity: Microeconomics* (1990): 85 - 124.

Hall, Bronwyn, "The Private and Social Returns to R&D: What Have We Learned?" unpublished paper, UC Berkeley, April 1995.

Hulten, Charles, and Frank Wykoff, "The Estimation of Economic Depreciation Using Vintage Asset Prices," *Journal of Econometrics* 15 (April 1981): 367 - 96.

Jovanovic, Boyan, and Yaw Nyarko. 1994".The Bayesian Foundations of Learning by Doing", New York University. NBER working paper no. 4739.(J-N 1994)

Jovanovic, Boyan, and Yaw Nyarko, "A Bayesian Learning Model Fitted to a Variety of Empirical Learning Curves", *Brookings Papers on Economic Activity* (Microeconomics Issue no.1), 1995.(J-N 1995a)

Jovanovic, Boyan, and Yaw Nyarko, "The Transfer of Human Capital", *Journal of Economic Dynamics and Control* 19 (June 1995):1033 - 1064 (J-N 1995b).

Lach, Saul, and Mark Schankerman, "Dynamics of R&D and Investment in the Scientific Sector", *Journal of Political Economy* 97, no. 4 (August 1989): 880 - 904.

Lichtenberg, Frank, and Donald Siegel, "The Impact of R&D Investment on Productivity - New Evidence Using Linked R&D-LRD Data", *Economic Inquiry* 29 (April 1991): 203 - 228.

Link, Albert, "Basic Research and Productivity Increase in Manufacturing: Additional Evidence", *American Economic Review* 71, no. 5 (December 1981): 1111 -2.

Mansfield, Edwin, "Basic Research and Productivity Increase in Manufacturing", American Economic Review 70, no. 5 (December 1980): 863 - 73.

Mansfield, Edwin, "Composition of R&D Expenditures: Relationship to Size of Firm, Concentration, and Innovative Output", *Review of Economics and Statistics* (1981): 610 - 615.

Miyazaki, Kumiko, "Search, Learning and Accumulation of Technological Competences; The Case of Optoelectronics", *Industrial and Corporate Change* 3, no. 3 (1994): 631 - 654.

National Science Foundation, *Methodology of Statistics on Research and Development*, Washington, D.C., 1959.

Nelson, Richard "Modelling the Connection in the Cross Section Between Technical Progress and R&D Intensity", *Rand Journal of Economics* 19, n. 3 (1988): 478-488

Pakes, Ariel, and Mark Schankerman, "The Rate of Obsolescence of Patents, Research Gestation Lags, and the Private Rate of Return to Research Resources", in *R&D, Patents, and Productivity*, Zvi Griliches, ed., University of Chicago Press, for NBER, 1984.

Oliner, Stephen, "New Evidence on the Retirement and Depreciation of Machine Tools", Working paper no. 147, Board of Governors, December 1993.

Trajtenberg, Manuel, Rebecca Henderson, and Adam Jaffe",Ivory Tower Versus Corporate Lab: An Empirical Study of Basic Research and Appropriability", NBER Working Paper no. 4146, August 1992.

6. University Patenting Amid Changing Incentives for Commercialization

Rebecca Henderson, Adam B. Jaffe, Manuel Trajtenberg*

Modern economies are characterized by a significant division of labor in the production of scientific and technical advance. In broad generalization, universities, other non-profit institutes and government laboratories pursue fundamental scientific and engineering research with little expectation of immediate commercial result. In contrast, private firms undertake applied research and development intended to improve the products and processes available for commercial use. Traditionally, these different missions have been largely carried out within incentive/reward systems that differ significantly between the two realms. University and government research is carried out largely with public funding, with the individual researchers receiving much of their "compensation" in the form of recognition within the profession, derived from publication of research results. Scientists and engineers working in firms are financed by those firms, and are more likely to receive financial rewards and less likely to receive wide public recognition for their discoveries.

The two research worlds have never been as distinct as this broad-brush portrait suggests, and there appear to be important trends underway that are blurring the distinctions between the two worlds and bringing elements of each's incentive and reward system to bear in the other's realm. An important aspect of this evolution, at least in the U.S., is the increasing role of universities in the commercialization of new technology. A crucial aspect of the increased emphasis on commercialization at U.S. universities was a statutory change initiated by the U.S. Congress in 1980. The Bayh-Dole Act and subsequent amendments removed most significant restrictions on the ability of universities to earn commercial royalties on inventions derived from publicly funded research.

The motivation for this change in the rules was a belief that potentially valuable commercial byproducts of publicly-funded research were going undeveloped because of a lack of clear property rights in these inventions. Ideas with *potential* commercial value that grow out of university research typically require significant subsequent investment to move them from the laboratory to the marketplace. If the underlying technical idea cannot be protected (because it derives from public research), then the university and its scientists have little incentive to try to find a commercial developer, and such potential developers have little or no incentive to invest the resources necessary to make the transition to commercialization. The

---

* Massachusetts Institute of Technology, Cambridge (USA) and NBER; Brandeis University, Waltham (USA) and NBER; Tel Aviv University and NBER

Bayh-Dole Act aimed to change these incentives, by allowing universities to license publicly-derived technologies to private firms, and to derive license royalties therefrom. Thus the Bayh-Dole Act provides an interesting opportunity to observe the extent to which the incentive/reward environment affects the way institutions behave. Understanding how this change in incentives has affected university behavior is important both for direct evaluation of the specific policy change, and more generally for understanding how incentives affect behavior related to the development of new technology.

There can be little doubt that the Bayh-Dole Act has affected university patenting behavior. In *1965*, just 96 U.S. patents were granted to 28 U.S. universities or related institutions. In 1992, almost 1500 patents were granted to over 150 U.S. universities or related institutions. This 15-fold increase in university patenting occurred over an interval in which total U.S. patenting increased less than 50 percent, and patents granted to U.S. inventors remained roughly constant. This dramatic increase in university patenting, much (but not all) of which has occurred post-Bayh-Dole, is widely cited as evidence of the success of Bayh-Dole. From a policy point of view, however, we would like to understand the extent to which the increased patenting represents a large increase in the universities' output of commercially valuable inventions, and the extent to which it represents simply increased filing of patent applications for a given number of new inventions.

This paper explores this issue in some detail, both as a phenomenon of interest in its own right and as a window into the changing role of universities as a source of technology for the private economy. A number of surveys and some detailed case study work has documented substantial shifis in the nature of the relationship between universities and the private sector (Blumenthal, 1986; Cohen, Florida and Goe, 1994; Argyres and Liebeskind, 1996). Here we focus on university patents, both because patents are a unique and highly visible method of "technology transfer" (Archibugi, 1992; Basberg, 1987; Boitani and Ciciotti, 1990; Schwartz, 1988; Trajtenberg, 1990a), and becanse their accessibility allows for a more comprehensive analysis than is possible with either surveys or case study work. Understanding the ways in which university patenting behavior has changed over the last twenty years is an important component of understanding the ways in which the relationship between universities and the private economy has changed (David, Mowery and Steinmueller, 1992; Mansfield, 1991; National Science Foundation, 1982; Pavitt, 1991).

We draw on a comprehensive database consisting of all patents assigned to universities or related institutions from 1965 until mid 1992; a one percent random sample of all U.S. patents granted over the same time period, and the complete set of all patents that cite either of these groups. Together these amount to a data base of over 100,000 patents that includes detailed information about each patent and about the linkages between them as revealed by patterns of patent citation.

Drawing on previous work, we construct measures of the "importance" and "generality" of university patents, and show that, averaged over the whole time period, university patents are both more important and more general than the average patent, but that this difference has been declining over time, so that by the late 1980s we cannot find significant differences between the university patent universe and the random sample of all patents. We show that this decline has occurred across the broad spectrum of technological areas, that it does not appear to be due to any shift in patenting away from more highly cited towards less highly cited fields or to any "truncation bias" resulting from our inability to observe future citations, and that it is not the result of the spread of patenting to smaller institutions. Instead it appears to reflect a lowering of the patent "threshold" at universities so that an increasing number of marginal patents are being filed. A major factor in the decline in the difference between universities and random sample patents is a change in the fraction of university patents receiving no citations. In the mid 1970s, a random patent was 5 times as likely to get zero citations as a university patent. By the mid 1980s this difference had essentially disappeared. Thus a large fraction of the overall increase in university patents is due to a significant increase in the fraction of uncited patents. This suggests that the observed increase in university patenting may reflect an increase in their "propensity to patent" rather than an increase in the output of "important" inventions.

The paper is organized into five sections. The following section describes in more detail the data we are using, and explains some of the institutional changes that appear to be related to the growth of patenting. Section III demonstrates the difference between university and other patents in the citation-based measures of importance, and the decline of that difference in the 1980s. Section IV explores possible explanations for that decline. Section V provides concluding observations.

6.1 The Growth of University Patents

6.1.1 The Basic Numbers

At first glance, a focus on university patents might seem a strange route to better understanding of the changing relationship between universities and the private sector. University patents are a small fraction of all patents, and since in general universities are in the business of the creation and free dissemination of knowledge "for its own sake," one cannot expect to learn about the full spectrum of university research from their patenting behavior. Yet university patents are informative. They reflect research that the university believes has direct commercial application, and thus changes in patenting behavior may reflect both changing motives within the university research community and shifts in the focus of university research from basic to more applied work.

This paper is part of a larger research project that exploits the declining cost of access to large quantities of patent data. In other work we have used patent data to show that spillovers are geographically localized (Jaffe, Trajtenberg and Henderson, 1993), that spillovers from university research are less likely to be geographically localized than privately funded research (Henderson, Jaffe and Trajtenberg, 1996) and to explore the degree to which citation based measures provide useful information about the scientific and economic impact of the idea captured in a patent (Trajtenberg, Henderson and Jaffe, 1996). Here on we draw on these data to observe the quantity of university patenting, as well as citation-based proxies for the "quality" of those patents, to examine how the quantity and quality have changed over time, and to compare both to the overall universe of U.S. patents.[1]

We have some information on all patents granted since 1965, and complete information on all patents granted since 1975.[2] Appendix (1) describes in detail the data that are used in this paper. We use four sets of patents: all patents assigned to universities that were granted between 1965 and mid-1992 (12,804 patents); a one-percent random sample of all U.S. patents[3] over the same period (19,535 patents); all patents after 1974 that cited the university patents (40,859 patents) and all patents after 1974 that cited the random sample patents (42,147 patents). For all of these patents we know the year of application,[4] the identity of the institution to which it is assigned, and the "patent class," a detailed technological classification provided by the patent office.

Figure 1 illustrates the dramatic increase in patenting we have already described. Panel A compares the rate of university patenting to all U.S. patents and to domestic U.S. patents. Panel B shows university patenting relative to university research, and an analogous ratio for the U.S. industrial sector. University patenting has not only increased, it has increased more rapidly than overall patenting and much more rapidly than domestic patenting, which is essentially flat until the late 1980s. In addition, university patenting has increased more rapidly than university research spending, causing the ratio of university patents to R&D to more than triple over the period. In contrast, while domestic patenting was roughly constant, industry research spending increased significantly, particularly between the mid-70s and the mid-80s, so that the ratio of domestic patents to domestic R&D nearly halved over the period. Said differently, universities' "propensity to patent," given their research effort, has been rising significantly at the same time that the overall

[1] In our earlier paper "The Bayh-Dole Act and Trends in University patenting 1965-1988" we documented the existence of this decline. However in that paper we were not able to control for problems such as truncation bias or shifts in citation patterns over time, and we were not able to explore the causes of the decline in any detail.

[2] There is no publicly available information on unsuccessful patent applications.

[3] By "U.S. patents" we mean patents granted by the U.S. patent office. By the end of this period, about half of such patents were granted to non-U.S. residents. About 1 percent of the patents assigned to U.S. universities were taken out by individuals who gave the Patent Office non-U.S. addresses.

[4] We prefer to date patents by the year of application rather than the year of grant, because that is when the inventor identified the existence of a new invention, and there are variable lags involved between application and grant date. Because of these lags, however, totals by date of application are incomplete for years approaching the 1992 data cutoff date, since some patents applied for at the end of the period were almost certainly still under review at the time we collected our data. Thus we terminate our time-sensitive analyses in 1988.

Figure 1

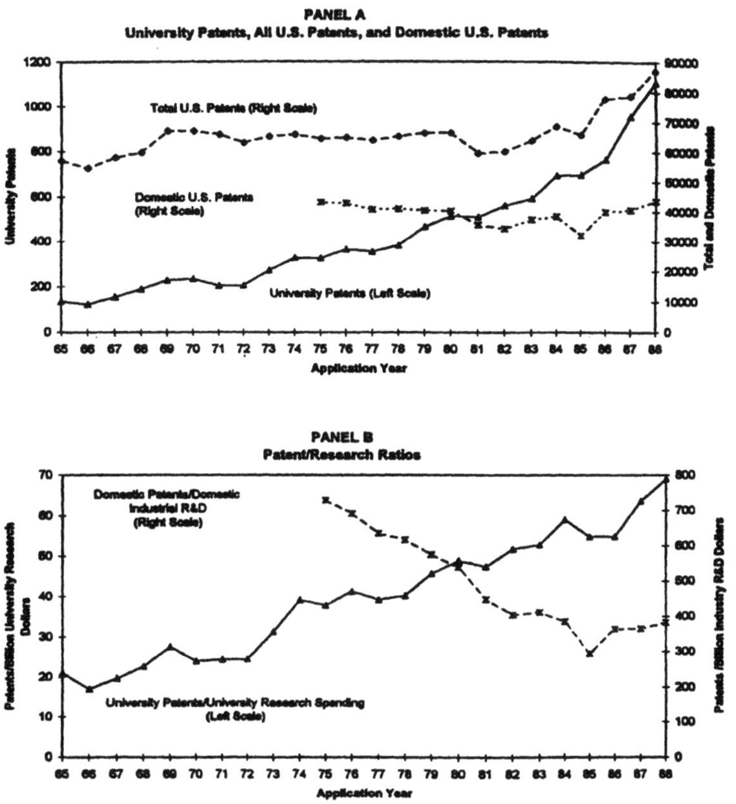

propensity to patent has been falling. Note that the increase in university patenting has been fairly continuous since the early 1970s. There is some evidence of an acceleration in the late 1980s, but this is a period in which both university research and overall patenting accelerate as well, making it difficult to assess its significance.

The increasing propensity of universities to patent is also evident in a significant increase in the number of universities and related institutions taking out patents. Whereas in 1965 about 30 universities obtained patents, in 1991 patents were granted to about 150 universities and related institutions. None the less, university patenting remains highly concentrated, with the top 20 institutions receiving about 70 percent of the total, and M.I.T., the most prolifically patenting institution, alone receiving about 8 percent. The top 10 institutions and their total patent grants for 1991 are shown in Table 1.

The increase in university patenting has not been uniform across the spectrum of technologies. To examine this issue, we define 5 broad classes of technologies, based on the much finer classification used by the patent office:[5]

[5] Full details of this classification are given in Jaffe, Adam B. (1986) "Technological Opportunity and Spillovers of R&D: Evidence from Firms' Patents, Profits and Market Value," *American Economic Review*, December 1986, pp 984-1001.

Table 1. Top Ten Institutions for University Patents, 1991

Institution	Patent Count
Massachusetts Institute of Technology	100
The University of California	91
The University of Texas	82
Stanford University	56
Wisconsin Alumni Research Foundation	44
University of Florida	43
Iowa State University Research Foundation Jnc	39
California Institute of Technology	32
University of Minnesota	30
Johns Hopkins University	26

- *Drugs and Medical Technologies ("Drugs/medical")*
- *Chemicals and Chemical Technologies Excluding Drztgs ("Chemical")*
- *Electronic, Electrical, Optical and Nuclear Technologies ("Electronics, etc.")*
- *Mechanical Technologies ("Mechanical") and*
- *All Other.*

Panel A of Figure 2 shows the breakdown of university patents into these fields over time; Panel B shows it for all patents.[6] The differences are dramatic, if not surprising. As of the end of the 1980s, Drug and Medical patents comprise about 35 percent of the university total, up from less than fifteen percent in 1965; chemical patents 25-30 percent; electronic and related patents 20-25 percent; mechanical patents 10-15 percent; and about 5 percent other. In contrast, overall patenting is 30-35 percent mechanical; 20-25 percent each for chemical and electronics, 10-15 percent other, and less than 10 percent drugs and medical. Thus universities are much more interested in drugs and medical technologies, and much less interested in mechanical technologies, than other inventors, and the difference has increased over time.

6.1.2 The Broader Context of Increased University Patenting

There are several possible explanations for this dramatic increase in patenting behavior. Changes in federal law affecting university patenting in 1980 and then again in 1984 made it significantly easier for universities to patent the results of federally funded research. Industry funding of university research has notably increased and at the same time there has been a substantial increase in organized university "technology transfer" or "licensing" offices. Since all three changes

[6] This and all subsequent analyses are based on our 1/100 random sample of all patents. Given the large number of such patents (over *500* per year), the composition by field of the sample is very likely to be close to the composition by field of the universe of all patents. Note that the random sample does *not* exclude university patents. Even at the end of the period, however, university patents are less than *2* percent of the total.

Figure 2. Panel A: University Patents by Broad Fields.

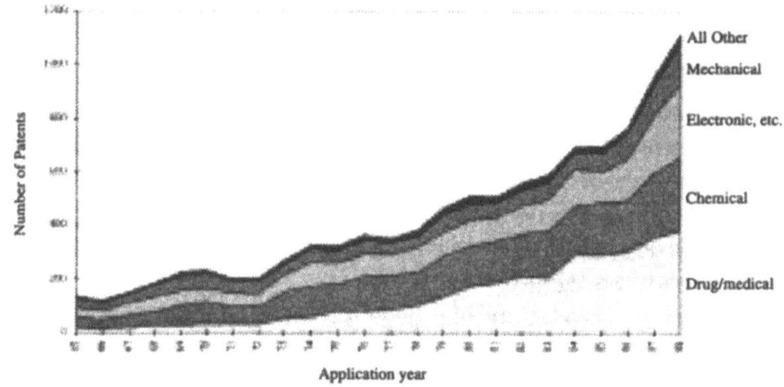

Figure 2. Panel B: All U.S. Patents by Broad Field.

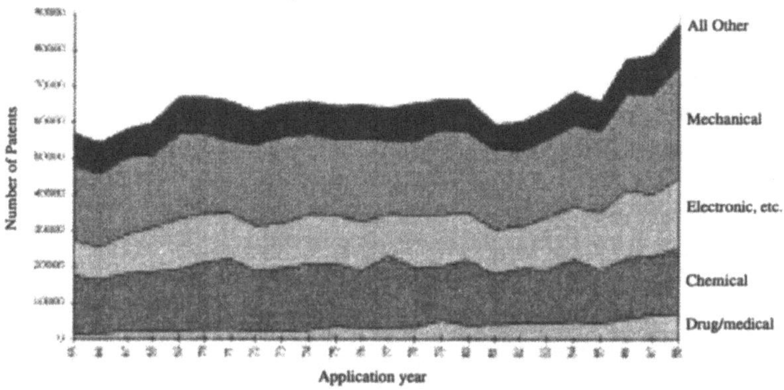

occurred roughly simultaneously their different effects cannot be easily separated, but it seems plausible that all three have played an important role in increasing the number of university patents.

Federal law affecting University patenting. Before 1980 the federal government had the right to claim all royalties or other income derived from patents resulting from federally funded research. Federally funded researchers could apply for patents, and could assign those patents to universities, but the exclusive property right associated with the invention remained with the government whether or not a patent was issued. The only way that a university could profit from federally-derived patents was to seek a title rights waiver from the funding agency. Some agencies provided such waivers, but their availability and the procedures that had to be followed varied from agency to agency, and were generally cumbersome. Since approximately 70 percent of university research during this period was funded by the federal government, this was a major barrier to widespread university patenting. University patents from this period thus represent either (1) the fruits of that portion of university research not funded by the federal government; (2) patents sought only for the public or professional prestige that they confer, rather than their commercial profit potential; or (3) inventions from federal research for which a specific title rights waiver was sought and received from the federal agency funding the research.

In 1980 Congress passed The Patent and Trademark Amendments of 1980 (Public Law 96-517), also known as the Bayh-Dole Act. The Bayh-Dole Act gave universities (and other non-profit institutions, as well as small businesses) the right to retain the property rights to inventions deriving from federally-funded research. The 1984 passage of public Law 98-620 expanded the rights of universities further, by removing certain restrictions contained in Bayh-Dole regarding the kinds of inventions that universities could own, and the right of universities to assign their property rights to other parties.

Both pieces of legislation may have reflected concern that the results of university research were not being effectively transferred to the private sector (U.S. General Accounting Office, 1987). If inventions often burst forth from federally funded research in a form that permitted immediate commercial application without significant development expenditure, continued federal possession of the associated property rights would be a reasonable policy stance to take toward new knowledge generated at public expense. However few inventions come out of basic research in a form that permits immediate commercial application, and significant additional investment is typically required to transform the results of university research into marketable products. The proponents of the Bayh-Dole Act reasoned that no private firm is likely to make such investments if the ultimate property right in the underlying invention is publicly held.

Thus since 1984, universities have had very broad rights to exploit inventions derived from their research, even if it is federally funded. They can charge royalties for the use of the patent, and they can assign the patent to a third party if they so desire. As a result, major research universities now typically have explicit policies requiring faculty and other researchers to assign patents deriving from on-campus research to the university, and specifying how any income deriving therefrom is to be divided among the institution, the researcher and research centers or departments.

An increase in organized university "technology'" offices. Though it is obviously difficult to separate the chicken from the egg, after the passage of Bayh-Dole there has been a dramatic increase in the scale and significance of the patenting and technology licensing function at universities. The Association of University Technology Administrators (AUTM) has recently begun conducting surveys of its members. The surveyed institutions[7] employed 767 full-time equivalent professional employees in technology transfer and licensing activities. In 1993, they received royalties totally about $375 million on about 4016 licensing agreements; more than 4000 additional active agreements were not currently generating revenue.[8]

---

[7] Survey responses came from 112 U.S. institutions that were granted 1169 patents in Fiscal Year 1992, compared to our data that indicates about *1500* patents were granted to over *150* institutions. Thus survey totals are lower bounds on the actual numbers.

[8] *The AUTM Licensing Survey. Fiscal Years 1991, 1992 and 1993.* AUTM categories

Jncreased industry funding of university research. Another factor related to the increase in university patenting is an increase in industry funding of university research from 2.6 percent in 1970, to 3.9 percent in 1980 and 7.1 percent in 1994.[9] This trend is presumably the result of both an increase in the supply of such funds, and in the "demand" in the sense of an increased desire by universities to do the kinds of applied research that interests industry. This desire might be driven either by the universities' increasing interest in applied research, *per se,* or by the perception that a decline in the rate of growth of federal support for university research has made industry funds an increasingly necessary source of support.

Thus the historical record provides us with a series of mutually reinforcing trends: An increased applied and commercial focus of universities, as evidenced by a rapidly increasing propensity to patent; a change in federal law that facilities university patenting; an expansion of the institutional ability and focus on patenting in the form of university technology offices; and an increased role for industry funding at the university. It is impossible to assign the roles of "cause" and "effect" to these different trends. The increase in university patenting predates the passage of Bayh-Dole, but continued exponential growth probably could not have been sustained without removal of the cumbersome barriers to patents from federal research. The increase in universities' institutional commitment to patenting, in the form of new and expanded licensing offices, would likely not have occurred if the impetus towards more commercial research, and the change in federal law had not occurred. But once created, these offices presumably facilitate the patent application process and thereby contribute to the increased patenting.. Finally, increased industry funding is probably partially a response to universities' increased interest in applied research, but it, in turn, increases the resources for these activities and thereby also supports increased patenting.

Table 2 illustrates this close correlation quantitatively. For the 113 universities reporting comprehensive data to AUTM it presents correlations across patenting rates, employees in the licensing office, invention disclosures, gross royalties and the level of industrial and publicly funded support. In this cross sectional data patenting rates are less correlated with levels of industry funding than with levels of public funding, disclosure rates or the size of the licensing office, suggesting that increased industry funding may be less important in driving patenting behavior than changes in the law and the expansion of technology licensing offices, but the high degree of serial correlation evident in the raw longitudinal data make it impossible to draw any firm conclusions as to the relative importance of these various factors.

---

included in the quoted totals are U.S. Universities, U.S. Hospitals, Third-Party Management Firms and Research Institutes. Excluded are government and Canadian universities. The royalty total has been adjusted to eliminate double counting that results from shared license agreements (personal communication, Ashley Stevens, AUTM).

[9] With federal funding at 60 to 70 percent of the total, the remainder is funded by state and local governments and institutions' own funds.

6.2 Characterizing University patenting

6.2.1 Citation-based Measures of Importance and Generality

There is evidence that the flow of technology out of universities contributes to technological innovation in the private sector (Jaffe, 1989), and a widespread belief that more effective transfer of technology from universities to the private sector would be beneficial to innovation and growth (U.S. GAO, 1987; National Academy of Sciences 1995). In this light, the rapid increase in university patenting appears to be a highly desirable trend, to the extent that it signals an increase in the successftil commercial application of university-derived technology. It is widely known, however, that patents vary tremendously in their importance, making it dangerous to draw conclusions about aggregate technology flows based on numbers of patents (Griliches, 1990). In this section, we look more carefully at the university patents, to understand better what the patent data do and do not say about increases in the flow of technology out of universities.

In an earlier paper (Trajtenberg, Henderson, and Jaffe, 1996), we used patent citation data to construct a variety of measures that we interpreted as capturing the importance or 'basicness" of the invention covered by a patent. This approach builds on earlier work that suggests that citation patterns are significantly correlated with both the economic and scientific importance of the invention they cover (Trajtenberg, 1990a, Carpenter and Narin, 1993). Implicit in this approach is a view of technology as an evolutionary process, in which the significance of any particular invention is evidenced, at least partly, by its role in stimulating and facilitating ftiture inventions. We assume that at least some of such future inventions will reference or cite the original invention in their patents, thereby

Table 2. Correlation Coefficients across key patent related variables, 1993 data

	FTEs for Licensing Activities	Disclosures	Gross Royalties	Industry support	Public support
Total US	0.88	0.91	0.71	0.69	0.82
Patents filed	(113)	(112)	(112)	(110)	(113)
FTEs		0.84	0.81	0.61	0.86
		(112)	(112)	(110)	(113)
Disclosures			0.72	0.66	0.83
			(112)	(109)	(112)
Royalties				0.53	0.71
				(109)	(112)
Industry support	0.64 (110)				

Source: The Autm Licensing Survey. Figures in parentheses are number of observations. The number of observations varies because not all universities participating in the survey provide comprehensive data.

making the number and character of citations received a valid indicator of the technological importance of an invention.[10]

The first citation-based measure that we introduced in the previous paper we called "importance." We define "importance" as:

$$\text{Importance}_i = \text{Nciting}_i + \lambda \sum_{j=1}^{\text{nciting}_i} \text{Nciting}_{i+1j}$$

where $0 < \lambda < 1$ is defined as an arbitrary "discount factor" that in the previous paper we set to *0.5*. In the absence of data about "second generation citations in the data set on which this paper relies, we set λ equal to zero, and measure importance simply by total citations received. We hypothesize that this measure captures the technological impact of an invention as reflected in the number and importance of its descendants (Carpenter and Narin, 1993; Trajtenberg, 1 990a).

The second citation based measure that we use is "generality." We hypothesize that patents that cover more "basic" research will be cited by work in a broader range of fields, and define "generality" as:

$$\text{General}_i = 1 - \sum_{k=1}^{N_i} \left(\frac{\text{Nciting}_{ik}}{\text{Nciting}_i}\right)^2$$

where k is the index of patent classes and N_i is the number of different patent classes to which the citing patents belong. Notice that $0 \leq \text{General} \leq 1$, and that higher values represent less concentration and hence more generality. In our previous paper we were able to show that both of these measures were reassuringly high for a number of patents that are known to have had a very significant impact on their field.

Citation-based measures of importance and generality are, to some extent, influenced by variations in citation practices across time and technological areas. They are also very influenced by the fact that when we count the citations seen to date of a patent issued in, for example, 1989, we are missing many more of the citations that it will ultimately receive than we are missing in our count of the citations of a patent issued in 1975. For these reasons, when comparing importance or generality it is necessary to control for time and technological field effects. The special difficulties raised by the truncation in time are discussed further below.

[10] Citations or references serve the legal function of delimiting the scope of patent protection, by identifying technological predecessors of the patented invention. Thus if patent 2 cites patent 1, it implies that patent 1 represents a piece of previously existing knowledge upon which patent 2 builds, and over which 2 cannot have a claim. The applicant has a legal duty to disclose any knowledge of the prior art, but the decision as to which patents to cite ultimately rests with the patent examiner, who is supposed to be an expert in the area and hence to be able to identify relevant prior art that the applicant misses or conceals. Trajtenberg (1990a and 1990b) showed that citation-weighted patents were a good proxy for the consumers' surplus generated by inventions. For more discussion of the value and limitations of citation data, see Trajtenberg, Henderson and Jaffe (1996)

6.2.2 Comparing University and Random Sample Patents

As a first step in exploring the degree to which the increase in university patenting rates reflects an increasing transfer of knowledge to the private sector we first explore the degree to which university patents are more important or more general than the random sample of patents and the degree to which this has changed over time.

Table 3 presents the results of regressions of our measures of importance and generality on a series of dummy variables for application years and technological areas, and dummy variables for whether or not the original patent was a university patent. These regressions are based on application years 1965-1988 for importance and 1975 to 1988 for generality.[11] Over the entire period, controlling for technological field effects and time effects, university patents received almost 25 percent more citations on average, and this difference is highly significant statistically. They were also about 15 percent more general, again a statistically significant difference. These overall averages conceal a moderate amount of variation across fields. The difference between university and random sample patents is largest in Electronics, etc., and Mechanical patents, and smallest in the Drug/medical area.[12] These results control for time effects, but they do not allow the university/random sample difference itself to vary over time. Results of regressions that allow each year-cohort of patents to have its own university/random sample difference are shown in Table 4 and again graphically in Figure 3. To make sure that the university/random sample difference is not due to the different technological focus of the two samples, the regressions reported in Table 4 replace the 5 technology field dummies used in Table 2 with 364 separate dummies for Patent Office patent classes. While the year-by-year differences are somewhat noisy, there is a clear overall trend: the university/random sample difference grew during the 1970s, reached a plateau during the period from about 1975 through about 1982, and fell significantly afier that. The regression results show that the differences between the two groups are statistically significant between 1970 (1975 for generality) and about 1982 or 1983; after that the two groups are not statistically different from one another in either generality or importance.

6.2.3 Robustness of the Apparent Decline

The results reported in Table Four appear to show that university patents were significantly more important and more general than corporate patents in the 1970s,

[11] The generality measure cannot be calculated for the pre-1975 patents because we lack information on the citing patents before 1975, and we terminate the analysis in 1988 because a significant fraction of 1989 applications might be granted after mid-1992 when our data end. Also, those granted in 1990 and 1991 would have very little time to receive citations.

[12] The differences across fields must, however, be interpreted with caution because of the changing composition over time. See discussion surrounding Figure 4, below.

but that this difference declined sharply in the early 1980s, so that by the mid-1980s at the latest there was, overall, no statistically significant difference between the two groups. In this section, we explore a number of possible explanations for this decline, focusing particularly on the degree to which this result may reflect truncation bias, or be an artifact of our use of citation data and the ways in which citation patterns have changed over time.

First, although the regressions reported in Table 4 control for field and time effects, they do not control for time-field interaction effects. For example, it could be that Drug patents have become increasingly less citation-intensive over time, relative to other fields. If so, then university patents (which are increasingly concentrated in the Drug/medical area) would appear to be increasingly less important in the sense of receiving fewer citations, because the regressions reported in Table 4 control only for the *average* level of citations in drug-related patent classes. To test for this possibility, as well as to examine whether the decline in university citation-intensity is concentrated in particular fields, we re-ran versions of the regressions of Table 4 *separately* for each of the 5 major fields, including application year and patent class dummies, and parameterizing the difference between university and random sample as a cubic function of time.

Table 3. Comparison of University and Random Sample Patents

	Importance 1965-1988 n=28313	Generality 1975-1988 n=14775
Random Sample Mean		
Drug/Medical	4.00	0.258
Chemical	3.87	0.296
Electronics, etc.	4.23	0.288
Mechanical	3.77	0.265
All Other	3.47	0.203
Overall University Difference, controlling for field	0.918	0.0452
	(0.072)	(0.0049)
University Difference by field:		
Drug/Medical	0.311	-0.0168
	(0.199)	(0.0135)
Chemical	0.416	0.0480
	(0.124)	(0.0087)
Electronics, etc.	1.718	0.0582
	(0.141)	(0.0094)
Mechanical	1.290	0.0740
	(0.153)	(0.0107)
All other	0.396	0.0148
	(0.255)	(0.0180)

Notes: Standard errors in parentheses
Differences are estimated controlling for application year effects

Figure 3

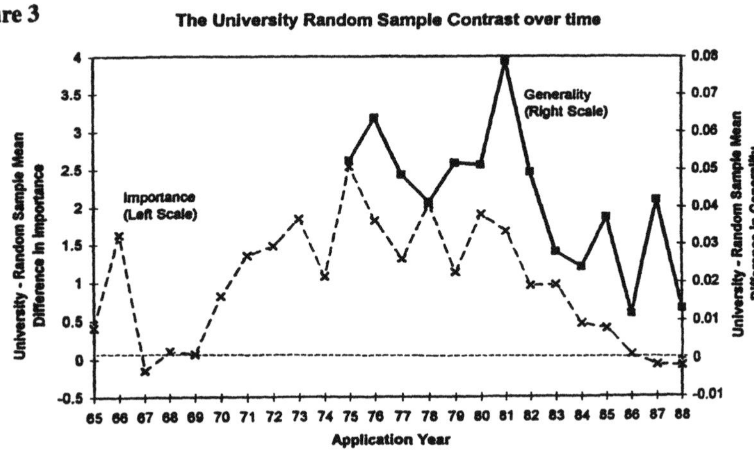

Figure 4 plots the predicted differences based on the cubic coefficients. While there are variations in timing and the magnitude of the peak, all of the areas show an initial rise, a peak somewhere between 1975 and 1981, and the disappearance of any university superiority by the second half of the 1980s.[13] It is interesting to note in this context that the Drug/medical field showed a small and insignificant overall university advantage in Table 2. Figure 4 shows that this is not the result of a consistent lack of difference; rather it results from the fact that the decline in advantage began early. This, combined with the fact that the number of university drug/medical patents is rising rapidly (so that the overall totals are dominated by recent patents) renders the overall difference insignificant. More generally, the decline in the university advantage is not a result of field effects and seems to occur across all of the fields.

The preceding regressions "control for" time effects, in the sense that we are comparing university patents applied for in a given year to random sample patents applied for in the same year. None the less, the proximity of the rapid decline in the university advantage to the end of the data period raises the issue of whether the decline might be an artifact of the truncation of the citation information in 1992. There are a number of reasons to suspect that such bias could be present. To see this, consider the distribution of citations over time from year of grant, i.e., the probability of receiving a citation in the year of grant, the next year, the second year, and so forth.

Assume first that the *shape* of this distribution for university patents is the same as for the random sample, but that in every year university patents receive proportionately more citations. The overall mean difference between the two would be the area between the two distribution curves, i.e., the cumulative difference. If we cut this off early in the distribution, the absolute difference that we have observed will be less than if we wait and look at a later date when the cumulative area between the two curves will be larger. This could conceivably

[13] Figure 4 shows results only for importance; the patterns for generality are similar.

Table 4. Comparison of University and Random Sample Patents Over Time

Dependent Variable	Importance (1)	Generality (2)
University - Random sample mean difference:		
1965	0.42	
	(0.29)	
1966	1.63 ***	
	(0.52)	
1967	-0.15	
	(0.41)	
1968	0.10	
	(0.35)	
1969	0.06	
	(0.44)	
1970	0.82 **	
	(0.42)	
1971	1.35 ***	
	(0.41)	
1972	1.48 ***	
	(0.41)	
1973	1.84 ***	
	(0.38)	
1974	1.08 ***	
	(0.35)	
1975	2.54 ***	0.053 ***
	(0.35) (0.019)	
1976	1.82 ***	0.065 ***
	(0.34)	(0.019)
1977	1.31 ***	0.048 ***
	(0.34)	(0.020)
1978	2.04 ***	0.040 **
	(0.34)	(0.019)
1979	1.13 **	0.052 ***
	(0.31)	(0.018)
1980	1.91 ***	0.051 ***
	(0.31)	(0.017)
1981	1.68 ***	0.080 ***
	(0.31)	(0.018)
1982	0.96 ***	0.051 ***
	(0.31)	(0.018)
1983	0.97 ***	0.028 *
	(0.30)	(0.017)
1984	0.47 *	0.024
	(0.28)	(0.017)
1985	0.40	0.037 **
	(0.28)	(0.017)
1986	0.06	0.013
	(0.27)	(0.017)
1987	-0.07	0.043 ***
	(0.25)	(0.017)
1988	-0.08	0.012
	(0.24)	(0.019)
Year Dummies	Sig.	Sig.
Patent class controls	Sig.	Sig.

Notes to the table

***, **, * Significant at the 1, 5 and 100% levels respectively.

explain the apparent disappearance of the difference between the two groups at the end of the observed period.

This hypothesis suggests a simple test: run the regression in logs, thereby capturing the proportionate difference between the two groups rather than the absolute difference. This requires eliminating from both groups those patents with zero citations. On this more limited data set, the log-citation regression produces results, (not reproduced here), that are broadly consistent with those reported above. The overall difference between the two groups in importance is about 15 percent; this overall difference conceals variation, with a high of about 30 percent in the mid to late 1970s, falling to insignificance by 1984. Thus the results cannot be explained by truncation of lag distributions, if the two distributions have the same shape.

A more subtle possible problem is that university patents may, on average, come later than those for private firms, meaning that the truncation has a more severe effect on them than on the random sample patents. In order to test for this possibility, we took all of the post 1974 patents[14] (both university and random sample) and created a data set with 1 observation for each year from the date of application of the cited patent until 1992, indicating the number of citations *in that year.* We then ran a regression with this per-year citation count as the dependent variable, and included as controls patent class and application year of the cited patent (as before), but now also including dummies for the *lag* between the

Figure 4

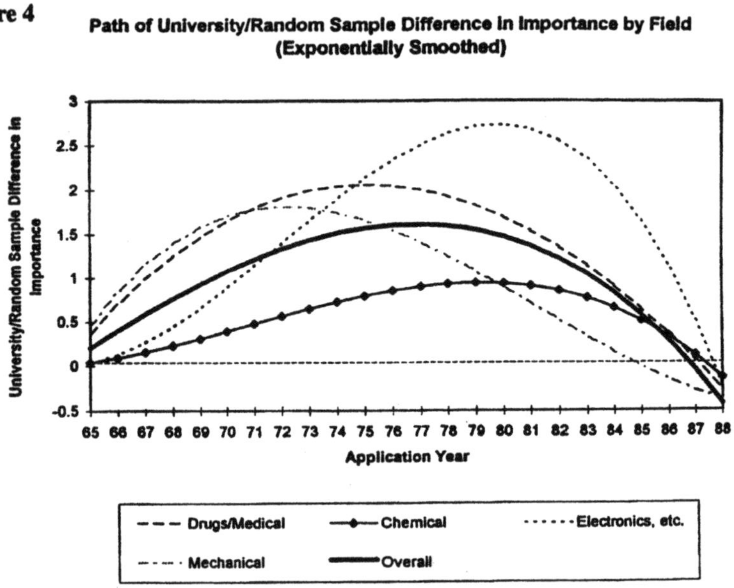

[14] For the patents before 1975, we know the total number of citations received (i.e. importance) but we do not know the identity of the citing patents, so we cannot decompose that total by year.

application year of the cited and citing patents, allowing different lag coefficients for the university and random sample cited patents.'[15] In addition to these controls, we allowed for a university/random sample difference in each application year Thus this more complicated regression estimates the difference between the average university and random sample patents in a given year, controlling for the predicted levels based on the years remaining to truncation and the average citation lag structure for each sample. The results of this exercise are very similar to the simpler ones reported earlier, with the university-corporate difference declining sharply around 1981 or 1982 and becoming statistically insignificant shortly thereafter.

In summary, then, university patents in all fields were more important and more general than average in the 1970s; this advantage disappeared in all fields by the mid-1980s; and this disappearance does not appear to be an artifact of truncation or of the way in which citation patterns have changed over time. In the next section we explore further the nature of the decline.

6.3 The Nature of the Decline

In this section we explore several potential issues with respect to the declining relative importance of university patents. First, we examine whether the change in the average relative importance is the result of a shift or a twist in the importance distribution — in particular whether it is driven by an increasing number of uncited university patents as the propensity of universities to patent has increased; second, we ask whether declining relative importance is likely to be the result of the *increasing* importance of non-university patents; and, finally, we examine whether the decline appears to be related to the spread of patenting to many more institutions.

Thinking about changes in the propensity to patent is facilitated by having a simple model of the decision to apply for a patent. This model is illustrated graphically in Figure 5. It seems reasonable to assume that inventions have a quality or value distribution, with many low-quality or low-value inventions, and few high-quality or high-value inventions. Patenting is a costly process, and conveys expected benefits that are increasing in the quality of the invention, but which also depend on the legal, competitive and institutional environment of the inventor. Assuming that inventors have some knowledge at the time of invention about the quality of their invention, this implies that there will be a "threshold" quality, above which a patent will be applied for, and below which it will not. Changes in the "propensity to patent" can be thought of as movements in this threshold. It follows that reductions in the cost of patenting (e.g, via elimination of the need to get funding agency approval for the pateant application) or increases in the benefits of patenting (e.g. if patents become more enforceable) will be

[15] There were 30 dummies altogether: a university and a random sample set, each with 1 for lags of *0* or 1, 1 for each value 2-14, and one for 15 or more.

Figure 5. An Increase in Patenting Due to an Increase in the Rate of Invention.

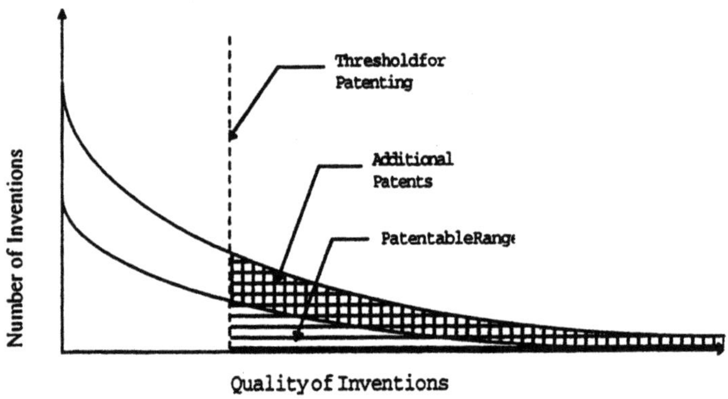

An Increase in Patenting Due to a Decrease in the Threshold for a Patent Application

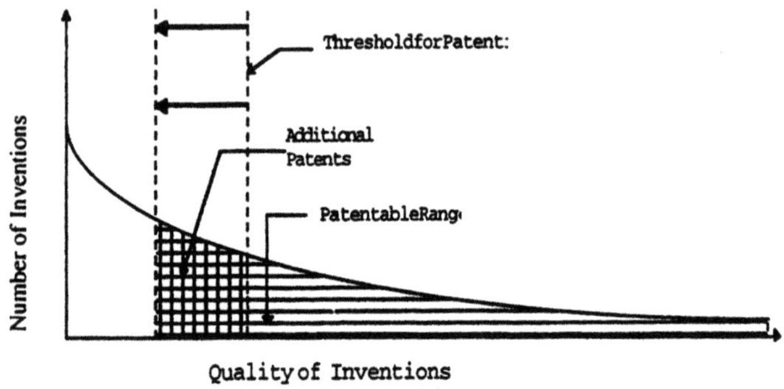

associated with a fall in the average quality of patented inventions, as low-quality inventions that were previously unpatented now join the patent distribution. This reasoning suggests that the fall in average university quality may simply be the result of the addition of many low-quality patents due to a reduction in the patenting threshold.

It also suggests that the change in relative importance is unlikely to be the result of increasing importance of non-university patents. As shown in Figure 1, the late 1980s were a time of increasing propensity to patent. That is, the overall patent/R&D ratio, which had been falling for most of this century, began to rise slightly. This is generally attributed to the creation of a special Court of Appeals for hearing patent cases, and the issuance of several decisions that have increased the perceived likelihood that patents will be enforced (Schwartz, 1988). Thus, the overall patenting threshold has probably been moving to the left, thereby, if anything, reducing the overall quality of patents.

Returning to the university patents, we can look at how the distributions over "quality" have shifted over time: if the increase in university patenting reflects an increase in the rate of patentable inventions, so that the top part of Figure 5 is most relevant, then we would expect to see little change in the general shape of the

Figure 6

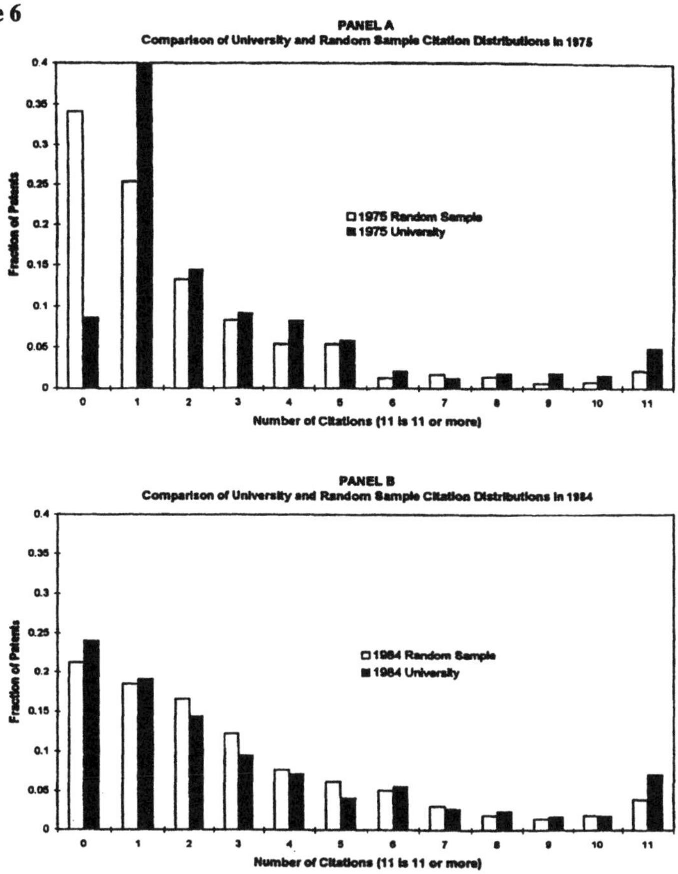

quality distribution of university patents over time. In contrast, if it reflects an increased propensity to patent, so that the bottom half of Figure 5 is more relevant, we would expect to see large increases in the fraction of university patents receiving very few citations.

Distinguishing between these effects over time is complicated by the difficulty of controlling for time and truncation effects. Figure 6 shows the distribution of patents across citation intensities for both the university and the random sample in 1975 and 1984, in each case looking only at those citations received in the first 8 years (to control for truncation effects), and showing also the distribution for the random sample (to control for time effects). The top panel shows that in 1975, the university distribution had fewer patents with zero citations, and more patents at virtually every other citation level. By contrast, in 1984, the university distribution actually has more density at zero than the random sample. In both years, the university distribution appears to have more density in the extreme right-hand tail.

Figure 7 looks specifically at the issue of zero-citation patents, but compares the university patents and the random sample from 1975 through 1987. To control for truncation we look at those patents that have zero citations in the first 5 years. Figure 7 shows that the decline in the university advantage flowing from a

Figure 7

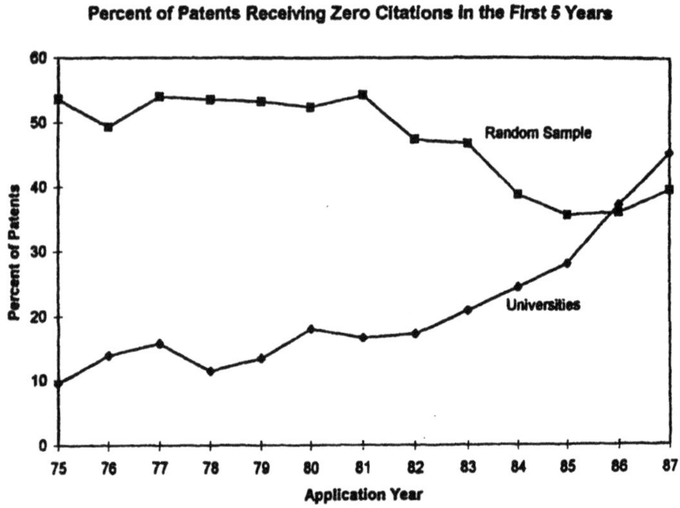

relatively low proportion of zero-citation patents has been dramatic. Hence there is strong evidence that part of the decline in average relative importance is an influx of low-importance patents, possibly associated with the effective reduction of the threshold for patent application by universities.

Identifying the decline with an increase in zero-citation patents does not, of course, explain what caused it. The "threshold" story is plausible but difficult to prove. A final aspect that we explored was to examine whether the decline might be associated with the spread of patenting beyond the handful of institutions that were active in the early years. For this to explain the decline in quality, it would have to be the case that the share of patents attributed to "large" institutions had declined. There are various ways to measure this. Figure 8 shows that such a decline has, in fact occurred: since 1965, the fraction of patents going to the top 4 institutions has fallen from about 50% to about 25%. The Herfindahl index of concentration across institutions has also declined, from about .1 in 1965 to about .04 in 1988. Interestingly, the citation-weighted Herfindahl has declined somewhat less than the unweighted Herfindahl, suggesting that the institutions getting more patents are getting more highly cited ones, and/or institutions getting better patents have not had the same decline in importance as other institutions.

We explored these possibilities directly by running a regression analogous to that underlying Figure 3, but in addition allowing the difference between university and random sample patents to differ not only over time, but also according to the "size" of the institution. To do this, we grouped all institutions that got any patents over the period into 3 categories: (1) those institutions in the top decile in terms of the number of successful patent applications in 1988;[16] *(2)*

[16] The distribution of patenting activities across universities is very stable over time, so that the choice of a particular year to divide the sample - in this case 1988 - seems unlikely to introduce any particular bias into the results.

Figure 8

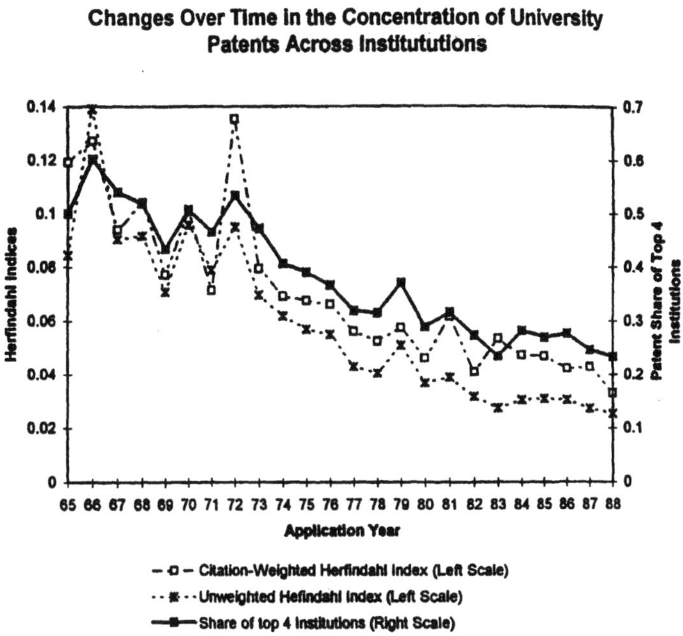

those institutions that got fewer patents than the top decile but more than the bottom quartile in 1988; and (3) those institutions that were in the bottom quartile in terms of patent total in 1988 plus those that had no successful applications in 1988 but received at least one patent from some other year. As in Figure 3, the regression used citations received as the dependent variable, and included controls for application years and patent classes.

The results are illustrated in Figure 9. The results show that, except possibly for a few years in the second half of the 1 970s, the bottom group of universities never produced patents that were statistically distinguishable from the random sample. At the same time, the 15 schools that comprise the top decile of institut ions had patents that were even more superior to the random sample than those of other universities, so that it is the case that on average those institutions that patent more are receiving patents of higher quality. None the less, this elite group has seen a decline in their relative quality since about 1983. Overall, then, the decline in the university average has two components: the very best institutions are indeed producing patents that are not as good (on average) as they were before; and, in addition, an increasing fraction of the university total is coming from smaller institutions, who have always produced patents that were not as highly cited as those from the larger institutions.

Figure 9

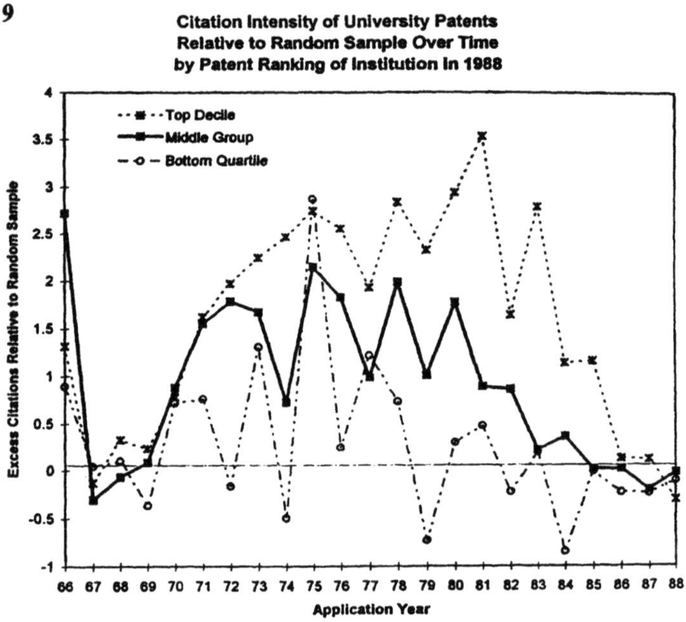

6.4 Conclusion

We have shown that the relative importance and generality of university patents has fallen at the same time as the sheer number of university patents has increased. This decrease does not appear to be a function of truncation bias or any shift in citation patterns over time. Rather it appears to be the result of a very rapid increase in the number of "low quality" patents being granted to universities.

Figure 10 provides a summary cut at this issue. It shows again the overall increase in patenting (the heavy middle line), juxtaposed with two contrasting components of that total. The bottom line is the number of patents from each year that have received no citations. It is virtually flat until the early 1980s, showing that the roughly 5-fold increase in overall patenting up until that time was not accompanied by much of an increase in the number of these low-importance patents. The dotted line at the top is the number of high-importance patents, dubbed "Winners" in the graph and plotted on the right-side axis. This is the number of patents that received more citations than the mean of the top 10 percent of random sample patents from the same year. This series increases *faster* than the overall total up until the early 1980s, implying that the proportion of very important patents was increasing over this period. From 1981 on, however, this series fluctuates up and down with no clear trend. Despite the approximate doubling in the total number of patents after 1980, there is no increase in the number of very important patents.

What are the policy implications of this result? From a theoretical perspective, the Bayh-Dole Act and the increase in industry funding had two distinct effects on university incentives. Both the incentive to perform research that could be expected to produce important commercial inventions, and the incentive to patent and license whatever commercial inventions were produced increased. Clearly, Bayh-Dole has been a success with respect to the second of these incentive effects. Both the rate of patenting and the extent of licensing have increased dramatically. In this context it is important to emphasize that even thought the body of uncited university patents that we have observed is probably less valuable *per patent* than previous university patents, these patents are not worthless in the aggregate. Some of these uncited patents are licensed and are commercially valuable. Before Bayh-Dole, they probably would not have been patented, would not have been licensed, and the invention would have been unlikely to generate commercial benefits. Thus the overall increase in the effective transfer of the technology generated by universities into the private sector has probably increased the social rate of return to the university research endeavor. In terms of the *magnitude* of this increase in technology transfer, however, our results suggest that the quality-weighted extent of technology transfer has risen less dramatically than the unadjusted patent totals would suggest.

In contrast to the impact on the *transfer* of technology, our results suggest that Bayh-Dole and related changes have not had a significant impact on the underlying rate of *generation* of commercially important inventions at universities. Universities either did not significantly shift their research efforts towards areas likely to produce commercial inventions, or, if they did, they did so unsuccessfully. It is unclear, of course, whether it would be socially desirable if universities shifted their research efforts towards commercial objectives. It is likely that the bulk of the economic benefits of university research come from

Figure 10

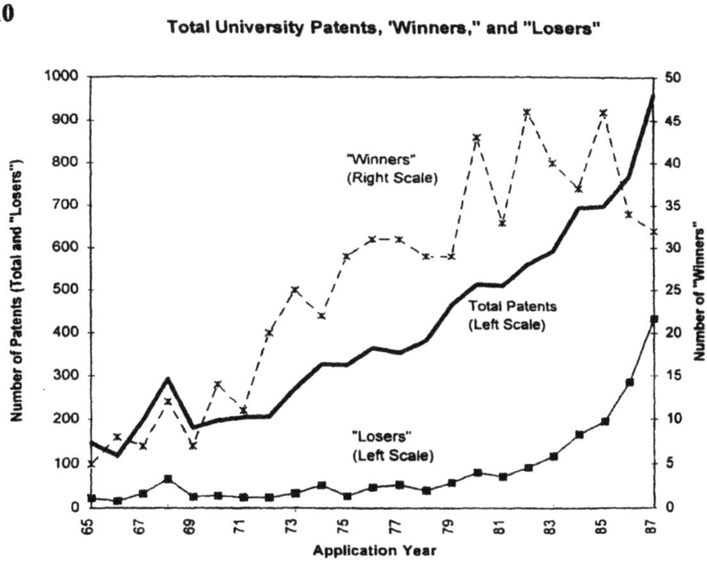

"Losers" are patents with no citations by the end of the period.
"Winners" are patents with more citations than the mean of the top 10 percent of Random Sample Patents from the same year.

inventions in the private sector that build upon the scientific and engineering base created by university research, rather than from commercial inventions generated directly by universities. In other words, if commercial inventions are inherently only a secondary product of university research, then it makes sense for policy to seek to ensure that those inventions that do appear are transferred to the private sector, but not to hope to increase significantly the rate at which university research directly generates commercial inventions. This appears to be what has occurred.

From a methodological perspective, our results show that it is possible to use citations to improve the usefulness of patent statistics as economic indicators. The economic usefulness of these widely available data has been limited by their perceived noisiness. Even in the time-series dimension, where cohort effects and truncation bias make citation comparisons difficult, the use of a reference group and careful controls for technology field allowed us to produce fairly clear results regarding the changing nature of university patents. This technique can be readily applied to other data, thereby greatly increasing the signal-to-noise ratio in patent data.

6.5 Appendix: The Data

A patent is a government-granted property right in the commercial use of a device or process (Schwartz, 1988). For a patent to be granted, the invention must be nontrivial, meaning that it would not appear obvious to a skilled practitioner of the relevant technology, and it must be useful, meaning it has potential commercial value. Ideas per se are not patentable. If a patent is granted, a public document is created containing extensive information about the inventor, her employer, and the technological antecedents of the invention. There is no public record of unsuccessful patent applications.

This paper examines the changing nature of university patenting using 3 large datasets of U.S. patents:

- a comprehensive or near comprehensive database of all patents assigned to universities or related institutions from 1965 until mid-1992 (13,024 patents);
- a one percent random sample of *all* U.S. patents granted over the same time period (19,528 patents);
- all patents from 1975 until mid-1992 that cirted *either* the university patents or the patents in the one percent random sample (60,215 patents).

We began by identifying all patents granted to universities and related institutions during the period 1965-1992.[17] To do this, we started with 243 patent assignees identified by the Office of Technology Assessment and Forecasting of the Patent Office in the mid 1980 as universities and related institutions. We supplemented this list by searching the patent database for all assignees containing the phrases "university," "research foundation," and the like, and then inspecting these by hand to determine if they were actually universities or related institutions. Altogether, we identified 657 assignees as universities. Many of these were minor spelling variations, typographical errors, or combinations of other institutions. After standardizing the institution names, we ended up with 244 distinct institutions that received over 13 thousand patents during the period.[18]

Our second step was to identify a 1/100 random sample of all U.S. patents over the same period.[19] This sample consists of almost 20 thousand patents. We have

[17] The commercial source from which we acquired these patent data contained grants through August of 1992. For the purpose of reporting totals for 1992, we have scaled up the totals through August by 12/8. Much of our analysis includes year dummy variables, or examines contrasts between university patents and other patents, in which cases the 1992 truncation does not create any bias.

[18] Some patents are assigned jointly to a firm and a university; for the purpose of the aggregate analyses these were considered university patents. Universities differ in the extent to which distinct schools or campuses patent under their own names or those of the parent. For example, patents from all campuses of the University of California are assigned to the Regents of the University of California; in other states different parts of the state system each hold patents separately.

[19] This was done by simply selecting all patents whose patent numbers ended in a distinct 2-digit sequence. Since patent numbers are assigned sequentially, this gives us every 100th patent issued.

assignee names for these patents as well, but the number of assignees is much larger and hence standardization would be a much more time-consuming task. We do not utilize the assignee information for the random sample patents in this paper.

The final step was to identify all subsequent patents citing both the university and the random sample patents. Because of data limitations, this was done in two different ways. For patents from 1965 through 1974, we know only the total ntimber of citations they received through 1991. For patents granted between 1975 and 1992, we know the actual patent numbers of all the citing patents, through August of 1992 (over 60 thousand patents). Thus for the later the patents, we have a slightly more complete *count* (through August 1992 instead of end of 1990), and we know more about the citing patents, including their patent classifications. Since we use the patent classifications of the citing patents for some of our analyses, we will for these purposes focus only on patents granted after 1975. The information from the patent record that we use in this research is as follows:

- the **date of grant** ("year") (The patent also contains the date of application, which for most purposes is a more appropriate means for dating the invention. However the information in our database on application dates is not complete, so we have limited our analysis to grant dates.)
- **the organization or organizations (if any) to which the inventor assigns the property right conveyed by the patent ("assignee").**
- **the primary classification of the patent** within the patent office's technological classification system ("class"). There are currently about 400 such classes. Examples of patent classes include "Measuring and Testing" and "Multicellular Living Organisms and Unmodified Parts Thereof."
- **the references to earlier patents** required by the patent examiner to recognize the existence of earlier related property rights, and thereby delimit the property right being conveyed in the current patent ("citations").

References

Archibugi, D. : "Patenting as an Indicator of Technological Innovation: A Review," *Science and Public Policy*, Vol.19, No.6, pp.357-368, 1992.

Argyres, Nicholas and Liebeskind, Julia "Privatizing the Intellectual Commons: Universities and the Commercialization of Biotechnology." Mimeo, University of Southern California, Spring 1996.

Association of University Technology Managers, *The AUTM Licensing Survey. Fiscal Years* 1993, 1992, 1991. Copyright 1994, The Association of University Technology Managers Inc, Norwalk Connecticut.

Basberg, B. : "Patents and the measurement of technological change: A survey of literature," *Research Policy*. Vol.16, pp.131-141, 1987.

Blumenthal, D. : "Academic-Industry Relationships in the Life Sciences (Extent, Consequences, and Management)," *JAMA*, Vol.268, 1986, pp.3344-49

Boitani, A. and E. Ciciotti: "Patents as Indicators of Innovative Performances at the Regional Level", in R. Cappellin and P. Nijkamp, eds., *The Spatial Context of Technological Development*, Aldershot Gower, p.139-163, 1990

Caballero, R. and A. Jaffe: "How High are the Giants' Shoulders: An Empirical Assessment of Knowledge Spillovers and Creative Destruction in a Model of Economic Growth," in 0. Blanchard and S. Fischer, eds., *NBER Macroeconomics Annual. Volume 8*, M.I.T. Press, 1993

Carpenter, M. and F. Narin.: "Citation Rates to Technologically Important Patents," *World Patent Information*, Vol.5, No.3, pp.180-185, 1993.

Cohen, Wesley, Richard Florida and W. Richard Goe (1994) "University-Industry Research Centers in the United States." Research Report, Center for Economic Development, Heinz School of Public Policy and Management, Carnegie-Melon University.

Dasgupta, P. and P. David. : "Information Disclosure and the Economics of Science and Technology," Chapter 16 in G. Feiwel (ed.), *Arrow and the Ascent of Modern Economic Theory*. NYU Press, 1987.

David, P.A., D. Mowery and W.E. Steimueller: "Analyzing the Economic Payoffs from Basic Research," *Economics of Innovation and New Technology*, Vol.2, No.1, pp.73-90, 1992.

Griliches, Z. (1990) "Patent statistics as economic indicators: A survey." *Journal of Economic Literature*, 28(4), 1661-1707.

Henderson, Rebecca, Adam Jaffe and Manuel Trajtenberg: "The Bayh-Dole Act and Trends in University patenting 1965-1988." Forthcoming in the proceedings of the Conference on University Goals, Institutional Mechanisms and the "Industrial Transferability" of Research, Stanford Center for Economic Policy Research.

Jaffe, Adam (1989) "Real Effects of Academic Research," *American Economic Review*, LXXIX, 957-70.

Jaffe, Adam, Manuel Trajtenberg, and Rebecca Henderson, "Geographic Localization of Knowledge Spillovers as Evidenced by Patent Citations," *Quarterly Journal of Economics*, 1993

Mansfield, Edwin, "Academic Research and Industrial Innovation," *Research Policy 20* (February 1991): 1-12.

Merton, D. (1973) in N.W. Starer (ed.) *The Sociologv of Science: Theoretical and Empirical Investigation* (Chicago, University of Chicago Press)

National Academic of Sciences *Allocating Federal Funds for Science and Technology*, National Academy Press, Washington DC, 1995.

National Science Foundation: *University Industry Research Relationships: Myths, Realities and Potentials.* U.S. Government Printing Office, 1982.

Pavitt, K. : "What Makes Basic Research Economically Useful," *Research Policy, 20,* 1991

Romer, Paul. 1986, "Increasing Returns and Long-Run Growth" *Journal of Political Economy,* Vol 94, No *5,* 1002-1037.

Romer, Paul. 1990, "Endogenous Technological Change" *Journal of Political Economy,* Vol 98, No *5,* S71 - S101.

Schwartz, H. F. : *Patent Law and Practice,* Federal Judicial Center, Washington, DC: 1988

Trajtenberg, M. : "A Penny for Your Quotes: Patent Citations and the Value of Innovations," *Rand Journal of Economics,* Vol. 21, No. 1, pp.172-187, 1990a.

Trajtenberg, M. : *Economic analysis of Product Innovation: The Case of CT Scanners,* Harvard University Press, 1 990b.

Trajtenberg, M., R. Henderson and A Jaffe: "University versus Corporate Patents: a Window on the Basicness of Invention." Forthcoming, *Economics of Innovation and New Technology.*

U.S. General Accounting Office, *Patent Policy: Recent Changes in Federal Law Considered Beneficial,* GAO/RCED-87-44, 1987

Zucker, Lynn, Michael Darby and Marilynn Brewer (1994) "Intellectual Capital and the Birth of U.S. Biotechnology Enterprises." NBER working paper no.4653, NBER Cambridge MA.

7. Communication Norms and the Collective Cognitive Performance of "Invisible Colleges"

Paul A. David*

7.1 Introduction and overview

Scientific research communities may be studied as social networks within which ideas or statements circulate, acquire validity as reliable knowledge, and are recombined to generate further new ideas. Social networks also form the locus for the transmission of tacit knowledge and skills requisite to the interpretation and operationalization of scientific statements. These extensive, yet informal structures of inter-personal knowledge-transactions have been referred to as constituting "invisible colleges". This paper develops an abstract and highly stylized account of the communications structure of an invisible college, and examines its

* All Souls College, Oxford and Stanford University
The development of this paper was encouraged initially by the comments of Patrick Cohendet and other participants at his CNET Workshop on the Economics of Networks, in Paris on 20th April 1995, where I described some very preliminary results. My stochastic formulation of the role of communications networks in science has benefited immeasurably from previous collaborative projects with Dominique Foray and Jean-Michel Dalle, applying Markov random field models in other contexts. The present version draws on material subsequently developed for a paper with Tom Flemming that was presented to the International Conference on Creation and Transfer of Knowledge: Institutions and Incentives, held in Castelgandolfo (Rome), 21-23 September 1995. Manuel Trajtenberg's insightful discussion, and the remarks of Adam Jaffe, Giorgio Navaretti, Jacques Thisse, and others on that occasion, have contributed substantially to improving the exposition. Elizabeth Johansson also helped, by diverting her time from more exciting activities to supply the answer to an applied probability problem required for the argument of section 4.2. I gratefully record, too, the probing questions and suggestions made by Albert Bolink, Sajeev Goyal, and Uskali Maki and others at the Economics and Philosophy of Science Seminar at Erasmus University, Rotterdam (13 November 1995); and the comments of Patrick Bolton and other participants of the Economic Theory Seminar of CENTER at the University of Tilburg (12 March 1996). Olivier Favereau's close critical discussion of the next draft of the paper, presented at the IMRI Workshop of the University of Paris-Dauphine (11 April 1996), contributed materially to its further improvement; so, too, did my conversations with Kenneth Arrow, Arie Rip, and other participants in meetings of the Seminar on the Economics of Information Diffusion convened at All Souls College, Oxford, during Trinity Term 1996. I take this occasion also to express my gratitude for the support that my work on the economics of science has received from the Renaissance Trust (UK). Those whose generous assistance has contributed to this paper must be fully exonerated from blame for such deficiencies as have persisted in the face of their best corrective efforts.

collective epistemological performance by employing concepts and results from Markov random field theory.

A central task in this undertaking is to characterize formally the relationship between the epistemological notion of scientific "closure", on the one hand, and the sociological notion of the organizational ethos and "norms" of invisible colleges and kindred scientific communities, on the other hand. The latter are taken to reflect themselves in micro-level stochastic behaviors affecting local social communications within a finite size system - which is constructed as the (global) network of (local) social communications networks. Using a particular stochastic structure (the "voter model" of Clifford and Sudbury (1973) and Holley and Liggett (1975)), a knowledge consensus, represented as a configuration of correlated belief orientations regarding a new scientific statement, is shown to be an emergent and path-dependent property of the global network. In that sense alone, however, "closure" is not held to be sufficient to establish a proposition's objective validity, nor is it tantamount to the stabilization of beliefs *over the long run.* I propose an approach to resolving the latter problem, by adopting a "weak relativist" position in regard to the objective "truth" or "falsity" of scientific statements, and embedding the stochastic process of consensus formation within a long-run evolutionary epistemological framework.

This analysis provides new and more concrete foundations for asserting that the degree to which researchers' behaviors typically adhere to some among the so-called "Mertonian norms", associated with the ethos of "open science", is of critical significance for the collective work performed in the cognitive domain by "invisible colleges." In so doing, it reveals the different implications for the invisible college's short-run and long-run collective epistemological performances, respectively, of the norms affecting the openness of internal communications, and those norms affecting the network's openness to entry of a socially heterogeneous membership.

As transparent as are the purposes and the logic of this paper, some background in the philosophy and sociology of science may be useful in order to situate it within the broader context of studies that have concerned themselves with the cognitive content of science and the intellectual and social organization of scientific communities.

A brief orientation of that sort is provided by the discussion in Section 1. There I have tried to indicate how the present contribution to "the new economics of science" fits within that larger enterprise; and also to notice its connections with other recent lines of work by economic theorists - specifically, those analyzing situations of strategic interdependence among agents that are embedded in local social networks, or whose information states can change only through word-of-mouth communications.

The organization of the paper's remaining sections follows in a straightforward manner the empirical motivations for, and analytical implications of the features of research networks that are caricatured by the model. Section 2 motivates the emphasis placed upon *local* communications through social networks among scientists, pointing out the distinct but complementary roles of tacit knowledge and codified knowledge in the conduct of research, and noticing the importance of personal contacts for the (social) transmission of tacit knowledge. The premised micromotives affecting the behavior of the individual researchers who choose to belong to these social networks is considered in Section 3. Section 4 discusses the nature of the cognitive transactions in which they are engaged. Starting with a highly stylized account of experimental research activity as an iterative process of Bayesian belief revision (section 4.1), I then provide (in section 4.2) an illustrative model in which conformity to the consensus of local peer belief emerges as a rational reputational strategy for individual researchers engaged in interpreting their experimental results. A brief digression is then required, in order to clarify (in section 4.3) where this formulation leaves us in reference to recent controversies between "scientific realists" and "relativists", concerning the social construction of scientific knowledge. Finding the extreme subjectivism of the radical social constructionists unsatisfying, I propose the importation of arguments from evolutionary epistemology as the basis for a plausible resolution, a conciliatory synthesis between the antithetical claims of the two camps in the on-going "culture wars." This clears the way for the core of the network model to be set out in Section 5: the apparatus of graph-theoretic representation of connected local networks of research units, forming an "invisible college" is introduced by section 5.1; the correspondence between the dynamic stochastic process of network interactions specified by the preceding sections, and the Markov random field model known as the "voter model", is described formally in section 5.2, and basic results are presented for the cases of network that can be represented as one- or two-dimensional connected graphs. Section 5.3 discusses those properties of the dynamics of consensus formation in relation to other theoretical results on critical properties of stochastic communication structures - specifically those deriving from the branch of probability known as *percolation theory*.

In section 6 it is shown that the relationship between micro-level behaviors of the research units, on the one hand, and the macro-level cognitive performance of the invisible college represented by the network of local networks, on the other hand, have a quite direct correspondence with the functionalist sociology of science in the tradition of Merton (1973). The significance of several key Mertonian norms is reconsidered in this light. Section 7 concludes the paper with a sketch of some directions in which it is possible readily to extend the analysis of the basic network model without extensive stochastic simulation exercises, namely, by examining the properties of the equivalent deterministic system. Consideration of the dynamics of endogenous network size adjustments, and the evolution of jointly determined network performance attributes is then quite tractable, as is the examination of the stability properties of growth paths for the "invisible college."

7.2 The larger context: Philosophy, sociology and the 'new economics of science'

The study of science as a cultural activity was for a long time fragmented into two domains. In one camp lived philosophers of many differing stripes, all of whom were preoccupied with the cognitive content of science - which is to say, with epistemological issues concerning the character of "scientific" statements and the way that these were arrived at.[1] With a few notable exceptions, philosophers and epistemologists of science have shared a primary interest in *normative* questions, dealing with the nature of the beliefs that people *ought* to hold about their surroundings. The classical approach in such endeavors sought to settle questions on an a prioristic basis, through solely conceptual arguments.[2] A complicating departure from the classical tradition in epistemology, part of a movement to "naturalize" epistemology and the philosophy of science that is closely associated with the work of the philosopher W.V.O. Quine (1953), undertook to inject into such accounts facts about the actual epistemic situations in which people find themselves. Quine (1969) coined the term "naturalized epistemology", arguing that the positivist distinction between contexts of "discovery" and "justification" could be "naturalistically" translated into questions of perceptual psychology and sociology of knowledge, respectively. The former involves studying people's behavioral dispositions to associate words with situations, whereas the latter is studied by examining the communication patterns by which such dispositions become stabilized for a community. In a robust sense, the present essay could be categorized as a contribution to the latter programme, launched from the theoretical perspective of the economic analysis of individuals situated in social communications networks.

To understand how the latter point of departure was arrived at, however, it is necessary to refer back to the second of the two domains into which the study of scientific activities was formerly divided. That comprised the territory inhabited by sociologists of science. Members of this clan, following the leadership of Robert K. Merton (1973), were absorbed in trying to understand the social context of modern science, the characteristic institutionalized modes of organization for this particular human activity, and the associated behavioral norms and cultural ethos which guided the participants' behaviors and permitted them to succeed collectively, if not always individually, in their pursuit of knowledge about the external world.[3] Although there were influential contributors to the history of science and technology who pointed to the role of external, material conditions

---

[1] See, e.g., Popper (1959, 1963), Quine (1962), Kuhn (1962), Latakos (1970).

[2] This picture of philosophy of science as a domain of concern completely disjoint from the sociology and history of science is deliberately overdrawn, for effect. See the more nuanced, but nonetheless compact account by Fuller (1994).

[3] See, e.g., Crane (1965), Zuckerman and Merton (1971), Cole and Cole (1967), Cole and Cole (1973), Cole (1978).

affecting the physical phenomena that scientists undertook to study, and the ways in which they sometimes went about it, the canonical investigations in the sociology of science focused upon generic behavioral and organizational problems that were represented to be independent of the specific substantive concerns of the scientific discipline in question, and orthogonal to its cognitive contents and claims at particular historical junctures.

Each of these philosophical and sociological tribes, appears collectively to have been disdainfully reluctant (and quite possibly also fearfully hesitant) to assert any claims to the territory staked out by the other; evidently, the two camps found it best to leave each other alone. In consequence of which, only occasionally, and rather tenuously, did their studies attempt to draw links between the ways in which communities of scientists were organized and rewarded, and the ways in which scientific statements were created and accepted within those communities.[4]

In the 1970's a new generation of sociologists took up the sociology of scientific knowledge - "SSK", as it came to be styled. They insisted, with ample reason, that the cognitive and the social dimensions of science should no longer remain compartmentalized as distinct, specialized fields of inquiry.[5] Instead, the subject matter had to be seen in reality to be inseparable. This was held to be so, because in "discourse" - within whose terms it was held possible to analyze everything - social and cognitive contexts are thoroughly inter-penetrating and mutually interactive. Left behind in the border-crossing rush of the SSK movement, however, were the old sociology of science's focii of attention, namely, the institutionalized reward systems, the social norms and the relationship of these to the organization of resource allocation within scientific communities. In the evolution of these academic transactions, as the case in other affairs, experience sometimes seems to reaffirm the aphorism of the French: *plus ça change, plus c'est la même chose*. The cognitive content of science and the macro-sociological, organizational structures of scientific communities, once again, could not be held simultaneously within the focus of scholarly attentions.

But, both Nature and the spirit of academic enterprise seem to abhor vacuums, even partial vacuums. On the terrain recently deserted by those who had gone off to build the new sociology of scientific knowledge, leaving behind only a small remnant of organizational sociologists, there soon appeared a vanguard of the "new economics of science." Encouraged by Dasgupta and David (1987, 1988, 1994), these newcomers, although still comparatively few in number, have gone about enthusiastically setting up camp to study the microeconomic resource allocation properties of science reward systems and scientific research organizations. To date, their programme has encompassed both theoretical and empirical analyses of these particular "knowledge producing and distributing

---

[4] This is a gloss on a very extensive literature, for fuller discussion of which see Ben-David (1990), Callon (1995) and Leydesdorff (1995).

[5] See Barnes (1974, 1977), Bloor (1976), Mulkay (1979), Latour and Woolgar (1979), Knorr-Cetina (1981).

activities", in which application has been found for some of the familiar tools and insights brought over from industrial organization economics, game theory, and applied econometrics.[6]

Economists, it must be said, are for the most part quite content to study the efficiency of resource allocation in the production of goods and services without stopping to inquire even superficially into the specific natures and concrete shapes of those commodities, much less entering into serious discussion of how they come to be differentiated in perception and design.[7] After all, it has been a signal achievement for their profession simply to have recognized and drawn out the implications that follow from the otherwise obtrusive fact that ideas, knowledge and information belong to a category of commodities that really are qualitatively different from traditional economic goods.[8] Considered from that angle, it is perhaps not surprising that the "new economics of science" found it most natural to start by reworking the area of organizational analysis originally ploughed by Mertonian sociology of science, looking at the implications of certain institutional arrangements for allocative efficiency in the production of generic information that acquires a certain measure of reliability; but not troubling itself over the nature of reliability in this context, nor the details of the way that attribute of information might be acquired, nor any of the other issues of socio-cognitive interaction that have occupied the sociology of scientific knowledge.

As there is still plenty left to do in the new economics of science along the trajectories that have already begun to be explored, perhaps it will be thought by some best to leave well enough alone. On the other side of the argument it must be said that to proceed by completely ignoring the foundation-level connections between the social organization of research and the nature of its information products is likely, sooner or later, to cause difficulties. Indeed, it has already been noticed that maintaining the focus exclusively upon the behavior of the individual researcher, a legacy from the approach of the older sociology of science, detracts attention from the study of collaborations; it encourages "representative agent" modeling of scientists, rather than analysis of role differentiation within larger working groups; it ultimately hinders the collection of empirical evidence about patterns of resource allocation within and between teams, consortia, and institutionalized research networks. Perhaps even more seriously, in terms of its policy animus, the individualist tradition tends to perpetuate the view that the cognitive performance of scientific communities is largely to be understood by

---

6 See, e.g., Arora, David and Gambardella (1997); Arora and Gambardella (1994, 1998); David (1994a, 1995, 1996); David and Foray (1995); David and Flemming (1995); David, Geuna and Steinmueller (1995); David, Mowery and Steinmueller (1992); Gambardella (1994); Trajtenberg, Henderson and Jaffee (1992). Some of the foregoing are noticed in the wider survey of the economics of science by Stephan (1996).

7 See Bacharach, Gambetta et al.(1994) for an exceptional departure from this tradition.

8 Indeed, an achievement that only came rather belatedly in the discipline's development, with the pioneering works of Arrow (1962, 1971), Hirschleifer (1971), and Marschak (1971).

simple aggregation of the independent behaviors of its constituent members; that is to say, by simple aggregation, rather than by the complex interactions among the agents who form a research community. That view, unfortunately, encourages a policy-disregard for the respects in which scientific researchers' collective epistemological performance in the long-run may be affected by specific internal organizational features of such communities, including the norms that govern the informational and other transactions among their members. But, these are only some among the considerations that have prompted making a start in this paper on the systematic economic analysis of the relationship between individual behaviors and the cognitive performance of research networks.

7.2.1 Social network approaches to the economics of knowledge production

Conceptualizing scientific research as a social phenomenon involving actions and transactions conducted within *networks* thus offers a quite natural and appealing direction in which to extend formal modeling approaches in the new economics of science. Among its potential virtues may be that of bringing the latter into closer contacts with other social science programmes. Both SSK, and the "sociology of translation" programme of Callon et al. (1989), have made illuminating use of "networks" both as metaphors and tools in many case studies, bibliometric analyses, and related empirical studies of science and technology.

But, the foregoing need not stand alone as warrant for the methodological weight that is placed upon the concept of a research network in this paper's treatment of the production of scientific knowledge. Exploring formal models of network dynamics that hinge upon the communication of knowledge, as is done here, also may have some "spillover" benefits for other areas of economics, where networks as entities for analysis are emerging with increasing frequency. A brief elaboration on the latter point will serve to establish the plausibility of this second claim.

It is undeniable that popular thought and expression have been encouraged in its use of network metaphors by the startlingly rapid and convergent advances in computer and telecommunications technologies, which recently have been transforming information industries and the users of their services. Whether in conscious or unconscious signification of the now trendier conceptualization of systems of production and social organization, economics too has taken up the study of *networks* - and not only in connection with the fashionable issues of telecommunication and computer network externalities and questions concerning compatibility standardization.[9] Social networks also are now treated formally in modelling many contexts involving strategic interdependence, where these structures are represented as conveying information, and forging mutual trust through repeated transactions - even though the connections among the players are highly localized and can be presumed to be effected without sophisticated technological supports.

---

9 See, e.g., the literature surveyed by David and Greenstein (1990), and David (1993b).

Several lines of inquiry in game theory also have converged upon local network structures, as the terrain for analysing the equilibrium properties of games characterized by strategic complementaries and interactive learning on the part of players. Interest has focused upon the strategic problem that arises when it is assumed that each player interacts directly with only some subset of the entire ensemble - those in the player's immediate "vicinity", and that the players are unable to adapt their behaviors to deal individualistically with each of their "neighbors"; in such circumstances every player must select a strategy that is uniform with regard to all their neighbors.[10] The conditions under which the dynamics of local interaction games of this kind will give rise to equilibria characterized by correlated beliefs or behaviors have been studied, both in deterministic decision frameworks, and in dynamic stochastic models. The latter make use of results from Markov random field theory to show how local network externalities can lead to *de facto* standardization of technologies and the spontaneous formation of conventions.[11] The present paper will be seen to be situated in that broader stream of the recent economics literature.

Whatever the causes,[12] it should be plain from the foregoing that work on the theoretical frontiers of economics has begun seriously to embrace some of the essential ideas about the roles of informal social networks that already are thoroughly familiar in the sociology of science (see, e.g., Luhmann (1984, 1990), Ziman (1984), chs. 4,5). Nevertheless, the functioning of such networks, their relationship to the systemic properties of larger scientific communities, and their part in the linkages between those entities and the surrounding technological and economic environment, are all topics that remain to be explored much more thoroughly. Progress in that direction, thus far, has been made through micro-level studies of "knowledge network" structures, and research of this kind recently has begun to engage the attention of economists as well as other social scientists.

Economists interested in the relationships between the worlds of academic research and technological innovation also have made a significant start of their own into socio-cognitive interactions, by using bibliometric techniques of co-

---

[10] See, e.g., Anderlini and Ianni (1993), Blume (1993), Ellison (1993), Bala and Goyal (1995), Morris (1996).

[11] See David (1988, 1992), Kirman (1992), David and Foray (1993, 1994), Dalle (1995), David, Foray and Dalle (1997), Ellison and Fudenberg (1995). The approach taken here will be seen to follow in the latter stochastic process vein.

[12] Perhaps the breath-taking advances in electronic communications networks have had a decisive influence in undermining economists' commitments to atomistic individualism in the representation of human behavior, where previous opportunities for direct observation had for so long proved unavailing; there certainly is a striking temporal coincidence between the fascinating developments in network technologies and the acknowledgement of the importance of modeling human agents as socially "embedded" - entangled in webs of recurring, and inter-personal and intra-organizational transactions with identifiable "others". Technological progress can, indeed, have many unexpected cultural repercussions!

citation analysis to measure the extent and the degree of localization in patterns of knowledge transfers ("spillovers") from university research laboratories to industry.[13] These initiatives in the new economics of science have built directly upon the quantitative foundations laid down by Price (1965), Narin (1976), van Raan (1988), and their followers in "scientometrics" who apply bibliometric methods to the study of cognitive structures in science. In that regard, this line of inquiry implicitly accepts the position maintained by the proponents of the SSK movement, and also by the sociologists in the "translation" school of Callon et al. (1989), who hold that social networks of research and knowledge dissemination have corresponding linkages in the cognitive domain; that they give rise there to counterpart "connected clusters of connected nodes" in co-citation networks formed among papers published in the scientific literature, patent applications, and other "inscriptions".

Beyond effecting these trans-disciplinary contacts, the extension of new economics of science in the indicated direction has something to offer that has hitherto eluded studies in the SSK tradition. In those and other explorations of the cognitive dimension of science, similarly based on networks and inscriptions and the artifacts (to which those inscriptions refer and are referred in turn) there is general difficulty in knowing how to resolve a larger picture from the microcosm of detailed filiations. One wants at the end of the day to know how the workings of these micro-level networks are connected to the properties of the macro-structure of scientific institutions; to the evolution of fields of scientific endeavors that become organized through the extension of social networks and their supporting apparatus of journals, conferences, professional societies, and suchlike. What characterizes social networks in science? How do the behaviors of the individual participants in scientific networks affect the ability of the collectivity to do its work, and does a network's cognitive-domain performance, that is to say, in creating and validating knowledge, affect its opportunities for growth? Questions of this sort crop up repeatedly on the critical "middle ground" between the older and newer sociologies of science. The new economics of science should have something to offer by way of answers. So, it is precisely in that potentially hazardous no-man's-land that the present contribution is meant to stand.

7.3 Tacit knowledge, local social networks and 'invisible colleges'

I shall begin with a characterization - a caricature, to be more precise - of the social context of communications among scientific researchers. Analysis of the economic logic of the academic science reward system concurs with the functionalist sociology tradition in studies of the cultural ethos of modern science, by laying stress upon the centrality of the norm of public disclosure of knowledge among those who belong to the Republic of Science (see Dasgupta and David

[13] For this, see the path-breaking work of Trajtenberg, Henderson and Jaffe (1992); Jaffe, Trajtenberg and Henderson (1993); Henderson, Jaffe and Trajtenberg (1995a, 1995b).

(1987, 1994)). Accordingly, we can agree with Ziman (1984: p. 58) that "the fundamental social institution of science is thus its system of *communication*". Much attention properly has been given to the conditions affecting formal communications conducted through books, journals, other archival "publications". But, for reasons that are briefly indicated below, these "broadcast" modes of communication are taken for granted and occupy a considerably less central position in the following analysis than do the informal and interactive transactions among the members of science communities.

Scientific communal interactions, specifically those among researchers acting *qua* scientists, tend to take place within the circle of those in the same research speciality, who thus form *invisible colleges*. The latter term has been applied in a variety of related ways, sometimes referring to informal (in the sense of non-institutionalized) clusters of scientists collaborating in specific, newly developing research frontiers; sometimes to more extensive collectivities within the same discipline, profession or research area, who are linked by a web of personal (non-anonymous) informational exchanges. The common features of invisible colleges in science are that they remain rather fluid as to membership and size, that internally they are not highly structured, and, especially in today's world of telecommunication technology and cheap air travel, neither are they localized institutionally, geographically, or nationally. Their members may communicate in personal conversations, by circulating working papers and pre-prints, via telephone conversations, and, now also by fax and email messages; whereas, in former times, they would have posted letters to one another.

Within these quite extensive communities whose membership numbers in the hundreds, there exist rather smaller and communicatively more compact relational entities. These are referred to here as *local social research networks*, or simply as "local networks". Although more tightly clustered, the latter social groupings are research "cliques", rather than organized teams (indeed, they may encompass some or all of the members of different project teams). They also tend to be rather more localized - possibly in the dimension of problem specialization, or in those of geographical and institutional proximity which afford frequent occasions for face-to-face communications, or in some combinations of all of the foregoing, and still other social and personal dimensions.

Within the more restricted ambit of a researcher's local network will be circulating many bits of crucial knowledge, about experimental procedures, equipment functioning, data analysis routines - all of which very often escape being codified and described with complete clarity in published accounts of research procedures and findings. Although the development and circulation of codified knowledge was traditionally taken to be of central interest in philosophical and sociological studies of science, the significance of non-codified, *tacit* forms of knowledge and their role even in scientific endeavors has recently been gaining more general appreciation.

Tacit knowledge, as conceptualized by Michael Polanyi (1966), refers to a fact of common perception that we are aware of certain objects without being focused on them. Lying outside the zone of conscious attention does not make them the less important, however; they form the context that makes focused perception possible, understandable, and productive. Tacit and codified knowledge should thus be viewed generally as complements, rather than substitutes in human cognitive processes. Like other human pursuits, scientific inquiry draws upon sets of skills and techniques that are acquired experientially and transferred by demonstration, by personal instruction and the provision of expert services. Knowledge of this sort may be highly precise and intricate, but it is most typically conveyed as a *gestalt*, and referred to by language and signs that is idiosyncratic, rather than being reduced to constituent elements and operations denoted by standard codes from which might be assembled programmes of implementation.[14]

The reasons for the persistence of scientific knowledge in uncodified states are various, but it is relevant here to note the argument that these reasons have less to do with epistemological problems arising in the intrinsic nature of the knowledge involved, than with the costs and incentives offered for rendering it in an explicit, codified form.[15] Nevertheless, it remains a fact that many "craft" aspects of scientific practice must be learned in modes of instruction akin to an "apprenticeship" by being afforded opportunities for first-hand observation of how they are done, leading to trials under the guidance and supervision of experts. Otherwise, something like the original process of acquiring mastery of such knowledge has to be repeated *ab initio*, guided and encouraged only by the belief that others have found this to be possible. A striking instance of the "craft knowledge" deployed in science is documented by Harry Collins' (1974) detailed and influential study of the construction of the TEA laser. Collins concluded that only those scientists who had themselves spent time in the laboratories where a properly functioning laser had been built were able to succeed in reproducing the apparatus at other laboratory sites.

The importance of this kind of "hands-on" experience in many laboratory and facility based research disciplines makes the problem of social communication of tacit knowledge especially germane for the cognitive work of those fields. But even in other fields where codification has been pushed to very high levels, such as pure mathematics, there are areas of advanced work where only initiates with immediate participatory experience in the research itself appear to be capable of

[14] In such a process of conscious analysis of knowledge developed first in tacit form, some things are likely to be "lost in translation". Consequently, following Dasgupta and David's (1994) usage, David and Foray (1995: 25) say that information is "knowledge that has been *reduced* and converted into messages that can be easily communicated among decision agents"; that "'Codification' of knowledge is a step in the process of *reduction* and conversion which renders the transmission, verification, storage and reproduction of information all the less costly". (Emphasis added.) It is equally appropriate to observe that knowledge is formed through the receipt, interpretation and synthetic integration of information, in the course of which new tacitly understood constructs may emerge, to which those possessing them can attach idiosyncratic meanings.

[15] Dasgupta and David (1994), and David and Foray (1995) advance arguments for this economic approach, stressing technical and institutional determinants of knowledge producing behaviors. Cowan and Foray's (1996) analysis develops this approach further.

following and checking the correctness of proofs developed by others, let alone producing new theorems.

In the modeling exercise undertaken in the following section, it is assumed for simplicity that codified knowledge alone can be effectively transmitted through non-network channels, that is, *broadcast* through a variety of public media that identify the authors of messages but are non-specific with respect to the identities of the recipients. But, no one-to-one correspondence is drawn here between the modalities for message transmissions containing codified information and tacit knowledge. Communications of the sort whose cognitive contents are uncodified (or incompletely codified, craft), and the codified scientific statements to which they relate, alike are taken to be emitted locally in the first instance, and to diffuse first within the immediate social network contexts in which they arise.

Both as a matter of formal logic, and in practical affairs, knowledge may be either disclosed to others or kept secret, regardless of whether it exists in codified form or remains tacit.[16] The view presented by the preceding discussion is that for the ideas contained in scientific statements to be understood and rendered operational, researchers must possess the complementary tacit cognitive associations. The local network within which tacit knowledge is shared, thus, need not be considered to be a strategic instrument whose primarily functional role is that of "capturing" and exploiting the benefits of tacit knowledge for its members. Indeed, I take just the opposite position and emphasize the benefits to the individual researchers of cooperative, reciprocated transactions in tacit knowledge possessed by specialists; their expertise enables those cooperating to increase the chances of solving complex, multi-step problems sooner than would be the case were they to work in isolation.[17]

In keeping with this, we may suppose that local networks are not autarkic; by having some members in common with other, similarly local social groups, they can be rendered inter-communicative. Thus, complementary packets of codified and tacit knowledge, along with explicit and implied conjectures about promising lines of scientific inquiry, eventually may percolate outward from particular local networks and so become diffused throughout the wider community of researchers that constitutes the "invisible college".

Even though codified information can be made accessible more quickly and cheaply through impersonal broadcast channels, the view of the world presented here is one in which researchers in order to utilize much of that information directly, must obtain the complementary tacit elements through their social network connections. That alone would make the conditions affecting the

---

[16] See David and Foray (1995), where three distinct dimensions are recognized as defining a space in which knowledge products can be located: the codified-tacit axis, the disclosure-secrecy axis, and the public -private property axis.

[17] This supposition is consistent with the arguments set out in the following section (3): *small* cooperative networks of information sharing can be supported - as an equilibrium of a game - among researchers, because some (enforceable) degree of cooperative behavior furthers their self-interests, even though they may be racing for priority.

operation of local social networks the factors proximately constraining the speed and extent of effective diffusion of new discoveries and inventions. Indeed, it could be argued that recognition of the critical role of tacit skills and learning mechanisms - which has been such a major theme in recent sociological research on "laboratory life"[18] - should be read as justifying an even more stark caricature of scientific communities or invisible colleges: nothing but an informally organized network of local social communications networks, or an interconnected set of "neighborhood clubs." But, it is neither useful nor necessary to go to the extreme of shutting one's eyes completely to the existence in modern scientific communities of a supporting apparatus of institutionalized procedures for publication via broadcast media, for the organization of global discussion fora, and the like. Indeed, it seems more plausible, especially, when one comes to try to formally characterize the way that new ideas are generated, to assume that at each moment in time the members of the invisible college are able to access the common archival knowledge-base containing scientific statements whose reliability is no longer a subject surrounded by doubt and active controversy.

7.4 Information-pooling, reputation and micro-motives for network membership

Contributors to the macro-sociology of scientific institutions have described the role of modern "invisible colleges" in channeling informal communications among scientists and forming the social substratum upon which the collegiate reputational reward systems of academic, open science are grounded.[19] Invisible colleges thus fulfill many functions. As seen by Derek de Solla Price,[20] they confer upon each scientist "status in the form of approbation from his peers, they confer prestige, and above all, they effectively solve a communication crisis by reducing a large group [such as the entire population of accredited scientists in a field as large as economics, or chemistry] to a small select one of the maximum size that

[18] See, e.g., Collins (1974), Latour and Woolgar (1979).

[19] I have argued elsewhere that the central problem of informational asymmetry in the late sixteenth century patronage system was being addressed by the formation of networks of correspondence and the flowering of "invisible colleges" among mathematicians and the experimentalists in the new science. See David (1995). Such networks, augmented by an expanding volume of printed pamphlets and treatises, provided an arena in which challenges could be issued, contests and competitions could be staged, and collegiate reputations could be both secured and widely broadcast. Challenges and contests among Europe's mathematicians were not an innovation of the seventeenth century, however. See e.g., Bell (1937), Boyer (1975), pp. 311-12, 341.

[20] Price (1976 :pp. 83ff) re-introduced and conceptually extended the seventeenth century term "invisible college", coined originally by Robert Boyle in referring to the small group of natural philosophers whose intellectual transactions with one another anticipated the formation of the Royal Society.

can be handled by interpersonal relationships".[21] But, insofar as these functions of an invisible college rest upon the members' sharing knowledge that might confer personal advantage if kept for their private use, subscribers to the theory of rational individual utility-maximizing behavior will want to know what induces researchers to seek admission to these "clubs"; and conform by-and-large with club rules, by sharing knowledge that they have not yet divulged and may never divulge by means of impersonal broadcast channels. What, then, do we suppose are the private incentives offered to the individual scientist, outside of an institutionalized setting in which such conduct an explicit has been made an explicit obligation (publish or perish!) of career advancement? Or is it necessary to presuppose the existence of some prior process of socialization into the role of "scientist", one that implants in the individual an acceptance, and perhaps even an active subscription to professional norms of conduct in such matters?

Two answers come to mind that avoid such presuppositions. First, there is the obvious "exchange value" of information that is implicit in Price's (1963, 1986) discussion - except that some of the knowledge involved may be "expertise" conveyed by direct demonstration rather than being reduced to "information." The division of intellectual labor confers advantages to those who need not work in isolation, but, instead, are able to draw upon the still-unpublished findings and tacit expertise of others in identifying or solving some particular problem. Access to such networks of assistance must in effect be "purchased up front" by proffers of worthwhile "material", or gained by accreditation, which holds out credible promises of a future reciprocation of valuable intellectual help received at the outset.

Second, a certain status is conferred by being accepted as a "correspondent" by already established members of the scientific community. This carries some "signaling value" for the individual, as a mark of recognition and approbation that can be displayed to advantage in transactions with third parties. The "price" extracted for elevation to membership in networks peopled by more prestigious scientists may be high - in terms either of attribution of credit for discoveries and techniques that otherwise could be claimed exclusively, or deflections of one's efforts to working on problems less centrally consequential for the advancement of one's own research program than to the work of others. Except where the reward system of the scientific community is characterized by strong personal patronage mechanisms rooted in teacher-student relationships, one should expect to find neophyte researchers, prompted by the first category of motives, entering into regular informal knowledge exchanges with disciplinary "peers" who are approximately on a par with them in achievements and reputational status.

---

[21] See Crane (1972) for an early elaboration of Price's view of invisible colleges as vehicles for diffusing scientific information, an approach that sociologists of scientific knowledge have criticized as neglecting the social interaction processes through which tacit knowledge is formed and conveyed. Merton [1986] discusses the relationships between the concepts of reference group, invisible college, and deviant behavior in science.

Unlike signaling value, the benefits of participating in a collegial research network (extending beyond the immediate group or team with which the individual is currently engaged) can be thought to have a substantive basis for both researchers and their patron-employers. This is so, whether or not the final disposition of the knowledge that is acquired will be public disclosure or private exploitation in some directly productive activity. Through the medium of their "network", the individual members have at their disposal assistance in discerning unrecognized anomalies and research opportunities, and help in locating solutions to unanticipated problems that tax their own or their research team's cognitive capabilities. The collectivity of the network thus operates somewhat like a "patent-pooling" arrangement without the patents; it is a loose coalition, in which some degree of non-compliance with the norm of mutual help will occur, and so yield an equilibrium in which the "common pool" is degraded by the private reservation of certain information. Nevertheless, access to help from "peers" remains a valuable asset for the individual.

To restate the thrust of the foregoing discussion, it is possible that cooperative behavior within a limited social sphere can emerge and be sustained without requiring the prior perfect socialization of researchers to conform (altruistically) to the norm of full disclosure and cooperation. This is a rather straightforward instance in which insights from the theory of repeated games are applicable to explaining cooperative behavior among potentially rivalrous researchers.[22] To sharpen this point, one can start simply by considering two researchers working towards the same scientific goal which involves the solution of two sub-problems, and suppose that each has solved one of the problems. Once each gets the other solution, it will be a matter of writing up the result and sending it off to a journal for publication, or reading the paper at a prestigious professional gathering, - the first to do so being awarded priority. Now suppose, further, that the write-up time is determined by a random process, and that if both get the two halves of the problem at the same moment, each will have the same (one half) probability of being the first of the pair to submit for publication. Whether the winner will be awarded priority will depend, however, on whether or not some other researcher has obtained the full solution and sent it off already. The question is: should the first researcher to get any part of the solution follow the strategy (S) of sharing that information with the other one, or should she adopt the strategy (W) of withholding?

If, without prior communication, they play the strategy pair (S,S) they can proceed immediately to the write-up stage; if they play (S,W) the second member of the pair will be able to proceed to the write-up, and the opposite will be true if they play (W,S). Should they both withhold (W,W), they must both spend further time working on the other problem. It is evident that if they are only going to be in this situation once, the "winner-takes-all" rule of priority will alone induce each of them to withhold, and they will end up (collectively, if not individually) at a

[22] The following parallels material developed in Dasgupta and David [1994].

relative disadvantage vis-a-vis other researchers who are hurrying to publish. If nobody else has the full solution yet, society also will have been forced to wait needlessly, because each member of the pair has a dominant (private) strategy of withholding what she knows. This game has the structure of a classic two-person "Prisoners' Dilemma."

It is well known, however, that an escape from the pessimal outcome and the attainment of the superior, cooperative equilibrium is possible under certain conditions: if the game is played repeatedly, if the future is not discounted too heavily, and if the players expect the other member of the pair to remember, and punish, their failure to cooperate by sharing. Indeed, there is a so-called "folk theorem" to that effect.[23]

Yet, that is not the end of the matter. As there are other researchers in the picture, we should really be considering an **n**-person game, involving the solution of an **m**-part problem, where **n** > **m**. Now the question of sharing information becomes one of sharing not only what you have learned yourself, but, also what you have been told by others. It obviously is advantageous to be in a group among whom information will be pooled, because that will give the group members a better chance of quickly acquiring all **m** parts of the puzzle and being the first to send it in for publication.[24] But, on the other hand, if there are individuals who

---

[23] In addition, the value in the future of developing and maintaining a good reputation for sharing has to be large in order to discipline the self-interested researcher into adhering to the sharing mode of behavior in the current period. If repetitive play comes to an end, or if the future is valued only slightly, cooperation will unravel from the distant terminal point in the game, right back to its inception. For a non-technical introduction to the literature on the repeated Prisoners' Dilemma and its broader implications, see Axelrod (1984). The "folk-theorem" of game theory holds that (if future payoffs are discounted by each player at a low rate) in the "super game" obtained by repeating a finite, two-person game indefinitely, any outcome that is individually rational can be implemented by a suitable choice among of the multiplicity of Nash equilibria that exist. See Rubinstein (1979), (1980), and Fudenberg and Maskin (1984).

[24] Campbell (1994 :p.143) attributes to Michael Polanyi the analogy of a jigsaw puzzle-working group. Polanyi suggested, as a functional justification for the norm of open information sharing in science, that it would be "optimal" for the collective solution of the puzzle that every member of the group see the whole board and the pieces that each of the others was holding. Such heuristic examples may suffice to make the point about the advantages of cooperation, but they cannot substitute for empirical investigation of the structure of communications channels in small (local) networks that will be most effective. Some work in the field of social psychology as been devoted to this question: Harold Guetzkow and Herbert Simon (1955), for example, reported experimental results in group problem-solving, connecting the members of six-person teams in three different forms of communication networks, following the patterns previously studied by Alex Bavelas (1950). These were (i) the "ring" structure, having as many channels as nodes (n=6), each node being linked to a node on each side; (ii) the "star" or hub-and-spoke structure, in which (n-1) = 5 channels suffice to link one node to every one of the others in binary couplings; (iii) the "fully" connected network, which requires n(n-1)/2 = 15 channels. Guetzkow and Simon found that when they distributed fragments of a poker deck among

behave opportunistically by exchanging what they have learned from one group for information from people outside that group, but do not share everything they know within their group, they can expect to do still better in their current race for priority of publication. However, because others would see that such "double-dealing" will be a tempting strategy, cooperation will be unlikely to emerge unless "double-dealers" (who disclose what you tell them to third parties, but don't share their knowledge full, with you) can be detected and punished. What is the form that retribution can take? Most straightforward will be punishment by exclusion from the circle of cooperators in the future; and even more severely, not only from the circle that had been "betrayed" but from any other such circle. This may be accomplished readily enough by publicizing "deviance" from the sharing norms of the group, thereby spoiling the deviator's reputation and destroying her acceptability among other groups.[25]

What, then, is the likelihood that this form of effective deterrence will be perceived and therefore induce cooperative behavior among self-interested individuals? If a group, i.e., "a research network" numbering **g** players ($\mathbf{g} \leq \mathbf{n}$) is large, identifying the source(s) of "leaks" of information and detecting instances of failure to share knowledge within it, will be the more difficult. It is worth remarking that the power of a large group to punish the typical deviator from its norms by ostracism tends to be enhanced by the higher probability that all those individuals with whom potential deviators will find it valuable to associate are situated within the group. In other words, the expected loss entailed in being an "outcast" is greater when there is only a fringe of outsiders with whom one can still associate. But, this consideration is offset by the greater difficulties the larger groups will encounter in detecting deviators. Smaller groups have an advantage on the latter count, and that advantage also enables them to compensate for their disadvantage on the former count. The more compelling is the evidence that a particular individual had engaged in a "betrayal of trust", the more widely damaging will be the reputational consequences for the person thus charged. Hence, unambiguous detection and attribution of deviations from recognized norms (regarding the disclosure and non-disclosure of information) augment the deterrent power of the threat of ostracism that can be wielded by any group that remains small - in relation to the total population of individuals with whom an excluded group-member could form new associations.

---

the participants and asked them to select the strongest poker hand from among the cards they held, the hub-and-spoke structure performed best, regardless of the ability of the person positioned at the hub; the ring was worst. Interestingly enough, in repeated play it was observed that the fully connected structure tended to evolve, through disuse of channels, into a hub-and-spoke pattern of communications. It may be noticed that the latter pattern is precisely the one that is assumed here by the stochastic opinion-polling process in the "voter model" of local social network interactions.

[25] See Greif (1989), and Milgrom, North and Weingast (1990) for analysis of repeated games of incomplete information that have this structure.

The foregoing suggests that small cooperative "networks" of information sharing can be supported among researchers, because cooperative behavior furthers their self-interest in the race for priority, and denial of access to pools of shared information would place them at a severe disadvantage vis-a-vis competitors.[26] Does this imply that the normative content of Merton's communalistic norm of disclosure is really redundant, and plays no essential role in fostering conditions of cooperation among citizens of the Republic of Science? Not so! For, it can be seen that networks of cooperative information sharing will be more likely to form spontaneously if the potential participants start by expecting others to cooperate, than if they expect "trust" to be betrayed; and cooperative patterns of behavior will be sustained longer if participants have reason to expect refusals to cooperate will be encountered only in retaliation for transgressions on their part.

Furthermore, detection of deviant behavior warranting punishment, and implementation of the retribution of ostracism from a particular network, will have more broadly damaging reputational consequences when the norms of behavior involved (i.e., the "custom" within the network in question) is common knowledge, and part of the shared socialization among all the potential members of networks. It is evident from this that even if the process of socialization among scientists were weak and quite imperfect, the common "culture of Science" makes it much more likely that the rule of priority will not tempt individuals to engage in opportunistic withholding of knowledge, and, instead will engage the self-interest of researchers in reinforcing adherence to the norm of disclosure - at least among those restricted circles of colleagues that form his or her local social network.

7.5 Cognitive communications: Beliefs, Bayesian learning, and scientific consensus

What is it that is being communicated in these transactions? For purposes of simplifying the following analysis, envisage two sorts of cognitive processes as taking place within the local social network structures. Information in the form of codified statements can be passed among the individuals (or among the teams, and other multi-person research organizations that institute communication policies binding upon their members), by the act of one of them sending a message or "sharing" a piece of knowledge, and the other receiving or "reading" it. The

[26] These "circles" or "networks" which informally facilitate the pooling of knowledge among distinct research entities on a restricted basis can exist as exceptions to both the dominant mode of "public knowledge" characterizing academic science, or the dominant mode of "private knowledge" characterizing industrial R&D. Thus, von Hippel (1988) and others have described how firms tacitly sanction covert exchanges of information (otherwise treated as proprietary and protected under the law as trade secrets) among their engineer-employees. Participants in these "information networks" who accepted money or remuneration other than in kind, most probably, would be dismissed and prosecuted for theft of trade secrets.

cognitive substance of the generic message-transaction comes in two parts. The first component contains or otherwise identifies a particular scientific *statement* about the natural world, or about the design of a measurement instrument or other artifact, an experimental procedure, or a logically connected "bundle" of such statements.[27] The message's second part conveys the sender's present state of belief as to the "reliability" or "unreliability" of the accompanying statements.[28] I do not speak here either of "truth", or of degrees of reliability. At any moment the researchers (acting either in a team organization or as a solo investigator) are of only one mind about the reliability of the particular statements they are emitting or transmitting, and so they mark them either as "acceptable", or "unacceptable". But their minds are open, in the sense that from moment to moment they can find cause to revise their labeling of the same statement(s).

This revision process is precisely where the cognitive content of the messages' second component (beliefs) comes into play. Two stories could be told now about how scientific inquiry is conducted in the world of this caricature. In both accounts the researchers are seen to be engaged in revising their orientations with regard to the reliability or unreliability of statements that have reached them in ways that permit them to be influenced, if not completely "persuaded" by the information they receive concerning the beliefs held by colleagues belonging to their local social network. They also are mutually compatible, so that for our purposes it does not seem that one must choose between them.

7.5.1 Beliefs and Bayesian experimental strategies

In the non-strategic story the researchers are assumed to be engaged in bounded Bayesian information processing, drawing observations from their own experiments, and the reports of similar observations made by others belonging to their network. Inasmuch as they are assumed to be following a common epistemological strategy (and referring to the same subjective probability thresholds when declaring a conjecture about an unobserved state of nature to be acceptable, or not) they view the *a priori* beliefs conveyed by neighbors as having been shaped by a Bayesian revision process. Therefore it would have been

---

[27] One may imagine that the cognitive content of such "statements" is produced by combining the texts of codified information already, or soon to be broadcast through public channels, and further comments of a qualifying , or clarifying sort resembling "marginalia" on those texts. The latter may be linguistically expressed, or conveyed by actions ('gestures"), and reflect the tacit knowledge of the sender - including particular, idiosyncratic renderings he or she may give to published "texts".

[28] It can be supposed that the channels of communication are multiplexed, and capable of bundling (or "packaging") cognitively interrelated propositions, so that it would be possible for them to handle more or less concurrently a flow of numerous messages of the foregoing sort about many distinct, and cognitively independent scientific statements. Each of those problematics could be assigned to its own "layer" over the network, and the resulting information-processing architecture thus would be enabled to simultaneously execute multiple consensus building routines in parallel.

influenced by both the pooling of beliefs communicated by other neighbors, as well as being occasionally updated by the formation of the likelihoods of their own experimental observations.

The expressed beliefs of local colleagues about the acceptability or non-reliability of propositions *cum* observations, are assumed in this story to weigh heavily. Remember that these beliefs have been communicated by the folk with whom they regularly are interacting. So, another appeal to a "folk theorem" provides some warrant to suppose that even if the individuals involved had not been socialized to report accurately their current scientific views, the prospect of extended repetitive play would have led them toward some Nash equilibrium characterized by "cooperative" rather than opportunistically deceitful strategies in this regard.[29]

But, it is now necessary to take on board some of the doubts that sociological, philosophical, and psychological studies have raised regarding the degree to which scientific progress occurs through the experimental invalidation of conjectural propositions, especially when strong supporting beliefs surround the proposition in question.[30] The 1970s were marked by the rising chorus of doubts among philosophers and historians of science about the basis for the view that experimental falsification of hypotheses played a crucial role directing the progress of scientific knowledge. This scepticism emerged from the sociological and psychological critiques advanced by Kuhn (1962/1970), Lakatos (1970) and Feyerabend (1975), and reinvigorated the logical problems that the earlier writings of Duhem and Quine posed for Popper's (1959) account of the process of scientific discovery. As Quine (1953:45) summed up the problem: "Any statement can be held true, come what may, if we make drastic enough adjustments elsewhere in the system". Thus, it might be argued that whereas the whole edifice of scientific theory was at risk of empirical refutation, particular theories or hypotheses could escape experimental falsification.[31]

Thomas S. Kuhn, in *The Structure of Scientific Revolutions* (1962/1970), held that major changes in scientific understanding occurred through shifts in paradigms, and that although competing paradigms each defined a context of "normal science" within which certain kinds of experiments played an important

---

[29] See David (1995) for further discussion of the emergence of cooperation from "repeated play" in this context. Since individuals are free to revise their beliefs, discovery of "misrepresentation" would not be quite straightforward; when challenged on such a matter, a researcher would have to acquit him- or herself by back-tracing the sequence of events that led from one belief orientation to its opposite.

[30] Little will be said here about some of the early doubts about validation processes in science that were raised by research on the social psychology of perception and observational reporting, but Campbell (1994) offers a useful review (from the social psychology perspective) of sociological and philosophical literature noticed here.

[31] See Franklin (1986:4, 106) for the latter reading; Harding (1976) for discussion of the so-called"Duhem-Quine problem" regarding the possibility of scientific refutation, and the original essays of P. Duhem and W. V. O. Quine.

role in the evaluation of particular theories, alternative paradigms, essentially, could not be put to empirical tests on common experimental grounds. Lakatos (1970:119) also contended that experimental evidence would not suffice to reject a theory: "...no experiment, experimental report, observation statement or well-corroborated low level falsifying hypothesis alone can lead to falsification. There is no falsification before the emergence of a better theory." Feyerabend (1975) went farther, portraying changes in scientific opinion as propaganda victories by one group of scientists over others, and denying the validity of the modern scientific enterprise's claims to possess a superior method of gaining knowledge through experimentation.

In this climate of reappraisal of the traditional emphasis placed upon "crucial experiments" in the progress of science since the 17th century, historians of science cast a skeptical eye upon putatively epic events of empirical refutation and found them to be largely "mythical". Not only were doubts expressed as to the very occurrence of the experiment whereby the authority of Aristotelian mechanics, supposedly, was undermined by Galileo's finding of essentially the same rate of fall of two unequal weights dropped from the Tower of Pisa; modern replication showed that the observable difference in the speeds of the objects over their 200 foot descent would have been too large to justify Galileo's reporting it to have been negligibly small.[32] Similarly, revisionist history of science now instructs us - through Worrall (1976) - that Newton's corpuscular theory of light was not overturned, and the wave theory established by the experiments of Thomas Young, since corpuscular explanations were available for both interference and diffraction.[33]

To be sure, counter-arguments can be mobilized against the extreme position that scientific changes are wholly theory-dominated; and that, rather than being driven by the force of experimental evidence, it is sociological forces and even political and economic considerations external to the logic of particular scientific questions that impinge in crucial ways upon the fortunes of contending theories.[34] There have been instances in which, clearly, experiments did prove "crucial" in over-turning a prevailing theoretical model. A striking modern case in point is that

[32] Franklin (1986:p.2, n.7), citing the replication study by Alder and Coulter (1978), goes on to point out that an Aristotelian could readily have modified the theory to accommodate the experimental data.

[33] Moreover, the early (pre-Fresnel) wave model could not account for the rectilinear propagation of light, which was as troublesome for that theory as interference was for the corpuscular model. See Worrall (1976), discussed by Franklin (1986:pp.2 n.8).

[34] Franklin (1986) reviews recent historical research on experimental physics (by Galison, Pickering and others), in addition to presenting his own case-studies of the role of experimental evidence. These include Millikan's 1913 "oil drop" experiments which established the quantitization of electrical charge and measured the electron's charge; and the experimental tests reported in 1957 papers by Wu et al., by Garwin et al., and by Friedman and Telegdi, which brought rapid acceptance of Lee and Yang's (1957) theory of parity nonconservation.

of the experimental results prompted by Lee and Yang's (1957) famous theoretical paper, suggesting that the theory of parity conservation (mirror symmetry) - which at the time was accepted by the physics community as characterizing the strong and electromagnetic interactions - might not hold for the weak interactions. Yet, the careful historical re-examination by Franklin (1986: Ch.1) concludes that this was perhaps an atypical case, and a really rather special set of circumstances afforded the experimental evidence the leverage they exerted.[35] The discovery of nonconservation of parity, therefore, may serve less to exemplify the "crucial" role of experiments, than to expose the special nature of the circumstances that would make it likely for experimental results to prove decisive in rapidly altering a scientific consensus.[36]

More generally, close examination of the ways in which runs of experimental physics results are generated, reveals how the latter can sometimes appear to imply parameter magnitudes that are conditioned by either theoretical expectations; or by the initial values that were obtained in contradiction of received theory, but, after extended replications, may undergo significant alterations causing them to converge to the theoretical expectations. Franklin (1986: 235-236) writes:

---

[35] The list itself is instructive: (1) the contending theories were mutually exclusive and exhaustive - they belonged to two qualitatively delimited classes, i.e., those that satisfied symmetry and those that did not, rather than simply making differing quantitative predictions; (2) a reinterpretation of the experimental results in the logical spirit of the Duhem-Quine gambit - contending that they did not establish the absence of symmetry specifically in the weak interactions because parity might not be conserved in the "background" conditions which the experiment presupposed - would have led to the even more radical rejection of symmetry in all the forces of nature; (3) while there was extensive experimental evidence consistent with parity conservation in the other forces, as Lee and Yang had pointed out, there was no such evidence about the weak interactions; (4) the operation of the apparatus that was used depended only upon the strong and the electromagnetic interactions, and so, while the experimental design was "theory-laden", the results could not be compromised by circular dependence upon the theory that they appeared to refute; (5) there already existed "framing conditions" of a sort that could be seen as providing a precedent for the partial violation of the parity conservation law - that is, for symmetry to be violated in only one class of interactions. This was the theory of the conservation of "strangeness", put forward by Gell-Mann and Mishijima during the immediately preceding period (1953-55). It accounted for certain well-established experimental observations by the conjecture that particles known as K meson and hyperons possessed a quantity called "strangeness" that was conserved in the strong, but not in the weak interactions. See Franklin (1986: 32-33, 39).

[36] Indeed, as Franklin (1986: Ch.2) describes, there were experimental results in the 1930's on the scattering of fast electrons by metals that, in retrospect, provided evidence for the nonconservation of parity in the weak interactions. The asymmetry findings were recognized at the time as anomalous because they were part of a larger body of experimental results that were at variance with the predictions of an other-wise well-corroborated theory of Dirac; but, because there was no theoretical context in which parity nonconservation evidence could be read, the latter became lost from view.

> "In any experiment, the sources of error, particularly systematic error, may be hidden and subtle. This is particularly true of the technically difficult experiments... The question of when to stop the search for sources of error is then very important. One psychologically plausible end point is when the result seems 'right' (i.e., in agreement with previous results)".

One explanation of the role of theoretical expectations in guiding experimental outcomes in the evaluation of physical constants, is that attributed to the physicist E.O. Lawrence:[37]

> "In any highly precise experimental arrangement there are initially many instrumental difficulties that lead to numerical results far from the accepted value of the quantity being measured. It is, in fact, just such wide divergences that are the best indication of instrumental errors of one kind or another. Accordingly the experimenter searches for the source or sources of such errors, and continues to search until he gets a result close to the accepted value. *Then he stops!* [Emphasis in original.] But it is quite possible that he has still overlooked some source of error that was present also in previous work. In this way one can account for the close agreement of several different results and also the possibility that all of them are in error by an unexpectedly large amount".

Such attitudes, and the experimental experience that they illuminate, constitute what has been referred to as "intellectual phase-locking" (Franklin 1986: 236). Should we then believe that experimentation cannot offer a check on theoretical consensus? Are agreements among scientists so powerful a force for conformity that they compel convergence of the experimental evidence to the prevailing "truths"? To assert this would be to ignore the mechanisms that are thought to be reliable in the long run to effect escape from serious experimental errors. The performance of quantitative measurements by "different" methods, and especially the repetition of measurements when new and more powerful experimental arrangements and instrumentation becomes available, are perhaps the most oft-cited mechanisms of auto-correction for "bandwagon effects" that would otherwise perpetuate experimental errors. But the reward system within which the experimenters work must be such as to repeatedly impel the search for new techniques, and their swift engagement when they do become available. In the case of an important magnitude, such as the fundamental physical constants, there have been indeed, many repetitions of experimental measurements. According to Franklin (1986:236):

> "The histories of such measurements, particularly those for *e*, the charge on the electron, and *c*, the speed of light, often do show such major changes when a new, and presumably better, technique is used."

---

[37] The reference in Franklin (1986: 239) for the following quotation is to R. Birge, "A Survey of the Systematic Evaluation of the Universal Physical Constants", *NC* [Suppl.], 6 (1957):51.

So, there is a selective element of built-in correction to the research process, because when experimental techniques change there is an opportunity to score a scientific "coup" by establishing a new and different value from the one previously accepted when the magnitude in question is "important" and widely relied upon. Alternatively, so long as there is some persisting acknowledged discrepancy between the theoretical expectation and previous experimental findings, there is hope of achieving recognition from their reconciliation in one way or another. These aspects of the science reward structure will thus set some bounds upon unintended tendencies towards conformity with prevailing theoretical expectations.

7.5.2 Conformity to local peer consensus as a reputational strategy

Introducing some further strategic considerations, however, may provide a direct rationale for the persuasive power that a prevailing consensus for, or against, a given proposition might exert in the formation of an individual researcher's reported beliefs about its validity. Unlike the foregoing arguments from the philosophy and history of experimental science, this story is meant to be more generically "economic" in its appeal. In the spirit of arguments suggested by Dasgupta and David (1994), we may suppose that a scientist working in a collegiate reputational reward system will consider the nearer-term reputational consequences of current actions (including expressions of scientific opinion), as well as considering long-term payoffs possibilities in the form of lasting fame for "having gotten it right". Whether one will be found to have been "right" in the judgement of the invisible college depends upon the alignment of one's recorded (or remembered) beliefs in relation to the consensus that existed at the time among members of the local network to whom those beliefs were disclosed; and also in relationship to the global consensus that may emerge within the invisible college as to the "truth" of the conjecture in question. It is quite possible to envisage a structure of "expected reputational payoffs" that makes it a dominant strategy to be found to have had beliefs conforming with those presently held by most of one's local network, even when the ethos of the lone scientific hero would accord maximum kudos and immortal glory to a researcher who had not conformed with local peer opinion, and whose beliefs eventually came to be shared by an overwhelming segment of the discipline at large.

Consider the following example, in which **c** denotes conformity with the preponderance of scientific opinion in one's network, and **d** denotes disagreement with that consensus. For expositional convenience we will consider the case where the prevailing consensus holds a particular statement to be "reliable/true", denoted as **R**. (One can treat symmetrically the case in which the consensus among the agent's reference groups holds the statement to be "unacceptable/false", not-R or, equivalently W.) The consensus that can emerge *eventually* in the global network, i.e., the limiting configuration of beliefs among members of the invisible college, may either hold the statement to be "Reliable Knowledge", denoted by **R**, or not, denoted by **W** ("Wrong"). But we suppose that this eventual determination cannot be known with certainty when the researcher is deciding which opinion to give on the matter. It is in that sense an unobserved "state of (social) nature". If the individual researcher treats the local network as the reference group whose esteem matters, we then have the following notation for possible states, to each of which there will correspond "reputational payoffs" for the representative researcher:

{**c**, **R**|**R**}= $\mathbf{b_1}$: being right, with the crowd;
{**c**, **R**|**W**}=$\mathbf{b_2}$: being wrong, with the crowd;
{**d**, **R**|**R**}= $\mathbf{b_3}$: being in a minority and wrong;
{**d**, **R**|**W**}=$\mathbf{b_4}$: being in a minority and right.

A suitable general payoff structure for an epistemic community is one that assigns greater value to an individual researcher who is found to be "in the right" than to the individual who embraced the "wrong" view. But it is also plausible that being wrong in a crowd will be deemed not as bad (for her subsequent reputational standing) as being found to have been wrong more-or-less on one's own; and being "lonely yet right" is deemed reputationally more glorious than being correct among a crowd of one's peers. This scheme of valuation of the outcomes of the game against (social) nature corresponds to the condition : $\mathbf{b_4}>\mathbf{b_1}>\mathbf{b_2}>\mathbf{b_3}$.

Now, let θ denote the individual's subjective probability assigned to the global consensus forming eventually on **R**. Where does this subjective probability come from? There will be some processes of consensus formation for which it is the case that the best probability estimate that the configuration of beliefs in the college at large eventually converges on **R** will be given by the distribution of beliefs prevailing at the initial point.[38] But, even if we have warrant for assuming such is the case, the preceding discussions of the role of local social networks gives grounds for assuming that the individual researcher's field of information on this matter is restricted to the state of beliefs among members of her immediate, local network. So, the proportion of the latter who presently hold say "R" when canvassed could thus provide the individual's probability estimate of the eventual "truth" towards which the consensus of the invisible college as a whole must converge, θ. Then, with a little algebra it follows the condition $[(\mathbf{b_1}-\mathbf{b_3})/(\mathbf{b_4}-\mathbf{b_2})]>[(1-\theta)/\theta]$ is sufficient for the strategy of conformity, **c**, to be the one that will maximize the individual researcher's expected reputational payoff. Conformity

[38] It will be seen below in section 5.2, that the stochastic process known as "the voter model" is one system for which this is true. As ***Property 3*** of the model states (for one- and two-dimensional network graphs) the system achieves perfect correlation of beliefs on one or the other of the binary options, and the probability that can be assigned ex ante to either of those outcomes corresponds to the initial distribution of beliefs. The "voter model" is constructed by symmetrically assigning to each node of the graph (each researcher) a belief revision strategy that reduces to probabilistically following the local consensus, and the illustrative micro-model provided here is consistent with that. In other words, the hypothesized Bayesian revision strategy is seen to be individually rational under the specified payoff structure and the restriction that ties the individual's subjective probability estimate to the distribution of beliefs in her local social network.

with the prevailing local network majority opinion implies that when the best estimate of θ is substituted into that condition, it simplifies to: $(b_1\text{-}b_3) > (b_4\text{-} b_2)$.

This says that for conformity to be generally rational, it will be sufficient for the term $(b_4\text{-} b_2)$, to approach a small positive value. In other words, deviance from the local peer consensus should not be too strongly rewarded even when it turns out eventually to be "justified", whereas being mistaken should not deal severe damage to one's reputation when one has done so in plenty of company.

Keeping matters simple, we may postulate that such conditions do obtain and go on to build upon the implied micro-level behaviors. We therefore suppose that all the researchers are symmetrically situated in their respective local networks, and that each is (rationally) adhering to a decision routine of the following sort: at random intervals in time the opinion-messages emitted by others in the local network are polled, and if there is unanimity among them, either on the reliability or the unreliability of a particular proposition, the researcher at the polling node will accept the local consensus and accordingly adapt the messages she emits concerning the state of her belief about the validity of the proposition in question. When a disagreement is found by polling within the local network, the researcher takes a quasi-Bayesian approach, and probabilistically forms an opinion about the true state of the natural world, by a procedure equivalent to tossing a coin whose loading mirrors the division of opinion among the set of her neighbors. In this procedure, implicitly, equal influence is accorded to the opinions of the local reference network members.

How does this work? Consider these two trivially simple illustrative situations:

(i) If the social network was composed of three members, then disagreement among those polled would take the form of an even split. So, the index-agent who was polling her "neighbors" would toss a fair coin (with a 1/2:1/2 loading) to arrive at an unequivocal position on the question under investigation.

(ii) Alternatively, if the social network had four members, disagreements among the index-agent's "neighbors" would take the form of 1:2 or 2:1 splits on the question of reliability. Accordingly, the index-agent would form an "opinion" on the question by tossing a coin loaded (1/3: 2/3) for reliability when the local consensus was against that view, and a different coin, with loading (2/3: 1/3) in favor of the statement's "reliability", when the reference group's balance of opinion was favorable.

The toss of the (local opinion-weighted) coin envisaged in this routine could be interpreted as a highly abstract representation of the conduct of an inherently ambiguous experiment or observational procedure, in which the "reading" of the outcomes would be strongly shaped by the experimenter's or observer's priors as to the validity of the hypothesis under examination. Only when opinion in the local peer group is quite evenly balanced would the testimony of the experiment exercise much leverage upon the experimenter's reported belief, and then, the conclusions that will be drawn from a "face value" reading of the results remains uncertain.

At this point two remarks may need to be entered concerning this schema. The first is directed to readers who have noticed that these hypothetical scientists cannot be very bright if they are using this cumbersome procedure for gathering and processing the information about beliefs prevailing in their local network. In situation (i) above, it evidently would be far simpler to equi-probably sample one of the two others in the network, and adopt his or her orientation on the question under consideration: were both in accord, it would make no difference to the outcome which of them was consulted; if they happened to disagree, the index-agent's orientation on the issue of "reliability" would be determined under either procedure by the toss of a coin with a 50:50 loading. A similar argument holds for situation (ii), in which the polling agent shares a network with three other researchers and, in the event of their disagreement, will have a one-third probability of polling a member whose views represent the 2:1 majority, and the same chance of drawing the minority belief. The outcome, in terms of the orientation of the polling agent who merely mimics that of a network member chosen (with equal probability) at random, is the same as that which would result from canvassing the entire local network and - in the absence of unanimity - forming an opinion probabilistically from the information obtained about the distribution of beliefs among its members. The same argument can be generalized to larger groups than the ones just considered, and the two procedures will be exactly equivalent in their probabilistic outcomes, so long as the rule for coin-tossing in the absence of unanimity is designed in the symmetric form that sets the coin-loadings to mirror the pattern of disagreement among other members of the social network. As soon will be seen, in Section 5, this formal equivalence of mechanisms turns out to be very useful. It simplifies analysis of the properties of the stochastic diffusion of belief statements throughout the extended community (the invisible college) which can be formed by the intersections, or overlappings of local social networks that share two-or-more members with some other social networks.

But, even though, for purposes of analysis it is strictly correct and convenient to treat the two mechanisms as equivalent, from another vantage point it may be held that these two procedures for forming an opinion really are not at all the same. This brings us to the second of the promised pair of remarks, a digression addressed to those who protest that the problems of theoretical framing and the ambiguities of complex experimental observations notwithstanding, the preceding description is a sheer travesty of science, and therefore ought not to be allowed to stand. Surely it is excessive to have minimized the persuasive force of modern science's cognitive content, to the point of portraying its practitioners as being simply engaged in efficient opinion sampling among their close colleagues. Worse yet is the caricature proposed in the latter reduced-form polling routine, in which the actors are quite passive, and mechanically copy the view of the colleagues with whom they have just made a (random) contact. Can it be thought reasonable to so completely disregard the claims of scientists to possess other, more reliable methods, such as direct observation and experimentation, which they deploy to arrive at "closure" in discussions about the validity of scientific statements?

7.5.3 Realists v. relativists: "Culture Wars" and an evolutionary reconciliation

One style of direct answer already has been given to this sort of protest. It comes from adherents of the so-called Edinburgh "Strong Programme" in the sociology of knowledge, following the seminal arguments of Bloor (1976). These "social constructionists" would maintain that, inasmuch as nothing distinguishes science from other domains of human action and insofar as scientists behave like other social actors, there is no cause to seek to explain agreement, disagreement, success and failure in science in *special* terms. A generic account of consensus formation, such as the mimetic process captured by the foregoing random polling model, might well suffice to make the general point.

Rather than having recourse to such an extreme rationale, however, it seems more satisfying to point out that it is only in the face of total unanimity among a close set of scientific colleagues that our researchers go along with "the accepted view". And here we can look to Kuhn (1962) and the philosophers and historians of experimental science who have documented the power of reigning scientific paradigms to withstand assaults by contradictory empirical observations. In the other and more frequent circumstances in which there is some disagreement, the researchers in our model adopt a position formed probabilistically, and so it may be imagined that there would be occasions (admittedly they will represent the rarer contingencies) when a contradiction of the weight of received opinion by the results of the researcher's own inquiries is sufficient for the former to be rejected.

Will this median stance be sufficient? Can it serve to stave off an outright refusal on the part of dogged critics of the "social construction" movement in the sociology and philosophy of science to suspend, for even a moment further, their disbelief in the foregoing "toy" model of a science research network? The current state of debate over the issues that have been raised is not entirely encouraging on this score. In a recent exchange in the *New York Review of Books*, for example, Gerald Geison (1996:68) characterized the unsympathetic review by Max Perutz of Geirson's book, *The Private Science of Louis Pasteur*, as another "salvo in the culture wars that now divide too many scientists from too many historians, philosophers, and sociologists of science." Perutz (1996:69) answered by reiterating his rejection of the "relativist's" view that knowledge, ultimately, is warranted by institutionalized communities whose acceptance of an empirical claim certifies it as true. The terms in which he did so suggest no imminent abatement of the ongoing hostilities:

> "These statements represent a fundamental misunderstanding of scientific knowledge. For example, the Second Law of Thermodynamics states that heat cannot be transferred from a cold to a warm body without performing work. This is neither an empirical claim, nor a social construction, nor a consensus by institutionalized science, but an inexorable law of nature based on the atomic constitution of matter. Scientific laws are different from social ones like 'extreme poverty breeds crime,' which may be called a consensus among liberal sociologists."

It is manifest that Perutz and other practicing scientists do not accept the point of the SSK inquiry into the way in which "law like statements" come to be accepted. Their notions of the progress of scientific knowledge are more than merely unreconstructedly Whiggish: "acceptance" does not enter into Perutz's vocabulary in connection with scientific statements; and the possibility that propositions may be accepted as being "true", only to be discarded at a later time for a "better truth", is not something that interests him. Such propositions, one suspects, would not be seen by Perutz to belong to Science, which he treats as being interesting as an epistemological category, and not as a cultural activity. Taking an essentially teleological position, Perutz regards "science" as knowledge upon which the processes of inquiry *eventually and inexorably* will converge. Thus, it is not the acceptance by contemporary peers, but by later generations of scientists that confirms the truth of a scientific statement. For Perutz, Pasteur's claim to greatness rests not only on having gotten a useful answer, but one that is "seen to be correct" in the light of subsequent experience:

> "[Pasteur's] greatness derives from a long series of fundamental discoveries which have stood the test of time and have much reduced human suffering....I accept the validity of Pasteur's work not merely because of his practical successes, but because his experimental results have never been falsified and their validity has been underpinned and given a strong theoretical basis by modern molecular biology. It is the hallmark of scientific genius to find the right answers even before the rationale for them becomes apparent."

Geison (1996: p.69) is quite justified, therefore, in characterizing this approach - which emphasizes the "end of the story" - as one that is "fundamentally *ahistorical* and inappropriate to any assessment of Pasteur's scientific achievements in the context of his own time". The last seven words of this sentence constitute its operative phrase, because it would seem that Perutz cannot see the sense of undertaking such an assessment except as a prediction of what will be the ultimate "judgement of history."

The idea that accepted scientific truths are at least partially matters of convention, social construction or consensus formation (that is, pieces of culture), seems virtually inescapable if one is seriously concerned to understand how scientific communities function in real time. At the same time, however, it does seem hard to deny that there is a point to Perutz's (1996) insistence upon taking a longer view; it is an important quality of "scientific inquiry" as a collective activity that it is able to escape *eventually*, even if not instantaneously, from the grip of consensus upon propositions that are falsified repeatedly by observational experience. Acknowledgement of this special quality, surely, is one of the things that gives studies of the social and economic organization of *scientific* communities a special, distinctive claim upon our interest. Otherwise, why would we have a sociology of scientific knowledge that was differentiated from that of bureaucratic knowledge, or legal knowledge, or theology? The Edinburgh "strong programme" for the sociology of scientific knowledge proposed that the mechanism of consensus formation should be one that was neutral with respect to the eventual conclusion as to the validity or falsity of the propositions. This was a useful departure, in asking sociologists of science to take notice of what they had learned of a generic nature about belief-formation in communities of specialized practitioners. It was a mistake, however, to suppose that such a programme by itself would be sufficient to generate a useful sociology of knowledge in modern *scientific* communities. This, surely is what Perutz (1996) means when referring to the Second Law of Thermodynamics as not merely a "social construction", but something objectively real about the atomic makeup of matter.

If that is so, then a resolution of the conflict between the "two cultures" in this regard would seem to lie within reach: we may think of the development of scientific knowledge as *an evolutionary process* - which is to say, it may usefully be conceptualized as involving mechanisms of selection to which alternative recurring propositions (ideas) will be submitted, as well as processes that serve to regenerate the diversity of candidate propositions. This formulation corresponds in a general way to the approach of the so-called "model" school of evolutionary epistemologists.[39] To introduce it within the confines of our overly simplified model, the invalidation and ultimate destruction of a consensus supporting "objectively unreliable" statements - i.e., statements that fail repeatedly to derive reinforcement from predictive successes in operational contexts - must be treated as probabilistic occurrences, rather than being foreordained. Probabilistic failures of this kind will be triggered by "accidents" in the interpretation of ambiguous observational evidence, that in effect lead researchers to state counter-consensual opinions, thereby injecting into the network of scientific communications some "seeds of doubt" about the currently reigning set of beliefs. Adoption of a Darwinian "selectionist" position, familiar enough from a variety of formulations to be found in the evolutionary epistemology literature following the work of Campbell (1965, 1974), would then allow us to imagine that such sparks of disagreement (to switch the metaphor) are more likely to be released frequently by the recurring collisions between research programmes that adhere to erroneous and inconsistent propositions, and the comparatively rigid boundaries set by the world of Nature. The key question then becomes one of whether or not these expressions of doubt and dissent will be quickly smothered by the weight of contrary opinion prevailing in the scientific community, or whether the conditions governing the propagation of ideas will permit them to escape rapid suppression. By surviving within some local networks, these minority opinions must raise the likelihood of their acceptance in a wider sphere, and so may ignite a spreading conflagration of revisionism that eventually reorients the configuration of beliefs throughout the community at large. So long as the latter outcome remains a possibility, then even on a purely chance basis, one may expect that through their

[39] For further discussion and introductory references to this literature in addition to Campbell (1965, 1974), and Hull (1988); see David and Sanderson (1996b) on the limitations of cognitive learning in such "evolutionary" processes.

more frequent collisions with obdurate Nature, the "objectively unreliable" propositions would tend over the long run to get selected out.

Thus we reasonably may hope to arrive at an evolutionary epistemological synthesis as a form of comprise between the camps in the "culture wars." Under the proposed terms of peace, both sides should agree that a scientific community can arrive via generic "social processes" at a consensus on the acceptability of certain statements about the material world. Also, there are some rules governing the way those statements are presented, and treated by members of the community, which have the effect of preserving the possibility that such a socially constructed "truth" will tend to be revised. There is, indeed, a long-run expectation that such social constructions will be discarded should they be found repeatedly to be difficult to square with other "truths" - especially those which possess a higher measure of "fit" with the logical implications and inferences that can be drawn from the available body of empirical observations.

7.6 From local stochastic social communications, to a model of global network behavior

We are at last in a position to draw together the strands of the preceding arguments and examine the properties of the stochastic communications model to which they lead. To simplify the analytical task. I shall need to put aside any further distracting questions, such as how local social communication networks among scientists come to be formed in the first place, and whether strategic or other considerations on the part of the researchers might enter into the process of their formation. Instead, these structures will be treated as exogenously given, multidimensional "neighborhoods" that are defined for each research scientist in the epistemic community under consideration.[40] The social boundaries of that larger research community, the *global network* which is the "invisible college", likewise must be taken to be exogenously delimited. Nevertheless, as shall be seen, the degree of coherence and agreement of its members on cognitive issues is an attribute of its disciplinary identity and intellectual connectedness that is in some important measure *endogenously* determined through the interactions of its members. The invisible college as conceived of here, therefore, is not an organizationally distinct entity; it is constituted of the ensemble of local social networks, through the chains of informal communication channels that connect the neighborhood of one researcher to the overlapping but not identically composed neighborhoods to which his or her immediate "correspondents" also belong.

[40] Although the possibility is not pursued here, it would appear feasible to treat the local social communication networks explicitly as coalitions, and, following the lead of Kirman, Oddou and Weber (1986), to model their endogenous formation - and possibly also their ramifying interconnections - by applying concepts and analytical techniques from the branch of probability known as random graph theory. See Bollobás (1979).

7.6.1 Graph-theoretic preliminaries

Some concrete spatial examples may help in visualizing structures of this kind. First, think of a two-dimensional neighborhood in which there are four researcher-nodes (or research organizations), situated at the crux and the three extremities of a Y; the index node of this assembly is at the crux, and is connected to the rest of the local social "neighborhood" by three communication channels. Two other such neighborhoods can be linked with this one, at the right and left upper branches, forming an arrangement like this: YYY. By now placing four more Y's at the four upper extremities of this structure, and positioning two more Y's below, so that their upper branches connect with the vertical's of the initial trio, a hexagonal honeycomb lattice can be extended in each direction, thus:

YYYY
YYY
YY

To take a different spatial arrangement, we might start from a five-member neighborhood, in which the index member has neighbors to the north, south, east and west. A similar connected graph can be joined to that arrangement, in which two members of the first neighborhood will be shared by the second, and so on, adding still others until a square lattice develops. To keep everything perfectly symmetrical and leave no nodes in boundary positions, these two-dimensional graphs can be wrapped around in both the horizontal and vertical directions, connecting the right side edges to nodes on the left, and top side to the bottom - and so forming a *torus*.

In choosing to work with graphical forms of this kind, we are explicitly specifying that the neighborhoods, or local social networks of each researcher are symmetrical with those of all the others, and that they remain fixed for the purposes of our analysis. Both assumptions are convenient, which is the most that can be said for making them. Clearly, some scientific researchers have larger social networks, and networks of higher dimensionality at their immediate disposal than to others, but we simplify the analysis greatly by making the ensemble of social networks symmetric and homogenous.[41] In actuality, of course, invisible colleges are not temporally stable, as they consist of the researchers who happen at the time to be engaged with a particular set of cognitively interrelated scientific questions. Such populations in part or in whole may abandon the research field that formerly gave it intellectual cohesion and structured its members' communicative interactions; or extended invisible colleges may

[41] Morris (1996), using mathematical tools other than those employed below, has shown that a number of the key properties of local interaction games, concerning the dynamic propagation of strategies chosen in particular locations, and the existence of equilibria of correlated equilibria, hold generally for a wide class of local (spatial) structures. On the other hand, the assumptions of symmetry and homogeneity are not wholly innocuous. For example, Bala and Goyal (1995) show that greater symmetry increases the speed of information diffusion in a local interactive learning game.

fragment into more compact areas of research specialization. The formal representation of an invisible college as *a network of localized social networks* abstracts from all these realistic complications, inasmuch as the fixed nature of the latter precludes researchers from migrating from one part of the larger network to another, or leaving the invisible college in question to participate in another.[42]

7.6.2 Markov random fields and the "voter model"

An undeniable attraction of the probabilistic routine for opinion formation set out in the preceding section is that it corresponds directly with the so-called "voter model", which was introduced in different contexts by Clifford and Sudbury (1973) and Holley and Liggett (1975), and is best known in the form elaborated by Harris (1978). The latter heuristic model is based on Markov random field theory, a branch of probability theory that has lent itself to application in the study of social networks (see Kinderman and Snell (1980)). More recently, it has been extended to the analysis of the dynamics of technological competitions in economic contexts that are characterized by the existence of local network externalities.[43]

Leaving aside technicalities, it is possible to set out this framework formally in a way that enables us to represent scientific communication and consensus formation in inter-linked local networks as a stochastic, path dependent process. Following the notation by Kinderman and Snell (1980), we may begin with the basic definitions relating to Markov random fields.

Let **G** = (**O**,**T**) be a connected graph, with **O**=$(o_1, o_2,...,o_n)$ the vertices (i.e. the set of nudes representing researchers, or research organizations) and **T**=$(t_1, t_2,...,t_m)$ the edges (i.e. the set of information transmission channels). For the moment, we restrict the discussion to connected graphs of social networks that are defined in one or two dimensions. A *configuration* **x** is an assignment of an element of the finite set S to each point of **O**. We denote this configuration by $\mathbf{x} = (x_o)$ where x_o is the element of **S** assigned to vertex o. If we let **S** = [*u*, *a*] represent assignments of the two possible opinion orientations regarding the reliability of a given scientific statement (*a* standing for "acceptably reliable", *u* for "unreliable"), a configuration would be an assignment of either o_u or o_a to each of the points in **O**. A *random field* p is a probability measure p(**x**) assigned to the set **X** of all configurations, such that p(**x**)>0 for all **x**. By the "neighbors" N(o) of the point o we shall mean the set of all points o' in **O** such that (o'o) is an edge. A random field p is called a *Markov random field* if:

$$p\{\mathbf{x}_o = s|\mathbf{x}_{O\text{-}o}\} = p\{\mathbf{x}_o = s|\mathbf{x}_{N(o)}\}.$$

[42] It will be noticed below (in section 7) that it is nonetheless possible within this framework to make allowance for newcomers to be inserted into the community - by attaching themselves, albeit independently, and randomly, to new social networks, or by occupying the places opened up in the network through the natural attrition of some of its members. But for the immediate purposes, such matters are left to one side.

[43] See, *inter alia*, David (1988, 1993b); David and Foray (1993).

That is, in trying to predict the value at o (either *u* or *a* in the example), given the values at all other points of **O**, we need know only the values assigned to the neighbors of o.

Assume now, following the "voter model", that associated with each point of a graph we have a researcher or research unit, and that with every such unit there is a reference set comprised of other units. The latter constitutes the neighborhood available for random polling at a given moment in exponential time, to select one a neighbor whose orientation the index node will mimic.

Now this process may be described informally as follows: each research unit selects an orientation in regard to the statement in question, *u* or *a*. At random points in time it will reassess its opinion. At these times it will commit to the choice *u* with a probability equal to the proportion of *u*-oriented research units in its reference set. The global process of migration between the alternative orientations of belief as to the reliability or unreliability of a given scientific statement is therefore a finite state continuous time Markov chain, with states being configurations of the form

$\mathbf{x} = (u, a, u, u, a, \ldots, u, a, u)$, where x(i) is the choice of research unit i.

A number of important properties of this well-studied process[44] may be immediately stated:

Property (1): It is evident on even the briefest consideration that the extremal states $\mathbf{x}^u = (u, u, u, \ldots, u, u, u)$ and $\mathbf{x}^a = (a, a, a, \ldots, a, a)$, in which there is a perfect correlation of beliefs throughout the population, constitute absorbing states for this system. Once such a state is entered, there can be no further change. The existence of a multiplicity (two) of absorbing states tells us plainly that *this process is essentially historical, in the sense of being non-ergodic* - it cannot invariably shake loose from all initial configurations.

Property (2): A somewhat less obvious proposition, also true, is that for any starting state $\mathbf{x}$ the chain eventually will end up in either $\mathbf{x}^u$ or $\mathbf{x}^a$. Thus, in the limit, *the process must become "locked-in" to one of its extremal solutions.* The system invariably does produce eventual "closure" on the scientific issues submitted to it.

Property (3): There exists a limiting probability distribution over the macrostates (opinion configurations) of the system which is non-continuous, such that, starting in $\mathbf{x}$, the probability that the chain will end in $\mathbf{x}^u$ is equal to the proportion of *u* in the initial configuration $\mathbf{x}$ (without regard to their position in the array); and the probability that it will end up in $\mathbf{x}^a$ is equal to the proportion of *a* in the initial configuration $\mathbf{x}$. Therefore, although subject to random influences, the nature of the asymptotic macrostate consensus in this system can be predicted (not with certainty, but probabilistically) from information on the initial configurations of opinions.

In the present application context what is the import of these results? A multiplicity of possible outcomes, eventual closure, predictability of the eventual outcome of scientific controversies on the basis of information concerning the

[44] See Kinderman and Snell (1980b), and the fuller discussion in David (1993b).

distribution of the researcher's initial opinion orientations? Obviously, one thing they signify is the possibility of formally connecting the social organization of science (affecting communications) with an important performance attribute of science communities in the cognitive domain "closure". Another rather pleasing formal reaffirmation of insights obtained via other routes is that "the details of history may matter" for the cognitive development of a scientific field. Further, in this light, the propensity of scientific communities to remark especially on instances in which new ideas have won eventual acceptance in the face of an initial consensus opposing them, is entirely understandable, because, at least in the near term, such cases would constitute the rarer contingencies.

Several technical qualifications must be appended to the foregoing conclusions about the voter model's properties, especially as these also admit of interesting interpretations in the present context. First, the property of complete closure, in the sense of "lock-in" to a consensus characterised by perfect unanimity, will not survive generalization of the framework to higher-dimensional graphs. It has been found that substantial but less-than-perfect correlations of opinions eventually would emerge in a community that was arranged with reference to three social dimensions. Plausibly enough, it may be surmised that as one goes to still higher dimensions in the social space, and as the community becomes in that sense "less compact", its ability to arrive even at a strong consensus would deteriorate further (see Kinderman and Snell (1980b).

Perhaps the recurring formation of disciplinary sub-specialities in science serves as a "social compacting process", the latent function of which is to preserve network performance in producing closure. On the other side of the coin, higher dimensional social networks may be seen to improve the chances of the survival of heterodox opinions within small minority clusters of researchers who, in effect, shield one another from the conformity-inducing pressure of exposure to the preponderance of belief prevailing in the epistemic community at large.[45]

A second point of qualification is that the properties of lock-in to closure, and predictability of the nature of the resolution, are ones that strictly hold only for *finite* populations. If the population of the network were to be constantly growing at a comparatively rapid rate - strictly, at a pace rapid enough to cause the introduction of newcomers (who are entering the field with randomly distributed beliefs about the scientific issues of the day) to overwhelm the pace of the process of random polling in the local social networks, then closure would no longer be assured. Moreover, under those conditions, the nature of the cognitive outcome would cease to be predictable on the basis of the system's initial configuration. It has been surmised by probability theorists that the dynamics of convergence to one or the other extremal (uniform consensus) configurations in a large finite system would approximate those of the infinite population case. Such systems continue to migrate back-and-forth between extremal states, albeit with transit times between them that are quite prolonged.

---

[45] The broader significance of this will be further remarked upon below, in Section 6.

Here, then, we may remark upon another respect in which the cognitive performance of scientific communities could be related to the organizational dimensions and dynamic attributes of invisible colleges. Those characteristics would certainly have to include the rate of entry of new members in relation to the speed of opinion-forming informational transactions within local social networks , and the "pre-entry orientation" of new recruits in reference to the prevailing distribution of scientific opinions held by those who currently constitute "the field". A substantial consensus can be expected to emerge even in the absence of strong pre-orientation as a criterion of eligibility for entry, once an invisible college "stabilizes" demographically - in the sense that its growth rate slows to the point that it is exceeded by the average internal opinion-polling rate. Here is where communications technology can help support the rapid growth of research communities that retain their epistemological functionality. A speed-up of the 'polling rate' will permit the mobilization of resources at a research frontier to proceed more rapidly, without jeopardizing the network's ability to reach closure on the new questions that it has taken up for investigation.

Yet another, and a quite important class of qualifications arises from closer consideration of the assumptions of the basic "voter model" with regard to the symmetry of communication behaviors on the part of the research-agents. These can be brought out now by turning to consider the properties of a somewhat different stochastic communications structure, in which not all the actors follow the same policy in regard to the informational transactions with other research units.

7.6.3 Imperfect communications, connectivity and critical percolation probabilities

The population of researchers portrayed by the "voter model" in the preceding sub-section is homogeneous, in the special sense that the structure of communication links among them is completely symmetrical. Moreover, it is assumed to take a rather special form that renders the transmission of influence among them completely deterministic, even though the timing of interactions between pairs of agents within the local network is randomly generated. All of the researchers are always sharing their opinions with all who ask, as all others in their social network stand ready to do. Putting aside the possibility of entry, the source of randomness in the revisions of beliefs within the population has to do not with whether or not particular researchers might be open to the influence of particular neighbors, but rather with the direction of the re-orientation of beliefs that such influences would bring about. The concepts and terminology of *percolation theory* provides a precise way of describing these specifications, and showing their relationship to a more general specification of the model.

The term *percolation* refers to the dual of a diffusion process. "Diffusion", to speak strictly, refers to the random movements of particles through an ordered, non-random medium - as in the case of the diffusion of molecules of salt in water, whereas the term "percolation" conjures up the image of droplets of water moving under the deterministic pull of gravity through a dis-ordered, random medium - such as filtration tank filled with sand and pebbles of different sizes. When the water, entering at some source sites, eventually finds its way into enough open channels to pass throughout, wetting the entirety of the interior surfaces, *complete percolation* is said to have taken place. It is from this that the mathematical statistics describing the properties of analogous processes has acquired the label "percolation" theory (see Grimmet (1989)).

Adapting the notation of Hammersley and Welsh (1980) to the Markov random field framework, let **G** be a graph in which some, none, or all of the edges may be directed. Thus, as before, **G** consists of a set of research units (corresponding to the graph's vertices or nodes), $\mathbf{O}=(o_1, o_2,...,o_n)$. These are connected by a set of (possibly directed) edges representing channels of social communications, $\mathbf{T}=(t_1, t_2,...,t_m)$. An *operative path* in **G** from an research unit o_1 to another research unit, o_n, is a finite sequence of this form:

$$\{t_{12}\, o_2\, t_{23}\, o_3 \ldots t_{[n-1]n}\, o_n\},$$

where t_{ij} denotes a relational line connecting o_i to o_j. The graph **G** is *connected* if for each pair of researchers o_i and o_j there is a path in **G** from o_i to o_j.

Now construct a *random maze* on **G**, as follows. Let each research node o of **G** be *open*, or ready with probability p_s to transmit messages that can influence any of its neighbors' opinions on the reliability of the statement at issue. Alternatively, it will be *closed* (unwilling to share its present knowledge on the question) with probability $q_s = 1- p_s$. Similarly, each line of inter-personal or inter-organizational communications t_{ij} may be thought of as potentially carrying messages that will be actually "read" with probability p_r, or fail to do so with probability $q_r = 1-p_r$. Furthermore, we shall assume all these events are to occur independently of each other.

An operative path, $D = \{ t_{12}\, o_2\, t_{23}\, o_3 \ldots t_{[n-1]n}\, o_n\}$ from o_1 to o_n is said to be "open" if all its communication links are functioning and all its research nodes are ready to "share" their knowledge-conclusions. Thus, the probability that the particular path *D* is operational in that sense is given by $(p_r p_s)^{n-1}$.

Figure 1. Critical boundary for percolation in infinite graph, G.

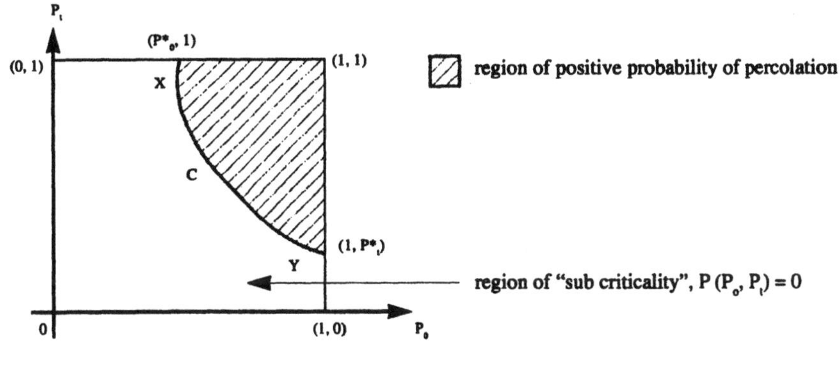

Let **Z** be some given set of "source" research units, from whom a particular "idea" or scientific statement emerges into G. The decisions to adopt that statement as reliable (or not) can flow along any open path from a source research unit and will then similarly reorient the other units on such a path ("wetting" them, to use the natural percolation metaphor). The *percolation probability* $P(p_r,p_s|\mathbf{Z},\mathbf{G})$ is the probability that **Z** can thus reorient some infinite set of nodes in **G**. We call the parameters p_r, p_s, respectively the probabilities of "reading" and "sending" or "sharing" information. In other words, in a large population, it can be expected that a proportion p_r are receptive to their neighbors' opinion on the reliability of a statement, while a proportion $1-p_r$ are unreceptive. The transactional lines of **G** connect neighboring pairs of researchers and the model supposes that a researcher already committed into a given scientific position has a chance p_s of "infecting" a neighbor, provided that the latter is receptive. Then $P(p_r,p_s|\mathbf{Z},\mathbf{G})$ is the probability that a belief initially established in the source research organizations **Z** will become adopted universally.

Suppose that **Z** and **G** are fixed, that **G** is infinite, and that we abbreviate

$$P(p_r,p_s|\mathbf{Z},\mathbf{G}) = P(p_r,p_s) = P$$

Then, clearly, the mixed percolation probability P is a non-decreasing function of p_r and p_s, and it follows that: $P(0,0) = P(1,0) = P(0,1) = 0$, while $P(1,1) = 1$. Consequently, $P_s(p) = P(p_s, 1)$ and $P_r(p) = P(1, p_r)$ are, respectively, called the *node percolation* and *connection percolation* probabilities of this system.

A fundamental mathematical property of the percolation process is that there exists some critical values of $p_r > p^*_r$ and $p_s > p^*_s$ beyond which there will be a positive probability that percolation occurs, and below which the percolation probability is zero (see Hammersley and Welsh (1980), Grimmett (1988)). In other words, the system undergoes a "phase transition" when these underlying critical probabilities are attained. There are corresponding critical values at which the node-percolation and edge-percolation probabilities, respectively, become positive. These define the endpoints of a region above which a "mixed-percolation process" (one for which it is not certain that either all nodes or all edges of the graph are open), will have positive probability of achieving complete percolation. Figure 1 illustrates the position and shape of this region. What these results from percolation theory tell us in the present context is that there is a minimum level of *persistently* communicative behavior that a finite size science network must maintain if ideas are to percolate within it, so that closure can be obtained. What significance can be attached to this fundamental property of percolation processes?

For a community of scientists to exist as a cognitively functioning entity, it has just been seen that there is a formal necessity of attaining some critical measures of "expected connectedness" through the average communication behaviors of its constituent members. A second pertinent result from percolation theory is represented by placement of the endpoints of the boundary shown in Figure 1: there is an important asymmetry between the effects upon network performance of reducing the representative agent's communication probabilities. Thus, a given proportional reduction of the mean probability of sending messages (node openness) will do more to degrade the percolation performance of the system than will the equiproportional reduction of the mean probability of a communication link being open.[46] In view of this, the invisible colleges' first condition for functionality, in the sense of its most exacting requirement, is that the network must maintain at least the critical level of openness in regard to the behavior of a "representative node", i.e., in the expected proportional composition of member "types" with respect to disclosure of their scientific knowledge.

7.7 A new perspective on the significance of the Mertonian "norms"

The representative researcher, of course, is a purely statistical construct in the percolation model framework - an average of nodes that are *permanently* open and those that are *permanently* closed. The fraction of those who, being closed, will never "share" (or never "write") what they have learned, therefore, must not be allowed to exceed the critical level ($1-p_s^*$) if the invisible college is to retain a positive probability of reaching closure on scientific questions. Thus, the "mix" of persisting behaviors is a critical matter for the system.

7.7.1 From the norm of disclosure to the probability of closure

What follows from the fact that the qualitative performance of the system undergoes this critical shift when the mixing fraction passes a specific level, but the precise value of that critical point is not likely to be known in advance, nor to become evident when the mean probability of node openness is progressively degraded - until it is too late? In such circumstances it would be sensible to protect the cognitive functionality of the global network by maintaining a "safety-first" policy of selectivity in regard to the recruitment of researchers to the college; in other words, impose some prior test of intrinsic propensity towards"openness" concerning what they will find and conclude in the course of their researches.

By the same token, the existence of strong and universal norms requiring cooperative behavior on the part of researchers, especially in disclosing what they learn, and a reward system that elicits such disclosures as the basis for establishing a collegial reputation, can be seen to serve as important bulwark protecting the invisible college's ability to deliver a clear consensus on the questions brought before it. Furthermore, inasmuch as $p_s^*>p_r^*$, in designing the incentives for individual behavior in such a system, it is reasonable from the viewpoint of assuring "connectivity" that assuring the consideration by others of one's own "findings" should take some measure of precedence over concerns about the

---

[46] See Hammersley and Welsh (1980), for proof of the generalized asymmetry theorem; David and Foray (1993, 1994) for its application in the social communication network context.

arrangements and facilities that affect the average propensity of network members to attend to each others' messages. "Publishing" should be made more important than "reading".

The foregoing observations thus, may be taken as a formal communications-theoretic rationale for the stress placed by Mertonian sociology upon the functional importance of the behavioral norm of openness among scientists, and also for the corresponding tendency of that literature to de-emphasize the significance of variations in the reliability of communications technologies upon the qualitative performance of scientific communities.

Still other implications of the Mertonian "norms" for the conduct of (non-proprietary) research can be brought into sharper focus by reconsidered their operation in the context of the stochastic models presented here. Viewed from the perspective of our earlier discussion of science as an evolutionary epistemological process, norms of "disinterestedness", "universality", and "disclosure" may be seen (in section 5.3) as crucial when taken together, precisely because they reinforce behaviors at the micro-level that permit the tendency for more "objectively reliable" forms of belief to emerge by selection from the collision between consensus and contradictory observations. It would appear that there has been an important failure of understanding among the relativists, the partisans of the "strong programme" in SSK, who dismiss these "norms" as a self-serving ideology that is betrayed by human failing and competitive stratagems adopted among the practitioners of science. It is naïve to suppose that the essential features of the qualitative performance of a mode of social organization will be lost if any one among its "norms" is violated by some members at some point in time. Any system of social organization that is so rigid and unrobust that it cannot tolerate a degree of deviance is not likely to survive for very long. Furthermore, it has been seen - from the properties of both the voter model, and the percolation model - that some degree of intermittent (random) interruption of communications, and even some margin of persistently non-communicative actors, are formally compatible with the ability of the global network eventually to arrive at closure.[47]

7.7.2 Reducing the hazards of premature closure: The "universalistic" norm and network dimensionality

No less deserving of criticism from this perspective are those champions of the objectivity of "the scientific method" who celebrate its power to uncover "the truth", yet fail to grasp the connection between the success of science and the nature of the social organizations that scientific communities represent. It is a too human failing of scientists who miss this point that they quite typically attribute the progress of scientific knowledge to the succession of individual "genius". It may be quite proper to share the doubts expressed by Perutz (1996), about the

[47] Trust in that capability, of course, is what has been presented as underpinning the rational microlevel strategies of the agents engaged in polling their respective networks in the case of the voter model.

philosophical representation of epistemological progress resulting from the dogged adherence to a formula-like "scientific method", and to deride as pure fantasy the picture of scientific researchers as just so many drones harnessed to the crank of a cleverly designed inferential engine, mechanically proceeding through cycle after cycle of theoretical hypothesis and empirical refutation until law-like propositions emerge, like so many sausages. Whether or not the process of inquiry at the individual, microcosmic level of the investigator is messy, irrational, driven by the insights of genius, or by human motives that are less than noble, really is not what matters. These are not the crucial aspects of the method of modern science upon which discussion should be focused. Instead, it should be more widely understood that there is an aspect of the "scientific method" for which it is true that the efficacy of the enterprise depends upon the subscription of most members of the research community. This is just the "method" or mode of *social* organization affecting the internal communications of invisible colleges.

It has been pointed out in the analysis of the voter model that in the case of networks that are arrayed in three or more dimensions, the collectivity loses the property of "lock-in" to *complete* closure. Further, it has been noticed that perfect correlation of beliefs throughout the global network constitutes an absorbing or "trapping" state for the system, so that compact, low dimensionality systems are exposed to the risk of being unable to free themselves from the grip of continually reinforced orthodoxy. The universalistic norm highlighted by Merton therefore may be seen to forestall the restriction of entry into network membership by selection criteria that otherwise would tend to render the social dimensionality of the resulting communications structure more compact, and for that reason the more likely to rapidly extinguish small pockets of dissenting opinion. The ability to shelter the views of dissenting minorities from the pressures for conformity with predominating peer beliefs, pressures such as are especially likely to exist in collegiate reputation-based reward systems, must thus be an important system attribute. It reduces the likelihood of premature consensus, and preserves the possibility of an evolution of beliefs away from those that persistently generate observational anomalies.

Thus, by enjoining members of the community to accept dissenting claims as worthy of examination, without regard for the economic, social, political or nationality status of the claimants; by insisting on disclosure as the condition for successful claims to the rewards that are attached to priority; by preventing a complete and enduring monopolization of knowledge and access to knowledge from protecting the consensus from challenge, these "norms" of the Republic of Science serve an important *epistemological* function. They allow signs of the collisions between the constructs of social communication and the constraints implied by the structures of the material worlds to be registered. They encourage the perception among individuals that such signs are to be interpreted as indicating opportunities for achieving recognition, and reward, rather than as areas in which they might stumble into heresies that will bring punishments.

We have now arrived, via a somewhat novel route, at a fresh appreciation for the critical balancing of the influences exercised upon individual behaviors by the institutionalized norms associated with the practice of "open science." Jointly, these tend to provide a mechanism that leads to closure, while protecting existing diversity of opinion from the generic social pressure for perfect conformity and uniformatization that otherwise would arbitrarily extinguish the process of consensus-revision. Consequently, there is a basis for supposing that the stabilization of beliefs among communities whose members adhere to such rules will tend to be indicative of the attainment of that greater degree of fitness with the unchanging features of the natural world -which specially qualifies the (scientific) statements involved to be described not merely as knowledge, and information, but as "reliable knowledge". More precisely, we should recognize that the stabilization of community opinion can occur not because "the truth" has been uncovered, but because the material and intellectual resources at the disposal of working scientists (for example, the sheer limitations of the resolution power of their telescopes, the magnification power of their microscopes, the sensitivity of their particle detectors and the computational power available for numerical analysis of data) have intervened; that it is perceived that although matters are not fully resolved, the opportunities for individuals to win recognition for contributions that will effect a revision of the consensus are a potentiality that for the time being remains unripe for exploitation.

7.8 Towards extensions of the model: Closure, creativity and the size of invisible colleges

A further range of insights is available if the analysis is focused on the performance of open science communities, all of whose members conform probabilistically to the norms of openness. Here we return to the world of transient random mazes represented by the voter model, where the communication probabilities pertain to the homogeneous behavior of the researchers in regard to their sending and receiving of messages. Under those conditions (see Cox (1989)), the expected length of time that is required for "closure" will be increased by all movements away from certainty that a randomly chosen node, and a randomly selected communications channel will be actually "open" when probabilistic polling occurs. The macro-level performance implications of alterations in the mean probability of node (and channel) openness remain. But what amounted to a critical qualitative change in the case of percolation through a permanent maze (in the case of the percolation model), is transformed into a continuous modification of the speed of attaining closure when we consider the case of *random mazes* (as in the voter model).

Exploring the dynamic implications these and related results about system performance in presumptively "open science" research networks is a task that must be left for the future. But, the direction of advance may be briefly indicated here. The basic model of a research network that has been presented in this essay can be elaborated quite easily to allow dynamic feedbacks, whereby the macro-level performance characteristics of the global network would affect the micro-level communication behaviors of the research-agents in their respective local social networks. In the most elementary implementation of this extension (by David and Flemming, 1996), the latter are just the homogeneous mean probabilities of sending and receiving information, or "sharing" and "reading". The resulting closed stochastic system can be transformed into an "equivalent deterministic system" by analyzing the *expected* performance characteristics of a science research network *viewed as an entity.* Pursuing the latter approach in initial explorations is especially convenient, as it dispenses with the need for extensive stochastic simulation experiments in order to establish the main qualitative aspects of the endogenous dynamics of network growth and performance.

In David and Flemming's (1996) implementation of the equivalent deterministic system modeling approach, the expected percolation time for establishing "closure" with regard to an idea, and the expected rate of "creativity" through recombinant generation of ideas, are each modeled as dependent upon endogenously determined features of the global network's performance. Key among these are the global network's size, and the mean probability that within local social networks all ideas will be shared freely (transmitted). Making use of a result due to Cox (1989) on the expected time to percolation on a three-dimensional torus, the mean speed of closure can be taken to be proportional to the product of the mean probabilities of sending, and of receiving messages, and inversely proportional to the cube of network size. The mean probability of a "new scientific proposition" being generated at any given moment somewhere in the connected network, or the invisible college's "creativity" rate, may be represented as a (positive) power function of the network's size, and of the probability that new ideas are "shared". The latter specification can be rationalized by appeal to findings in cognitive psychology and the social psychology of science, and in the history of invention. These point to the importance of "recombinant creativity", and the sensitivity ofthe latter phenomenon upon the size and diversity of the set of ideas and techniques that are available for permutation and recombination by members of an epistemic community.[48] In this context, emphasis is not placed upon the determinants of individual creativity, and there is no assumption as to the automaticity of the cumulative process through which new ideas are generated. Rather, like "closure", "creativity" is an emergent property of social networks that have certain features, among which "size" and "openness" are critical.

When the global network size can be varied exogenously, a variety of networks displaying a wide range of performance attributes all turn out to be equally sustainable equilibrium solutions. If the "creativity" effect of greater network size is strong enough to overcome the latter's effect in slowing the expected speed of

[48] See, e.g., the contributions in Shadish and Fuller (1994), Usher (1982), especially Ch.IV. Weitzman (1995) builds a model of endogenous economic growth around the core concept of "recombinant innovation", taking the process of hybridization of existing plant varieties to obtain new and improved strains as a metaphor for the process through which "knowledge can build upon itself in a cumulative interactive process" involving the combination and recombination of ideas

closure, the larger networks belonging to this family generally will be those which are consistent with a greater (probability) measure of "openness" at the micro-behavioral level. Hence, by virtue of both their size and (the assumed diversity of the ideas held by their members)the higher probability that the network's members will communicate those ideas openly, the bigger networks are able also to sustain higher levels of systemic "creativity".

A fully dynamic equivalent deterministic system also may be obtained by taking the further step of modeling the influence of those latter (average) performance features upon the ability of the network to maintain its size, and even to grow by recruiting members from a surrounding eligible population. This entails the formation of expectations about the implications of the overall network's performance for the benefits that potential members would derive by joining it, in comparison with those obtainable in competing employments. It is possible to consider either situations in which the latter opportunity costs are exogenously set, or "closed" models, in which they are determined endogenously by forces related to the performance of the network in question.[49]

There are at least two useful aspects of the demonstration that such simple formal models can generate results that reinforce the intuitions and generalize the insights gained through careful and detailed empirical research into both the sociology of scientific communities and the institutional and cultural contexts in which research is conducted. First, it suggests that future work in this vein may yield some more subtle insights into the connections between micro-behaviors and macro-phenomena that are of interest for science policy formulation. Second, there is a potentially significant heuristic benefit: formal exercises of this kind do open up an attractive avenue that should entice economists proficient in the requisite modeling skills into learning more about the social and intellectual organization of scientific research communities. From that empirical base economics can move on towards deploying their talents in increasingly sophisticated dynamic analyses of the production and distribution of scientific knowledge by networks of researchers.

---

[49] See David and Flemming (1996). It is even possible to extend the model to analyze the dynamics of competition between two networks that can draw recruits from a common pool of researchers, but this is left for future work.

References

Adler, C. G. and B. Coulter (1978), "Galileo and the Tower of Pisa Experiment", *American Journal of Physics*, 46: 199-201.

Alba, R. (1973), "A graph-theoretic definition of a sociometric clique", *Journal of Mathematical Sociology*, Vol. 1.

Anderlini, L., and A. Ianni (1995), "Path dependence and learning from neighbours", forthcoming in *Games and Economic Behavior.*

Arora, A., P.A. David and A. Gambardella (1997), "Reputation and competence in publicly funded science: estimating the effects on research group productivity". (A revised version of this paper presented at the National Bureau of Economic Research Productivity Workshop, is fothcoming in *Annales d'Economie et de Statistiques,* Numero Exceptionelle, ed. J. Mairesse.

Arora, A., and A. Gambardella (1998), "Public policy towards science: picking stars or spreading the wealth?", forthcoming in *Revue d'Economie Industrielle*

Arora, A., and A. Gambardella (1994), "The changing technology of technological change: general and abstract knowledge and the division of innovative labour", *Research Policy* 23, 523-32.

Arrow, K.J. (1962), "Economic welfare and the allocation of resources for inventions", in R.R. Nelson (ed.), *The rate and direction of inventive activity: economic and social factors* (Princeton, NJ: Princeton University Press).

Arrow, K.J. (1971), "Political and economic evaluation of social effects and externalities", in M. Intrilligator (ed.), *Frontiers of quantitative economics*, Contributions to Economic Analysis No. 71 (Amsterdam: North-Holland).

Bacharach, M., D. Gambetta et al. (1994), "The economics of salience: A research proposal", Unpublished paper, Oxford Institute of Economics and Statistics.

Bala, V., and S. Goyal (1995), "Learning from neighbors", Econometric Institute Report 9549, Erasmus University, Rotterdam, September.

Barnes, B. (1974), *Scientific Knowledge and Sociological Theory*, London: Routledge and Kegan Paul.

Barnes, B. (1977), *Interests and the Growth of Knowledge*, London: Routledge and Kegan Paul.

Bloor, D. (1976), *Knowledge and Social Imagery* ,London: Routledge & Kegan Paul.

Blume, L.E. (1993), "The statistical mechanics of strategic interaction", *Games and Economic Behavior,* 4: 378-424.

Bollobás, M. (1979), *Graph Theory: An Introductory Course.* New York: Springer-Verlag.

Callon, M. (1995) "Four models for the dynamics of science", in S. Jasanoff, G.E. Markle, J.C. Petersen and T. Pinch (eds), *Handbook of science and technology studies* (London: Sage Publications).

Callon, M., and J.-P. Courtial (1989), *Co-word analysis: a tool for the evaluation of public research policy* (Paris: Ecole Nationale Supérieure des Mines).

Campbell, D.T. (1965), "Variation and selective retention in socio-cultural evolution", in H.R. Barringer, G.I. Blanksten and R.W. Mack (eds), *Social change in developing areas* (Cambridge, MA: Schenkman Press).

Campbell, D.T. (1974), "Evolutionary epistemology", in P.A. Schilpp (ed.), *The philosophy of Karl R. Popper* (LaSalle: Open Court).

Campbell, D. T. (1994), "The Social Psychology of Scientific Validity: An Epistemological Perspective and a Personalized History", Ch.2, in *The Social Psychology of Science,* New York: The Guilford Press.

Caudill, Maureen and Charles Butler (1990), *Naturally intelligent systems* (Cambridge, MA: MIT Press).

Clifford, P., and A. Sudbury (1973), "A model for spatial conflict", *Biometrika* 60, 581-8.

Cole, J. and S. Cole (1973), *Social Stratification in Science*, Chicago: Chicago University Press.

Cole, S. (1978), "Scientific Reward Systems: A Comparative Analysis", in R. A. Jones, ed., *Research in Sociology of Knowledge, Sciences and Art*, Greenwich, Conn.: JAI Press, pp. 167-190.

Cole, S. and Cole J. (1967), "Scientific Output and Recognition", *American Sociological Review*, 32:.377-390.

Collins, H.M. (1974), "The TEA set: tacit knowledge and scientific networks", *Science studies*, 4: 165-86.

Cowan, R. and D. Foray (1996), "The Economics of Codification and the Diffusion of Knowledge", Unpublished working paper: IMRI, University of Paris-Dauphine, (March).

Cox, J. T. (1989), "Expected percolation times for the voter model on a d-dimensional torus", *Annals of Probability*, 17 (4): 1333-1366.

Crane, D. (1965), "Scientists at Major and Minor Universities: A Study in Productivity and Recognition", *American Sociological Review*, 30:.699-714.

Dalle, J.-M. (1995), "Dynamiques d'adoption, coordination et diversite: la diffusion des standards technologiques", *Revue Économiques*, (July).

Dasgupta, P., and P.A. David (1987), "Information disclosure and the economics of science and technology", in G. Feiwel (ed.), *Arrow and the ascent of modern economic theory*, New York: New York University Press.

Dasgupta, P., and P.A. David (1988), "Priority, secrecy, patents and the economic organization of science and technology", CEPR Publication No. 127, Stanford University (March).

Dasgupta, P. and P.A. David (1994), "Toward a new economics of science", *Research Policy* 23: 487-521.

David, P.A. (1988), "Path-dependence: putting the past into the future of economics", *Institute for Mathematical Studies in the Social Sciences Technical Report No. 533*, Stanford University, (November).

David, P.A. (1993a), "Historical economics in the long run: some implications of path dependence", in G.D. Snooks (ed.), *Historical analysis in economics* (London: Routledge).

David, P.A. (1993b), "Path dependence and predictability in dynamic systems with local network externalities: a paradigm for historical economics", in D. Foray and C. Freeman (eds), *Technology and the wealth of nations* (London: Pinter Publishers).

David, P.A. (1994a), "Positive feedbacks and research productivity in science: reopening another black box", in O. Grandstrand (ed), Economics of *Technology* (Amsterdam: North-Holland).

David, P.A. (1994b), "Les standards des technologies de l'information, les normes de communication et l'État: un problème de biens publics", *Analyse économique des conventions*, ed. A. Orléan (Paris: Presses Universitaires de France).

David, P. A. (1995), "Reputation and Agency in the Historical Emergence of the Institutions of 'Open Science'", Paper presented to the National Academy of Sciences Colloquium on the Economics of Science and Technology, held at the Beckman Center, U.C. Irvine, 20-21 October 1995.

David, P.A. (1996a), "Science reorganized? Post-modern visions of research and the curse of success", in *Measuring R&D Impact - Proceedings of the Second International Symposium on Research Funding*, (Ottawa: NSERC of Canada).

David, P. A. and T. Flemming (1995 & 1996), "Communications, creativity and research network dynamics: An elementary economic model of scientific communities", Paper presented to the International Conference on Creation and Transfer of Knowledge, held at Castelgandolfo (Rome), September 1995; revised for presentation to the All Souls College Seminar on the Economics of Information Diffusion, convened in Oxford, May 1996.

David, P.A., and D. Foray (1993), "Percolation structures, Markov random fields and the economics of EDI standards diffusion", in G. Pogorel (ed.), *Global telecommunication strategies and technological change* (Amsterdam: Elsevier).

David, P.A., and D. Foray (1994), "Dynamics of competitive technology diffusion through local network structures: the case of EDI document standards", in *Evolutionary economics and chaos theory: new developments in technology studies*, L. Leysendorff and P. van den Besselaar, eds. (London: Pinter Publishers).

David, P.A., and D. Foray (1995), "Accessing and expanding the knowledge-base in science and technology", *STI Review -Science, Technology and Industry, 16,* (OECD: Paris).

David, P. A., D. Foray and J.-M. Dalle (1997), "Marshallian externalities and the emergence and spatial stability of technological enclaves", forthcoming in *Economics of Innovation and New Technologies*, vol.6 (Special issue on *"The Economics of Localized Technological Changes"*, ed A. Antonelli).

David, P.A., A. Geuna, and W.E. Steinmueller (1995), "Additionality as a principle of European R&D funding", Report for the STOA programme of the European Parliament, MERIT Research Memorandum 2/95/012, Maastricht.

David, P.A., and S. Greenstein (1990), "The economics of compatibility standards: an introduction to recent research", *Economics of Innovation and New Technology* 1 (1 & 2), Fall 1990, 3-42.

David, P. A., D. Mowery, and W.E. Steinmueller (1992), "Analyzing the payoffs from basic research", *Economics of Innovation and New Technology*, Vol. 2, No. 4.

David, P. A., and W. C. Sanderson (1997), "Making use of treacherous advice: cognitive learning, Bayesian adaptation and the tenacity of unreliable knowledge", in *Frontiers of the New Institutional Economics*, J. V. C. Nye and J. N. Drobak, eds. (San Diego, CA: Academic Press.)

Ellison, G. (1993), "Learning, local interaction, and coordination", *Econometrica*, 61: 1047-1071.

Ellison, G., and D. Fudenberg (1995), "Word-of-mouth communication and social learning", *Quarterly Journal of Economics* 110 (1): pp. 93-126.

Feyerabend, P. (1975), *Against Method* (London: Humanities Press).

Franklin, A. (1986), *The Neglect of Experiment* (Cambridge: Cambridge University Press).

Fuller, S. (1994), "A Guide to the Philosophy and Sociology of Science for Social Psychology of Science", Ch. 17 in *The Social Psychology of Science,* New York: The Guilford Press.

Gambardella, A. (1994), *Science and innovation,* Cambridge: Cambridge University Press.

Geison , G. L. (1996) "Pasteur and the Culture Wars: An Exchange, *New York Review of Books*, XLII(6), April 4: pp.68-69.

Grimmett, G. (1989), *Percolation* (New York: Springer Verlag).

Hammersley, J.M., and D.J. Welsh (1980), "Percolation theory and its ramification", *Contemporary Physics* 21 (6).

Harding, S. (1976), *Can Theories Be Refuted?* (Dordrecht: D. Reidel).

Harris, T.E. (1978), "Additive set-valued Markov processes and percolation methods", *Annals of Probability* 6.

Henderson, R., A. Jaffe and M. Trajtenberg (1995a), "Universities as source of commercial technology: a detailed analysis of university patenting 1965-1988", National Bureau of Economic Research Working Paper No. 5068, March.

Henderson, R., A. Jaffe and M. Trajtenberg (1995b), "The Bayh-Dole Act and trends in university patenting 1965-1988", paper presented to the Conference on University Goals, Institutional Mechanisms and the 'Industrial Transferability' of Research, March (Stanford: Stanford Center for Economic Policy Research); revised August.

Hirshleifer, J. (1971), "The private and social value of information and the reward for inventive activity", *American Economic Review* 61, 561-74.

Holley, R., and T. Liggett (1975), "Ergodic theorems for weakly interacting systems and the voter model", *Annals of Probability* 3, 643-63.

Hull,, D. (ed.) (1988), *Science as a process* (Chicago: University of Chicago Press).

Imai, K. and Y. Baba (1989), "Systemic innovation and cross-border networks", paper presented to the International Seminar on the Contributions of Science and Technology to Economic Growth, June (Paris: OECD Division of Science, Technology, and Industry).

Jaffe, A.B., M. Trajtenberg and R. Henderson (1993), "Geographic localization of knowledge spillovers as evidenced by patent citations", *Quarterly Journal of Economics*.

Kindermann, R., and J.L. Snell (1980a), "On the relation between Markov random fields and social networks", *Journal of Mathematical Sociology* 7.

Kindermann, R., and J.L. Snell (1980b), *Markov random fields and their applications: Contemporary mathematics, Vol. 1* (American Mathematical Society).

Kirman, A. (1993), "Ants, Rationality and Recruitment", *Quarterly Journal of Economics,108:137-156.*

Kirman, A., C. Oddou, and S. Weber (1986), "Stochastic Communication and Coalition Formation", *Econometrica,* 54(1): 129-138.

Knorr-Cetina, K. (1981), *The Manufacture of Knowledge: An Essay on the Constructivist and Contextual Nature of Science,* Oxford: Pergamon Press.

Kuhn, T. S. (1962/1970), *The Structure of Scientific Revolutions,* First/Second Edition (Chicago: University of Chicago Press).

Lakatos, I. (1970), "Falsification and the methodology of scientific research programmes", in I. Lakatos and A. Musgrave (eds), *Criticism and the growth of knowledge* (Cambridge: Cambridge University Press).

Latour, B., and S. Woolgar (1979), *Laboratory life* (Beverly Hills: Sage Publications).

Leydesdorff, Loet (1995), *The challenge of scientometrics: the development, measurement, and self-organization of scientific communities* (Leiden: DSWO Press).

Luhmann, N. (1984), *Soziale systeme. Grundrisz einer allgemeinen Theorie* (Frankfurt a.M.: Suhrkamp).

Luhmann, N. (1990), *Die Wissenschaft der Gesellschaft* (Frankfurt a.M.: Suhrkamp).

Marschak, J. (1971), "The economics of information", in M. Intrilligator (ed.), *Frontiers of quantitative economics,* Contributions to Economic Analysis No. 71 (Amsterdam: North-Holland).

Merton, R.K. (1973), in N.W. Storer (ed.), *The sociology of science: theoretical and empirical investigations* (Chicago: University of Chicago Press).

Morris, S. (1996), "Strategic Behavior with General Local Interaction", University of Pennsylvania Department of Economics Working Paper. March.

Mulkay, M. (1979), *Science and the Sociology of Knowledge,* London: George Allen and Unwin.

Narin, F. (1976), *Evaluative bibliometrics* (Cherry Hill, NJ: Computer Horizons, Inc.).

Perutz, M.F. (1996), "Pasteur and the Culture Wars: An Exchange, *New York Review of Books*, XLII(6), April 4: pp.68-69.

Polanyi, M. (1966), *TheTacit Dimension* (London: Routledge and Kegan Paul).
Popper, K. R. (1959), *Logic of scientific discovery* (London: Hutchinson).
Popper, K. R. (1963), *Conjectures and refutations* (London: Routledge & Kegan Paul).
Price, D. J . de Solla (1965), "Networks of scientific papers", *Science* 149, 510-15.
Price, D. J. de Solla (1986), *Little science, big science and beyond* (New York: Columbia University Press).
Quine, W. V. O. (1953), "Two Dogmas of Empiricism", in *From A Logical Point of View*, (Cambridge, Mass.: Harvard University Press).
Quine, W. V. O. (1962), "Carnap and logical truth", in *Logic and language: studies dedicated to Professor Rudolf Carnap on the occasion of his seventieth birthday* (Dordrecht: Reidel).
Quine, W. V. O. (1969), *Ontological Relativity*, New York: Columbia University Press.
Schelling, T. (1971), "Dynamic models of segregation", *Journal of Mathematical Sociology* 1.
Shadish, W. And S. Fuller, eds., *The Social Psychology of Science*, New York: The Guilford Press.
Stephan, P. (1996), "The Economics of Science", *Journal of Economic Literature*, XXXIV (3), September: 1199-1262.
Trajtenberg, M., R. Henderson and A. B. Jaffe (1992), "Ivory tower versus corporate lab: an empirical study of basic research and appropriability", National Bureau of Economic Research Working Paper No. 4146, August.
Usher, Abbott Payson (1982), *A History of Mechanical Invention*, Revised Edition, New York: Dover Publications.
van Rann, A.F.J., ed. (1988), *Handbook of quantitative studies of science and technology* (Amsterdam: Elsevier).
Weitzman, M. L. (1995), "Recombinant Growth", Harvard University Economics Department Working Paper. March.
Worrall, J. (1976), "Thomas Young and the 'Refutation' of Newtonian Optics: A Case-study in the Interaction of Philosophy of Science and History of Science", in *Method and Appraisal in the Physical Sciences*, ed. C. Howson, (Cambridge: Cambridge University Press).
Ziman, J. (1984), *An introduction to science studies: the philosophical and social aspects of science and technology* (Cambridge: Cambridge University Press).
Zuckerman, H. (1977), *The Scientific Elite: Nobel Laureates in the United States*, New York:The Free Press.
Zuckerman, H. and R. K. Merton (1971), "Institutionalization and Patterns of Evaluation in Science", in R K. Merton,*The sociology of science: theoretical and empirical investigations*, N.W. Storer, ed., Chicago: University of Chicago Press, 1973.

PART II

Transmission and Diffusion of Knowledge

8. Literacy and the Diffusion of Knowledge across Cultures and Times

Jack Goody*

My contribution to this discussion differs from the substantial papers written by others. But it does arise out of an empirical background, in trying to understand the world as it is and as it was, and it needs to be viewed as a contribution to cultural history, even economic history, in a broad sense. The empirical background relates firstly to the kind of changes that are taking place in Africa and other parts of the world (preferably not the Third World, a misleading concept) as the result of a major addition to the communication system in the shape of *writing* and its consequences for what I call the technology of the intellect and what is here referred to as the creation and transfer of knowledge.

Problems of transferring and creating knowledge are of course basic to all cultures. They took an enormous step forward with the development of language which many would see as the defining criterion of Homo sapiens. That was at an indeterminate period in the past, which some have associated with the Upper Paleolithic, others with extensive tool making. From this moment non-genetic information flowed within and between human groups in an extraordinary manner - as witness the similarity in tool making all over the world in various early prehistoric periods.

This transfer was undoubtedly subject to attempts to limit the spread of the useful arts to neighbouring groups, combined with incentives so to do. What these were we can only guess. Undoubtedly one mechanism promoting transfer was the institution of exogamy, a rule enforcing out-marriage, so that women moved from one group to another, thus establishing not only affinal but kin relations, because every child would have his or her father in one group and his mother's brother in another. In other words there existed both constraints and incentives but also leakages.

The process of communication made another quantum jump with the invention of writing (c. 3000 BCE), that is, only some 5000 years ago, during what the historian Gordon Childe called the Urban Revolution.

What is the most striking feature of knowledge in modern societies? I ask this question while sitting in the great circular Reading Room of the British Museum, opened in 1857 and about to close, each cranny of which is stuffed with books, their readers and often their writers. The answer rests on the shelves and on the desks. It is the storage of information in books (or other written form), its retrieval

* St. Johns' College, Cambridge University, Cambridge (U.K.).

by reading and its supplementation by writing.[1] The contrast with oral societies, those without writing, is striking. There storage rests largely upon the individual memory, although artifacts (including some graphic ones) may provide a marginal additional means by which some information is handed down.

The problem with the memory storage of information is its limited capacity, its fading (forgetting) and the nature of its transmissability. Whereas books are infinite in number, ensure virtually permanent storage and provide for indirect transmission. That is to say, while in oral cultures all transmission is face-to-face and requires presence in space and time, written information can be communicated on material objects that can be transferred over both space and time. We can therefore gain access not only to the words of our cousins who migrated to America but to the thoughts of our distant ancestors as well. So while oral cultures are highly localised, written ones are more inclusive and in the long run global. Let me give one example. In 1960 the population of Black Africa was roughly 250 million and there were some 4000 languages. With roughly the same population the USA had officially one. In Africa all these, except for a few lingua franca, were languages of limited circulation which had emerged as variants in the course of the creative (and differentiating) play of cultural intercourse. With the adoption of modern education under the European colonial regimes, the languages of schooling and government have been drastically reduced, with the new independent nations adapting a tongue of major circulation, partly to permit the consolidation of national government but also to gain access to written information already stored in books. With the advent of computors and international transport, the tendency is for technical languages to be further reduced, to one in certain contexts. These processes of nation building and globalisation give rise to a diglossic situation with co-existing technical and maternal languages.

Such a shift has important effects on cognitive processes, those that control our ability to understand the world. The written past obviously differs from the oral in providing an archive of a more permanent, 'objective' kind, that is, one situated outside the memory internal to individuals. The external storage of knowledge means that the past can influence the future to a much greater extent. Equally writing increases our capacities to organise for the future, for example, by repeatedly recording natural events to enable predictions to take place, or by permitting budgetting in a more precise manner. Of course all is not gain with writing, which may set out in too permanent a form ideas of the 'men of old' which are better forgotten, or which may inhibit as well as promote social change by freezing inappropriate laws that would be quietly eliminated in a purely oral culture. The same is true of creativity; writing draws a line between those who recite (Shakespeare, for example) and those who compose in a manner, whereas in purely oral cultures every recitation is an act of creation.

---

[1] In this context I am disregarding the computer, which I see technically as an extension of rather than a substitute for literacy.

What are the economic implications of this shift in the modes of knowledge production and diffusion? Firstly, the complexity of the skills of reading and writing, not to mention the sheer increase in knowledge stored in that way, means that some individuals have to be removed from primary production to enable them to gain sufficient mastery of these. Until the latter part of the nineteenth century, with few exceptions only a relatively limited proportion of children were segregated in this fashion, the others following their mothers and fathers in their daily activities and learning in a direct, concrete rather than in a mediated, abstract way. As 'education', that is, school learning involving the '3 Rs', became more widespread, an increasing percentage of the child population was removed from domestic activities, including production, and handed over to an ever growing body of specialists, the teaching profession, up to at least the age of 14, more likely 16, and with 'further education' to 22 or later.

Thus the acquisition of written learning changed radically the division of labour as well as the nature of household production; it also meant that children of the poor had greater difficulty in gaining such an education since their parents had to make a proportionally greater sacrifice to dispense with their domestic activities. Indeed it is significant that widespread education of the labouring classes was largely coincidental with the development of work in factories and offices. There family labour had at first been prevalent, children accompanying parents to the mill or the mine. But subsequently the pressures of social reform and of overall economic forces removed them from the productive process, and indeed from any other employment. The disappearance of jobs for children meant that they were available to take part in full-time education, which eventually became a legal obligation, relieving parents of their day-time minding duties and leaving them freer to go out to work. That reduced the immediate economic importance of children and, if parents were to maintain or raise their standard of living, reduced their numbers too.

Secondly, the production of more decontextualised written knowledge entailed the specialisation not simply of the teachers but of the creators of knowledge. These had to be supported, whether by court or temple, by noble or merchant, by state or community. Specialisation, in the shape of withdrawal from primary production, was required both for those learning and those teaching how to read and write. But it was even more necessary for the production and advancement of written knowledge in whatever field. 'Hence it was in Egypt', remarked Aristotle, 'that the mathematical arts were first developed, for there the priestly class was set apart as a leisured one'.[2] While literacy was certainly not the only factor to contribute to the establishment of a leisure class, it required specialists removed from other demands who were spared the particular type of effort involved in manual work.[3] They filled what were largely privileged positions (even if as scribes they were not the highest in the land and sometimes attracted a measure of

---

[2] Aristotle *Metaphysics*, 981B, 23-25.

[3] On the formation of intellectual classes, see Gramsci.

contempt) which they were likely to preserve by laying down restrictions to entry, or in some cases even to access to the corpus of written knowledge. The latter was often true of Religions of the Book (or written religions generally) where the establishment of a defined canon was one the aspects of institutionalisation. Secular knowledge was usually more open but its circulation and production was nevertheless often subject to control or censorship in one form or other. In order to preserve their privileged position, scribes had to succumb to or even engineered such restrictions. But in any form of communication there is always a balance of gains and losses between preserving one's control over the information and extending one's audience or one's clientele. That situation may lead to a certain mystification of literacy, giving rise to special forms of writing, as on doctor's prescriptions, or of knowledge that has to be mediated by oral transmission through a specialist, as earlier with the Catholic Bible, with Hindu prayers or even with psycho-analysis.

As well as the production of knowledge, there was the question of its reproduction. Each manuscript had, by definition, to be copied by hand, a long and arduous process; it has been estimated that in the West it required six month's manpower to produce an average medieval text.[4] Whether that was done in temple scriptoria or in secular workshops, the cost of unit texts in pre-printing cultures was very high and required a great imput of manual labour, a situation that was radically changed with the invention of printing which introduced mass production into knowledge systems. Printing led to a considerable extension of the reading public but prices remained relatively high and it was not until the nineteenth century with the invention of the rotary press that books and newspapers became items of mass consumption, leading to an increase in the speed and range of the circulation of information, which in turn quickened the pace of the creation of knowledge as well as of fiction in all its forms.

The extent to which knowledge in oral cultures is embedded in specific situations is not often appreciated by those used to dealing with knowledge that has been abstracted in texts. I do not wish to imply that in such cultures all is concrete, lacking theoretical constructs. As Horton has argued, that is certainly not true of religious concepts. But knowledge systems take quite a different turn when information is consigned to books and partially decontextualised. Transmission is no longer face-to-face but rather hand-to-eye. The knowledge exists outside specific individual social relationships and out there in libraries or in data bases. Oral knowledge is inevitably more immediate whereas written is more mediated, often the product of a deeper process of self-reflection since one has a greater opportunity to consider the words of others or of oneself.

The shift from purely oral to written cultures clearly has economic implications, as well as causes. We only find writing in the Bronze Age cultures that had undergone the Urban Revolution. That in turn was based upon the shift to intensive cultivation involving the harnessing of animal energy to production, the

---

4 Chaucer's Clerk had an immense library of thirty books.

control of water and the development of a variety of crafts. We cannot really make the assessment of the advantages of writing on a purely individualistic level. For the first 5000 years that writing existed, the skills were only acquired by a minority of any particular population. One may argue that the others were deliberately excluded by a kind of restrictive practice in order to provide greater benefits to the practitioners, as is the case with a number of specialist occupations. But it is also the case that there was only room for a certain number of scribes in a pre-industrial economy, and that literacy was of limited value to the agricultural majority. Of little value, that is, to the other individuals. But to the community as a whole it was essential to promote and accumulate knowledge, to advance the economy and indeed to develop the whole socio-cultural system. Such advances required literacy only among a limited proportion of the population. The bulk could remain 'uneducated', that is to say, illiterate. Only a minority needed to acquire the skills in order to shift the whole culture to a different level in terms of the results of written communication. While it may have been advantageous to a particular individual to acquire this 'new' technique as well as the knowledge that was incorporated in texts (though this did not remain altogether hidden from the illiterate), it would have been largely uneconomic for more than a certain number to learn these skills.

That remained the case, with certain minor exceptions, until the Industrial Revolution and the advent of mass democracy in large-scale societies. Even so, the impulse to spread literacy was not directly economic; it did not add much to the skills of mill workers though it could be a great advantage to their children in the context of social mobility. The major social pressure for the extension of literacy came, as in other written civilisations, from religious sources. The skill of reading was required in order to gain first-hand knowledge of Holy Scriptures which constituted one of the first major products of written cultures and their Religions of the Book. Moreover, except in Ancient Greece and China, the training in and transmission of the literate tradition was usually assigned to sacred rather than to secular authorities, even when the training was for secular ends. In Europe, direct access to religious texts at first received greater encouragement in Protestant than in Catholic circles and produced higher levels of literacy. In the late eighteenth century the Sunday School movement in England had a significant influence in extending the reading public. However, as the result of the Counter-Reformation, the situation also changed in Catholic countries and church schools expanded the ability to read and write under all Christian denominations.

There had been two opposed tendencies with regard to Christian literacy in Europe, which I suggest were to be found in the past in other cultures such as India, where literacy was at some periods limited to Brahmins alone and at others made available to all twice-born castes. The same problem of extending the influence of the word of God but limiting its transmission to those trained to act as gatekeepers was present elsewhere. In medieval Catholicism direct access to the scriptures was open only to the priest who was responsible for transmitting the knowledge orally or visually to the rest of the population. Literacy was restricted;

literati were virtually the equivalent of *clerici* and could be saved from capital punishment for the most heinous crimes by the reading (though sometimes by recitation!) of the 'neck verse', which established their privileged position. The situation contained built-in contradictions which were exploited by the Protestants. For they embraced direct access to the scriptures by the people, and hence became involved in translating the Bible (from a 'dead' language, Latin) into the vernacular and encouraging the practice of reading and writing at schools so that children could acquire the techniques (as well as other knowledge). Just as one had direct unmediated access to God, so too one did to His Word. At the same period the advent of printing made books cheaper commodities, though even in the eighteenth century the price of novels for example was high in comparison to incomes, and hence the popularity of lending libraries among the middle classes. The mass production and consumption of reading material was by and large accomplished only with the rotary presses and expansion of education in the nineteenth century.

An additional secular reason for the acquisition of these skills did occur in nineteenth-century Europe. With the introduction of mass democracy in the shape of broadly based assemblies and voting procedures, many social reformers insisted that the bulk of the population should get to know how to read and write, not simply to make their mark on a ballot paper but to enable them to be informed about what was going on in the country in whose government they were now directly participating. That was a central theme in the works of John Stuart Mill. Only now, when industrial production took over, when domestically-centred employment became of less significance, was it possible to make the children of the poor attend school, though at an expense to working class families since they had to forgo the proportionately greater returns on the labour power of their children in the hope that they would 'better themselves' later on.

The fact that there was little national *economic* pressure towards universal literacy continues to be true in large regions of the world today. Just as for the first five thousand years of written cultures only a minority could read and write, while on a societal level the remainder enjoyed some (but not all) of the advantages, so too in developing countries, especially in the new nations, it might have made sounder economic sense not to provide universal schooling at once but to allow the shift from traditional oral cultures to take place more slowly. However one of the first acts the (educated) leaders of new nations did was to try to pursue the aim, earlier confined to religious groups, of getting every one to read and write, on the political grounds that a democracy needed the circulation of information that the printed word could provide; only that way could a transformation of the country be effected and modernisation attained. That was partly true, even if many of these nations did not long remain democracies and the economic requirement was not proven, or even plausible. Nevertheless education took a high proportion of national budgets which could otherwise have contributed more immediately to aspects of development. The results have been mixed. In Africa, literacy has given the educated greater mobility. Many seek better-paid work in developed countries, even if those jobs are less qualified than they could have expected at home. The loss to their own country is two-fold: firstly, the disappearance of qualified personel who could have contributed towards development; secondly, the loss of the cost of training which the community has largely funded. The gains are personal: preferred living conditions for one's conjugal family and possibly some return to those who remain behind, especially as a link in a chain migration. From Indians working abroad in the Middle East, the repatriation of funds to the home country is significant; that does not seem to be the same for Africa. Indeed it is doubtful whether the gains to the developed (receiver) country are offset by the amount of aid transferred to the developing (exporting) one.

What are the incentives to transfer literate knowledge? They are clearly different from those in oral cultures where the process is virtually always face-to-face and hence carried out in the context of specific social relationships with transmission remaining under the control of the transmitter. In literate societies that restriction no longer applies. One scatters one's seed on the water, to be gathered by friend and foe alike; after publication one loses personal control of the process, which takes place in the absence of a specific social relationship. The incentives are of a totally different order; the return depends mainly on the income from the sale of books or of copy-right.

That is not the case with knowledge in the process of being created and published (and publication is the witness of creation, hence the importance of dated communications and letters to journals such as *Nature*). At that stage the process is closer to that obtaining in oral cultures where friendships, enmities, chance encounters, face-to-face relationships play a central role. It is those events that constitute the excitement and interest of an account such as the discovery of DNA in *The Double Helix* by James Watson. In that story, the blocking of the flow of information to competitors, the acquisition of knowledge of what others were doing (as in the incident of the letter from Linus Pauling to his son, Peter), the conspiratorial sessions in *The Bun Shop*, a public house, these were intrinsic features of the creation of knowledge until the moment when all was revealed in a public announcement. Publication is then judged by the numbers of copies sold, the reputation made, the people influenced, which involves not only the acquisition of prestige and the more material returns of royalties, but also the possibility of getting a better job; that is one of the roles of the citation index in the USA. In addition literate publication provides a measure of individual longevity, even immortality of a kind, that few can ever acquire in an oral culture where cultural products soon lose their individual signature. Without the name on the tombstone, who knows who is buried there?

A central feature of literate knowledge is the fact that it forms a largely public depository of information which can be built on by others. That is not to say that incremental accumulation of a non-genetic kind cannot take place in oral cultures. If this was not possible, there would be no cultural evolution, for example, from hunting and gathering towards agriculture. But just as language has made that possible, so writing has enormously speeded up the process of creation and transfer since the new method of storage encourages the possibility of incremental accretion. One reads the work of someone unknown personally; one criticises, adds to and reorganises that knowledge. Hence the great significance of listing behaviour in early literate societies as well as the interest in compendia, encyclopaedias, *summae* of knowledge, in Greece, in Song China, in medieval Europe (largely religious knowledge). Hence the importance of quotes and of the limited plagiarism that is involved in the process of accumulation since one starts where someone else has left off. Hence the stress on learning about (what other people have done, especially in scholarly discourse) before one can add one's own pebble to the cairn.

We may look at the diffusion of this knowledge in terms of two processes, the intergenerational and the intragenerational. The first involves the communication of information from adults to children, either in the largely domestic environment of oral cultures or in a largely external setting, as with the schools of literate cultures. This particular process is one by which a culture is maintained by transmitting its knowledge to the new generation; it involves in the main the passing down of already canonised knowledge, such as is encapsulated in text books. A great degree of verbatim memorisation is involved on the part of the child. The incentives to the teachers are salary and personal satisfaction; to the child, it is acquiring good marks, becoming an adult, getting a job, but its rewards, actual and potential, do not always compensate for the heavy burden of these tasks and some get turned off, especially in those environments where alternative pursuits are availabe that do not require writing (farm work in the country, truancy and street society in the towns).

In Africa where there were no schools (which does not mean of course to say there was no education in a wider sense), the internal transfer of knowledge involved transmission within domestic groups. The only schools teaching writing were Muslim schools which with few exceptions taught what I call a restricted literacy in that it was largely restricted to religious topics. That was also the case in India (though not in China). The introduction of schools and universities is probably the most radical instrument of social change in Africa and elsewhere. Even if introduced by religious sects, it has been a major influence towards secularisation.

The second axis of diffusion is between adults and is concerned partly with the acquisition of existing knowledge, partly with its creation and partly with the transfer of information in the course of daily life. Here the problem of incentives takes a different turn. Communication within the generations is usually much less authoritarian than intergenerational transmission (in early Mesopotamian schools the teacher was 'Big Brother'), taking place in situations that are more open to negotiation and hence more subject to specific incentives.

In the case of unpublished information it involves not only the diffusion but also the retention of knowledge. What gets published (printed) and what remains restricted is a critical aspect of knowledge in literate societies, since once it is published control has been lost, lost not only to the particular culture or country but to the world at large. An example is Nicolas Appert's invention of the bottling (later canning) process in France in response to an appeal of the Directoire in 1795 for contributions towards solving nutritional problems created by the war situation and the British blockade, though the citation he received when getting his award refers to the advantage of his new invention for sea-voyages, for hospitals and for the domestic economy. Appert had been a chef before his proposal was given a series of public tests in Brest in 1804, after which he opened a bottling factory near Paris. Five years later he was awarded the prize of 12,000 francs by a committee that included Guy-Lussac and Parmentier, on condition he deliver a description of the process, in 200 copies, printed at his own expense. This description he published in 1810, under the title of 'Le Livre de tous les ménages...' which proclaimed bottling as a general aid to domestic life.

Appert invested his prize money in further production and research, founding the house of Appert in 1812. But the secret of bottling was out. Despite the war an English translation of his work appeared as soon as 1811, an American edition a year later. Others had learnt of the process earlier. In London Donkin set up a factory in Bermondsey, whose products were ordered by the Navy. From London the technique was transferred across the Atlantic first to New Orleans in 1817 by a man who had worked in the British industry. From there he made his way to Boston in 1819 and the American canning industry was born. Appert himself made nothing out of this extension of his invention; he had published the results and in any case there was no copyright, so diffusion across cultures could take place very rapidly, in a manner that followed the trend of many of the achievements of the Industrial Revolution, which were rarely confined for long to one culture.

The transfer of technical knowledge is closely dependent upon the ability of the host community to accept it. Much of the transfer of technology to Africa nowadays (and to other countries) has been military technology, in which the main aim has been to sell the product at great price without transferring the production. Indeed in some cases of advanced technology restraints are placed on the export of the product itself and all information about it. It has long been thus. European domination of the rest of the world depended, in Cipolla's phrase, on Guns and Sails. Adapting the Chinese invention of gunpowder, the gun played a major role in the domination of Africa and it was for that reason that the Pope forbade their export to Africa. In fact leakage took place since there were always renegades (especially Protestants) who were willing to engage in this trade with the 'enemy'. In Arabia, Malabar and Japan they sold guns which were then copied. Unlike Africa which never had the required steel technology, they already had the right kind of technical background.

As an example of a very different kind, one may look at the diffusion of sexual knowledge. Writing certainly brought a change for we find evidence of Roman and Greek textbooks on the subject offering information on different positions in intercourse, as well as on courting behaviour. There is some suggestion that Ovid was referring obliquely to such works in *The Art of the Love*. Such textbooks also existed in India (for example, the Khama Sutra), in China (as described by van Gulick) and in Japan where it was the custom to leave one by the bedside of a new bride, thus spreading information which she would not otherwise easily have obtained except in a garbled form from her peers. In India instructions in proper sexual behaviour were mixed with instructions on manners more generally. Given the price of manuscript books (because of the difficulty of reproduction) the circulation of such such works was mainly among the rich.

The situation in Europe changed with the advent of printing, mechanised writing, which produced multiple copies, especially of diagrams, much more cheaply. It encouraged authorship as well as readership. Porter remarks 'A genre that received special significance in the early modern period, notably toward 1700, was printed sexual advice literature, a subset of the enormous dissemination of know-how on matters from medicine to manners, horsemanship to housekeeping, gushing from the presses in the post-Gutenberg era'.[5] The most popular book on this subject was Dr Nicolo Venetti's *Tableau de l'amour conjugal*, published in 1696. It was much translated including into English in 1703 as *The Mysteries of Conjugal Love Revisited* and subsequently went into many editions. It combined medical and folk knowledge in the manner of many works of the period, which was one that led to the Enlightenment with its expansion of popular and scientific knowledge that played its part, among other things, in the events leading to the Industrial Revolution.

Literacy radically extends the possibility of accumulating and transmitting knowledge in space-time. As such it changes the nature of social interaction, including cognition both internal and external to the individual, in ways that have profound implications for economic history and organisation. At the same time it alters the market conditions for knowledge. A change in the mode of production (to capitalism) is inconceivable without a change in the mode of communication (to literacy), at least for a percentage of the population. This change had many implications for the development of specialisms, for the allocation of the labour of children as well as of adults, for the whole organisation of the priorities of domestic life. In the modern economy of the West those without literate skills have few niches, as compared with the situation in newly developing countries where investment in universal literacy may have few immediate economic benefits at the national level. However one such benefit is important. The impact of colonial development in Africa and elsewhere had a profound effect on public health as well as on political organisation; mortality, especially of infants, was drastically reduced. Consequently population increased on a dramatic scale. One of the few social factors capable of reducing rather than increasing the marital fertility of women appears to be their incorporation in school education, in other words the achievement of high levels of female literacy. In this context at least widespread literacy becomes an important socio-economic factor in world strategies, even in societies where other immediate gains may be few.

It is often asserted that in the past literacy has had a gender bias. It is true that women have been largely excluded from schools and universities, as well as from creative activities. But such exclusion exists in oral cultures in a yet more absolute fashion. In a face-to-face gathering of a ceremonial kind, it is relatively easy to separate men and women so as to prevent one group from acquiring the

---

[5] Porter 1994:138.

knowledge of the other. With the written word, such exclusion is less easy. While schooling may be confined to the male sex, the acquisition of the skills of reading and writing may be acquired at home. Some men like King Alfred believed in educating both sons and daughters, which was in keeping with the general aim of avanced stratified systems, I have argued, of maintaining the social position of children of both sexes. In other cases (some twenty per cent of families) a man might have daughters but no sons and wish to pass down his skills to them. So that even in societies where women were formally excluded from education, we have examples of female writers, Sappho in Greece, Lady Genji in Japan, the many women poets of Tamil India, the female troubadours of medieval Europe. In other words the potentialities for democratisation were present within literacy; the underprivileged may gain access to the means of communication before they do to the means of production and indeed may use them to subvert the existing social order.

I have touched upon some of the more direct economic consequences in terms of the division of labour which clearly had to be related to primary production and the acquisition of surplus. It is a question of withdrawing teachers and children from other activities on a major scale (and it is one of the major aims on which developing countries spend resources - on universal education).

The development of education systems has nearly always taken place in close connection with the needs of the economy and of society more generally - only a proportion being education. The late nineteenth century saw the development of mass educated systems in the West which were broadly in line with technology. What has happened in developing countries has been the adoption of an ideological approach to universal education (for democratic reasons) which has been grossly out of line with economic developments. Hence Africans are educated to fill US economic needs rather than their own.

The pressure to restrict literacy was in the long run countered by the pressure to extend it and to exploit fully the new medium by attracting wider participation, which resulted in evening out the possibilities of acquiring knowledge and the attached social positions, in other words, in a greater degree of social equality.

References

Aristotle [1957] *Metaphysics*. Oxford

Chaucer, G. [1913] *The Canterbury Tales*. London

Childe, G. 1942 *What happened in history?* Harmondsworth, Middlesex

Eisenstein, E. L. 1979 *The press as an agent of change: communications and cultural transformations in early modern Europe. 2 vols.* Cambridge

Goody, J. 1977 *The domestication of the savage mind.* Cambridge

Goody, J. 1986 *The logic of writing and the organization of society.* Cambridge

Goody, J. 1987 *The interface between the written and the oral.* Cambridge

Horton, R. 1967 African traditional thought and Western science. *Africa* 37: 50-71, 155-87

Porter, R. 1994 The literature of sexual advice before 1800. In Porter, R. and Teisch, M. (eds) *Sexual knowledge, sexual science: the history of attitudes to sexuality.* Cambridge

9. Market Failures, Education and Macroeconomics

Giuseppe Bertola, Daniele Coen Pirani*

This paper surveys the burgeoning macroeconomic literature which emphasizes the role of education in models of growth and distribution, and reviews traditional partial equilibrium approaches to the economics of education. Contemporary models of growth recognize that knowledge is accumulated under a variety of market imperfections, but view the macroeconomic role of human capital accumulation as similar and complementary to that of physical investment. The complex microeconomics of education, however, suggests that education does more than increase a worker's productive potential. We argue that such aspects need to be taken into account by macroeconomic models of growth and distribution.

Section 1 organizes a review of recent contributions to the theory of macroeconomic growth and income distribution around the three crucial roles of "human capital" in such models. First, allowance for other accumulated factors makes it possible to model endogenous growth even when physical capital accumulation encounters decreasing returns to scale. Second, education is perhaps more strongly influenced by capital-market imperfections (such as self-financing constraints and externalities) than other forms of investment, hence it is particularly well-suited to the distributional perspective of many recent contributions. Finally, and again in light of realistic market imperfections, explicit modeling of education gives an important role to policy interventions and to political interactions.

Section 2 outlines the microeconomic foundations of such macroeconomic work, focusing particularly on the (perhaps unwarranted) distinction between human and physical capital as pertains to financial market imperfections. We also highlight other microeconomic features of education provisions, which have not received as much attention in recent literature. Human-capital accumulation increases a given worker's productive potential (as physical capital also would) but, in the incomplete-information settings which rationalize financial-market

* Università di Torino; University of Rochester
Paper presented at the Fondazione Eni Enrico Mattei/The Beijer International Institute of Ecological Economics International Conference on Creation and Transfer of Knowledge: Institutions and Incentives, Castelgandolfo (Rome), 21-23 September, 1995. We are grateful to the discussant and other conference participants for helpful comments, and to F.E.E.M. for financial support.

imperfections, education also serves the purpose of *screening* labor - i.e., that of allocating heterogeneous individuals to different occupations, and of determining the rate at which they should be compensated.

The "screening" role of education has distinctive implications for both economic development and income distribution. Section 3 illustrates the relevant issues by sketching a simple model of information provision in a macroeconomic setting where information as to labor's qualifications and skills is imperfect, and may in principle be improved by schooling; Section 4 concludes.

9.1 The macroeconomic role of education

The standard Solow (1956) neo-classical model of growth focuses on physical capital accumulation. By contrast, contemporary growth theory assigns an important role to accumulation of human (or knowledge) capital, at both the aggregate and the individual levels. Increased attention to human capital accumulation is certainly justified on grounds of realism, but the macroeconomic literature's shift of emphasis from physical to human capital is also motivated by theoretical and methodological considerations. We organize our brief survey in this Section around three of them.

9.1.1 Compensating decreasing returns to scale

A first, obvious motivation for modeling human capital in an aggregate setting is that the long-rum growth of the economy may be endogenously explained, even when returns to physical investment alone are decreasing, if balanced accumulation of both human and physical capital affords constant returns at the aggregate level.

Consider the standard neo-classical aggregate production function relating aggregate output Y_t to K_t, the aggregate stock of physical capital, and L, the economy's workforce,

$$(1) \qquad Y_t = F(K_t, A_t L)$$

where A_t indexes labor-augmenting technological progress. Disregarding depreciation, we can write

$$(2) \qquad K_{t+1} - K_t = Y_t - C_t$$

if C_t denotes aggregate consumption at time t, which may be determined by optimal individual decisions (in infinite-horizon, as in Cass (1965), or overlapping-generations settings) or by a simple proportional-savings rule of thumb as in Solow (1956).

"Neo-classical" growth models assume that the function $F(\cdot, \cdot)$ has constant returns to scale in its two arguments, so that it is feasible to compensate private factors of production according to their true marginal productivities $F_K > 0$ and $F_L > 0$. Suppose that not only $F_{KK}(K, \cdot) < 0$ for all K, but also that

(3) $\lim_{K \to \infty} F_K(K, \cdot) = 0$

i.e., that the marginal productivity of the accumulated input decreases to zero as the economy's stock of capital grows and the production function's second argument remains constant. Then, per-capita income asymptotically grows like A_t and, to model unceasing growth, the labor-productivity index A_t must increase over time so that more efficient use of existing and new capital goods offsets the tendency of the marginal productivity of physical capital to decrease.

Technology has public-good nature in the model, since A_t is non-rival and nonexcludable if we view it as a proper factor of production but maintain the hypothesis that only L and K_t are compensated. This makes it problematic to endogenize its evolution in a standard framework of complete and competitive markets. Models of "endogenous" growth take two distinct approaches to this problem. The neo-Schumpeterian approach of Romer (1990), Grossman and Helpman (1991), and others gives up the assumption of perfect competition: in the resulting model, growth is driven by technological change as the result of costly R&D investments by monopolistic firms, and indices of technological advancement (such as the number of existing varieties, or the quality level of the most advanced product) have much the same reduced-from role as physical capital K in equations (1) and (2) above. Human capital, when explicitly modeled, need not be an indefinitely increasing input in this strand of literature, and can be a private (rival) input in the production of both patents and industrial goods (see Grossman and Helpman, 1991, Sec.5.2).

The models of Lucas (1988) and Rebelo (1991), by contrast, maintain the neoclassical assumption of perfect competition, and assign an important role to human capital: building on earlier work by Arrow (1962) and Uzawa (1965), the effective labor force variable A_tL is replaced by a human-capital aggregate H_t in equation (1), to obtain

(4) $Y_t = F(\phi K_t, \psi H_t)$

where ϕK_t and ψH_t respectively represent physical and human capital devoted to the production of tangible output (consumption goods and physical capital).

Importantly, however, H_t is neither an *exogenously* provided good nor a *pure* public good, like A_t was in the basic model of equation (1): economic theory can endogenously explain long-run growth if it can model the provision of (e.g.) education. Endogenous growth models such as that of Mulligan and Sala-i-Martin (1993) assume that an economy's educational level can be increased only at the cast of a lower production of consumption and/or physical capital goods: inputs K_t and H_t are diverted from the production of Y_t towards that of human capital according to an accumulation function in the form

$$\text{(5)} \qquad H_{t+i} - H_t = G((1 - \phi) K_t, (1 - \psi) H_t),$$

where G $(\cdot, \cdot)$ has constant returns to scale to allow long-run growth to be endogenous, and ϕ and ψ (the shares of physical and human capital employed in the two sectors) are endogenously determined by first order conditions of static and dynamic efficiency.

Lucas (1988) considers the extreme case where the only input in human capital production is human capital itself, so that at the aggregate level the accumulation equation (5) takes the linear form

$$H_{t+i} - H_t = \delta(1 - \psi) H_t$$

where δ is a productivity parameter; note that in this case the rate of return to both human and physical capital investment is pinned down by the productivity δ of existing human capital in the production of new human capital: the economy's growth rate is, then, unrelated to the parameters of the aggregate production function $F(\cdot, \cdot)$.

In these models, the speed of growth need not be inversely related to *physical* capital intensity. Economies where the physical capital/output ratio is low need not grow faster than those where it is high: the former may have higher human capital/output ratios, hence a more productive labor force. In essence, human capital plays the some role as "disembodied" technology in earlier models, namely that of offsetting physical capital's decreasing marginal productivity.

As in Rebelo (1991), human capital can be treated as a pure private good, to obtain a perfectly competitive model of long-run growth. Lucas's seminal 1988 paper, however, lets *private* accumulation of human capital produce positive externalities $S(H)$, increasing the productivity of all the economy's agents and of aggregate capital. The aggregate production function

$$F(\phi K, \psi H, S(H))$$

is then characterized by increasing returns to scale in K, H and $S(H)$ with $S'(H) > 0$: as in Arrow (1962) and Romer (1986), a decentralized competitive equilibrium exists because each agent takes as given the aggregate spillover $S(H)$.

Since (by construction) agents do not consider their investment's positive spillover on the other agents'productivity or human capital, the *laissez faire* equilibrium is in general inefficient: the private marginal productivity of human capital investment is $F_H(\cdot)$ and, being smaller than its *social* counterpart $F_H(\cdot) + F_S(\cdot) S'(\cdot) > F_H(\cdot)$, leads to inefficiently low investment in equilibrium.

9.1.2 Income distribution and education in models of growth

The endogenous-growth perspective often leads to abandonment of the neoclassical assumption of overall constant returns to scale and representative-agent decentralization in complete competitive markets. Accordingly, theoretical models need to explicitly address distributional and policy issues, and human capital accumulation has an important role in both respects.

9.1.2.1 Human capital and imperfect financial markets

The growth literature reviewed above considers a representative agent, so it cannot address distributive issues. If investment in education is self-financed, or local externalities enter in the human capital production function, then it is possible to explain diverse patterns of savings behavior and income distribution, and many recent growth models do so in an overlapping-generations context (among others, see Loury, 1981; Glomm and Ravikumar,1992; Galor and Zeira, 1993; Ljungqvist, 1993; and Checchi *et al.*, 1994).

The financial markets relevant to human-capital accumulation can be imperfect in at least three different (and interrelated) respects:

(i) an agent cannot borrow an unlimited amount of capital at the interest rate which he receives when he is a lender rather than a borrower: he is constrained in the amount of capital he can borrow, or the interested rate on borrowed capital is larger than that on loans;

(ii) A parent cannot borrow on his children's behalf leaving them a negative bequest, i.e. a debt they have to pay out of their future income;

(iii) When the outcome of a given investment in education is random, the risk associated with the investment is non-insurable.

Imperfection (ii) can be viewed as a limit case of case of imperfection (i), where the interest rate on loans spanning different generations is prohibitively high for the borrower. As perfect insurance would "freeze" distribution at some initial point, imperfection (iii) is essential if theory aims at modeling ongoing renewal of inequality but it does not provide additional insights. For simplicity, then, we may outline the common setup of the above mentioned models assuming that investment in education must be self-financed, but is not risky-so that only imperfection (ii) is at work.

Let the economy be populated by a continuum of families of unit measure. At time t, family i is composed by an old member with income y^i_t and a young one who is not productive, does not consume, and acquires education through schooling. Each agent lives for two periods, and his objective function is additive in his own consumption c^i_t and his offspring's income:

$$(6) \qquad U = u(c^i_t) + g(y^i_{t+1})$$

where $u(\cdot)$ and $g(\cdot)$ are both concave in their arguments and $u'(0) = g'(0) = +\infty$.

Abstracting for now from other determinants of educational achievement and income, we may focus on educational expenditures at the family level. If a parent invests I in his child's education, the latter obtains human capital $h(I)$, where $h(\cdot)$ is concave in I, i.e., there are decreasing returns to scale to human capital investments. The child's income y^i_{t+1} is then proportional to his human capital h^i_{t+1}. Denoting with $D_t(\cdot)$ the distribution of income at time t, and writing the aggregate production function in the form:

(7) $Y_t = AH_t$, with $H_t = \int_0^{+\infty} hD_t(h)\,dh,$

we obtain $y^i_{t+1} = h^i_{t+1}$ if the wage rate A is normalized to unity.

If financial markets are imperfect, each parent has to trade his own consumption off his child's in the budget constraint

$$y^i_t = c^i_t + I^i_t$$

The parent's resources are allocated to investment in human capital investment up to the point where the marginal rate of substitution between his own consumption and the offspring's income is equal to the marginal productivity of the investment:

$$h'(I^i_t) = \frac{\partial u / \partial c^i_t}{\partial g / \partial I^i_t}$$

This condition implicitly defines an optimal expenditure level. If $u(\cdot)$ (the parent's utility from his own consumption) is strictly concave, then educational expenditure is an increasing function of total available resources,

$$I^i_t = I(y^i_t), \qquad I'(\cdot) > 0$$

i.e., education is a normal good for the parent.

In models of this type, imperfect human capital markets link a parent's and a child's income: poor families will invest less than rich ones in their offspring, so that the latter will remain poorer than the former. Formally, we may insert the investment equation $I^i_t = I(y^i_t)$ in $y^i_{t+1} = h(I^i_t)$, to obtain:

(8) $y^i_{t+1} = h(I(y^i_t)) = k(y^i_t)$

with $k'(\cdot) > 0$ and $k(0) = 0$. The transition equation (8) describes the dynamics of intergenerational social mobility in the economy.

Given the initial income distribution, families characterized by different initial incomes may or may not converge asymptotically to the some income level. Concavity of the human-capital production function $h(\cdot)$ and of the objective functions $u(\cdot)$ and $g(\cdot)$ may imply that $k''(y) < 0$: the crucial condition for limited persistence of inequality is that returns to accumulation (in terms of accumulated resources or bequests) be so strongly decreasing as to ultimately cancel initial heterogeneity, or that

(9) $\lim_{y \to \infty} k'(y) = 0.$

This conditions is more than superficially related to that in (3): just as an asymptotically positive return to aggregate accumulation would make it possible for an economy to determine its own rate of long-run growth, so ultimately nondecreasing bequests would allow a dinasty's income to diverge endlessly from others'. Loury (1981) assumes that condition (9) holds, and establishes the existence and uniqueness of a non-degenerate invariant income distribution for this economy when agents' human capital depends not only on parents' investments, but also on stochastic i.i.d. shocks.[1]

Galor and Zeira (1993) show that Loury's convergence result may break down and multiple equilibria may arise if the human capital production function is not concave, but is characterized by threshold effects, so that returns to additional education are highest for intermediate levels of I. Then, the recursive equation $y^i_{t+1} = k\,(y^{it})$ may have two non-trivial fixed points, y^* and $y^{**} < y^*$. The proportion $\int_0^{y^{**}} D_0\,(y)dy$ of families whose initial income is below y^{**} will converge to a zero-income stationary state, while the complementary fraction of rich families will converge to y^*. Thus, if the investment in human capital is income-constrained at an individual level and if the economy's aggregate production depends on the accumulated stock of human capital, income distribution affects aggregate production in both the short and long rum. Let $y_t = h_t$ be the percapita production function of the economy described above; the average stock of human capital is $h_t = \int_0^{+\infty} \kappa(y) - D_{t-1}\,(y)\,dy$ where $D_{t-1}\,(y)$ changes over time as the economy grows. Asymptotically

$$(10) \qquad y = (1 - \int_0^{y^{**}} D_0\,(y)\,dy)\,y^*$$

as the income of poor families converges to zero. Then, for given initial average income a more equal initial income distribution will result in a greater long run average income.

From a normative point of view, because of capital markets imperfections, the return to an individual investment in human capital differs from its social counterpart if the human capital production function is characterized by decreasing returns to scale: under self-financing constraints the allocation of investment is based on individual returns while a first best allocation of investment resources would equalize returns across individuals.[2]

9.1.2.2 Social capital and income distribution

Some recent overlapping-generations models of education and growth with heterogeneous agents concentrate on social capital spillovers instead of (human

---

[1] In Loury's economy, agents are heterogeneous in terms of abilities, and these are unknown to parents at the time in which their investment takes place. See also Bertola (1995) and references therein for an analysis of stochastic wealth dynamics in infinite-horizon settings with imperfect insurance and precautionary savings.

[2] *"... The most important consequence (of on imperfect loan market) is the inadequate development of the human resources of the children of poor families - which I would judge, is one of the most serious inefficiencies of the American Economy today"* [Arthur Okun, 1975, cited by Loury, 1981).

capital) financial markets imperfections to address distributive issues (see, for example, Bènabou, 1994a and 1994b; Durlauf, 1992 and 1994; de Bartolome,1990).[3] In these models individuals aggregate themselves in different neighborhoods or local communities. The interesting feature of this literature is that the individual choice of a neighborhood is endogenous to the model and affected by economic considerations. Each individual, endowed with a certain level of human capital or income and altruistic versus his child's human capital, bases his rational choice of community on housing prices and on the local level of social capital.

To bring out the basic insight of this literature, we may ignore for simplicity the determinants of private expenditures on education as modeled by the contributions reviewed above. If the only educational inputs are family and social spillovers (rather than actual expenditure or opportunity costs), then issues of capital market incompleteness or imperfection are irrelevant, and the human capital production function reflects a rational investment decision only inasmuch as agents, choosing the community in which they live, also choose a certain social capital level.

Each neighborhood j is characterized by some level of social capital $h^{j,i}_t$ (for example crime, norms of behavior, role models, networking contacts, peer effects in school, etc.) which aggregates the human capital levels of the agents living in the community. The human capital production function of an agent belonging to a dynasty i living in community j can be written as

$$(11) \qquad h^{j,i}_{t+1} = G\,(h^{j,i}_t, H^j_{t+1})$$

where $h^{j,i}_{t+1}$ represents his educational attainment, or human capital, and $h^{j,i}_t$ the positive externality produced by the parent's education on the child's one.[4] The aggregate production function is equation (7).

Employing Bènabou's terminology these models potentially feature two kinds of equilibria, *stratification* and *integration*; assume, for simplicity, that the economy is composed by only two neighborhoods 1 and 2, each of dimension 1/2 and two types of agents A and B each of dimension 1/2 with $h^A_t > h^B_t$. In a stratified equilibrium community 1 is entirely composed by agents of type A and community 2 by agents of type B, with $h^A_t = H^1_t > H^2_t = h^B_t$, while in an integrated equilibrium each community is composed by agents A and B in equal proportions so that $H^1_t = H^2_t = H_t/2$.

If stratification is supported through house rental, as in Bènabou (1994) and de Bartolome (1990), a stratified equilibrium emerges when rich families are willing

---

[3] We ignore, for simplicity, the role of locally financed public expenditures on education, which is mostly relevant in the United States.

[4] Human-capital production may be a function of innate ability as well as of the father's resources: Checchi et al.'s (1994) theoretical and empirical work focuses on the extent to which high-ability individuals can obtain appropriate schooling in the presence of financial-market imperfections. Their model, accordingly, combines features of those reviewed in this and in the previous subsections.

and able to bid more than poor families for land in a rich community. There are many potential determinants of stratification (see Bènabou, 1994a): ignoring capital markets imperfections, the simplest one is given by complementarities between the parent's investment in the child's education h^i_t and community social capital H^j_t, i.e., $G_{hH}(\cdot) > 0$; in this case human-capital-rich families are more sensitive to a neighborhood's social capital, and are willing to bid more for land than human capital poor families.

Models of this type can explain the persistence of human capital (and income) inequality: if a segregated equilibrium prevails, then in the next generation human capital will be more unequally distributed with respect to an integrated equilibrium.

The evolution of *intra*-generational inequality depends on the returns to heterogeneous inputs in $G(\cdot,\cdot)$; if they are decreasing, then different income levels asymptotically converge: for example Tamura (1991)[5] shows that if an integrated equilibrium prevails and $G(\cdot,\cdot)$ is a constant returns to scale Cobb-Douglas,

$$G(h_t, H) = h^{\beta}_i (H/2)^{1-\beta}$$

with $i = A, B$, then, since $h^B_t < H/2 < h^A_t$ B-type agents have a greater rate of return to human capital investment than A-type agents. Hence, the human capital stock of B-type agents grows faster than that of A-type agents until their human capital and income levels converge.

In these models, the *laissez faire* equilibrium may be inefficient because of the presence of a spillover in the human capital production function. In the simple case presented above, two forces working in opposite directions determine whether stratification is efficient or not: on the one hand, assuming that $G_{hH}(\cdot) < 0$, a better social environment is more productive for educated A type families; on the other hand, if returns to social capital are decreasing i.e. if $G_{HH}(\cdot) < 0$, an additional unit of social capital is more productive where social capital is scarce. If the first effect outweighs the second, then segregation is efficient; otherwise, it is inefficient. Thus, while in case of human capital financial markets imperfections the *laissez faire* equilibrium is inefficient under decreasing returns to human capital investment, in this case the trade-off between efficiency and equality may still be relevant.

9.1.3 Education policies and political economy

The *laissez faire* equilibrium is not Pareto-optimal when market failures affect human capital investments. The modeling features outlined above, namely inefficiency of the *laissez faire* equilibrium due to aggregate and social capital

---

[5] In Tamura's model agents choose the amount of human capital (income) that they leave to their children when a human capital financial market does not exist; nonetheless, we have found more interesting to include his model in this paragraph rather than in the previous one, because of its convergence properties due to the aggregate externality H_t.

spillovers and/or misallocation of investments due to imperfect financial markets, rationalize efficiency-enhancing public intervention and give an important role to political interactions.

Consider the case of imperfect human capital financial markets in which education must be self-financed and private expenditures on education do not equalize rates of return among agents. The interesting feature of the literature on human capital investment and public intervention is that in these models politics is realistically an endogenous variable, being chosen by the median voter under the standard hypothesis of single-peaked preferences over the policy variable; interestingly, as Stiglitz (1974) has pointed out, a realistic economy characterized by the co-existence of both public and private school systems is likely to feature non-single peaked preferences over the quality of public schools: in this case a politico-economic equilibrium may not exist.

Consequently, the literature has concentrated either on the political analysis of the quality of public schools (Saint-Paul and Verdier, 1993) or on the choice between a private and a public school system (Glomm and Ravikumar, 1992 and Bènabou, 1995); other models (Perotti, 1993 and Fernandez and Rogerson, 1994) consider the case in which public intervention takes the form of income redistribution among agents rather than schooling expenditures' equalization. What remains unexplained, in these models, is the choice of policy instruments which agents have to take as given.

The simplest model of public efficiency-enhancing redistributive intervention is due to Glomm and Ravikumar (1992); their assumption that the utility function in equation (6) is additive and logarithmic implies that agents vote unanimously for a constant tax rate which is equal to the private propensity to invest in education in the *laissez faire* equilibrium: thus, ignoring tax-induced distortions and family spillovers, public policy simply redistributes the aggregate *laissez faire* expenditures, efficiently equalizing rates of returns among agents. If the median voter is poorer than the average, public education is introduced and maintained until income distribution becomes degenerate.

Saint Paul and Verdier (1993) consider a simple linear non-overlapping generations economy which grows by accumulating human capital. Human capital is partly inherited by children from their parents and partly publicly provided; the rate of growth of the economy is increasing in the tax rate if taxation is non-distortionary. Their basic point is that if poor agents prefer more public expenditures on education than rich ones, the economy grows faster when the distribution of income is more egalitanan: as income dispersion shrinks because of the redistributive effect of public expenditures, the median agent prefers smaller tax rates, political support for public education diminishes, and this leads to slower growth.

In Glomm and Ravikumar (1992) agents whose income is below the mean support the introduction of a public school system; in their model, a public school system increases the educational attainment of the poor more than it harms that of the rich, realizing a net efficiency gain; nonetheless, there is no immediate gain for the average agent stemming from expenditures redistribution because per-capita public expenditures are equal to the taxes he has to pay. Thus, the greater the distance between median and average income, the greater the political support for a public education system. Bènabou (1995) points out that if a public education system realizes an efficiency gain for the average agent, then the relationship between inequality and political support for public education is not monotonic as in Glomm and Ravikumar, but U-shaped: at low levels of inequality, an increase of incomes' dispersion reduces public support for redistribution while at high levels of inequality the standard redistribution effect prevails. If some agents have more influence than others in the political system, then the economy could feature multiple steady states: at relatively low levels of inequality the political equilibrium entails high efficiency-enhancing redistribution; this would reduce incomes' dispersion and increase the political support for more redistribution etc.; at relatively high levels of inequality, nonetheless, the political support for redistribution could be so low that a private and not a public school system is adopted: this political equilibrium would increase inequality, reduce efficiency and persist over time.

Perotti (1993) and Fernandez and Rogerson (1994) consider the case in which agents vote on the extent of income redistribution as a means to overcome the self-financing constraints which they face when capital markets are imperfect.

Perotti (1993) models the case in which the investment in human capital by an agent increases the productivity of all the other agents, as, for example, happens in Lucas (1988). This positive externality could provide the median voter with an incentive to redistribute less than he would in case of perfect appropriability of educational investments' returns: in a poor economy, for example, if there are indivisibilities in human capital investments, high redistribution could make the rich class - the only one which, ex-ante, can afford to buy human capital - so poor that it would not be able to invest in education.

The key point is that the median voter has to trade-off the short-run gains stemming from income redistribution with the long-run ones due to increased future productivity and income: the basic choice facing each individual in the economy is that between redistributing (towards oneself) the shares of a given pie or increasing its size, and it is crucial to the politico-economic approach that the perspective of the swing (median) voter on this tradeoff in general does not coincide with that of a hypothetical representative (mean) agent. As in other models like Bertola (1993), Alesina and Rodrik (1994) and Persson and Tabellini (1994), economies in which income distribution is relatively more equal are characterized by less redistribution and more growth. In a rich economy, conversely, Perotti's model features characteristics similar to Saint-Paul and Verdier (1993) and Glomm and Ravikumar (1992): the redistribution which is needed for increasing the economy's efficiency takes place if income distribution is sufficiently unequal.

Finally, Fernandez and Rogerson (1994) point out that educational public expenditures may redistribute income from poor to rich agents because every individual has to contribute to public expenditures in education, but only a subset of the economy's individuals receive post-compulsory education attending secondary schools and universities if, for example, poor individuals face higher opportunity costs of acquiring higher education. A realistic subsidy scheme can induce rich and medium classes to vote for the maximum tax rate that prevents poorer individuals from investing, so that the poor finance the other classes' investments without any benefit accruing to them.

9.2 Microeconomics of education

Turning to the microfoundations of the macroeconomic models reviewed in the previous section, we note at the outset that all of them adopt Becker's approach to education as human capital, i.e., "the stock of skills and productive knowledge embodied in people" (*The New Palgrave Dictionary of Economics*, p.682). This microeconomic paradigm, developed by Mincer (1958), Schultz (1961) and Becker (1975), builds upon an almost perfect analogy between physical and human capital: just as an entrepreneur incurs costs to upgrade or enlarge his existing stock of plant and equipment in order to increase his sales and profits, so an individual invests in education-incurring both direct and opportunity costs-to increase his stock of productive skills in order to command a higher wage in the future.

The internal rate of return on thusly defined "human capital" investments can be calculated, just like the internal rate of return of physical capital investments, and rational agents invest in both physical and human capital up to the point where their marginal rates of return equal the opportunity cast of investment. This analogy is fully accepted by the macroeconomic literature reviewed so for, hence it is far more interesting -from the perspective of this work- to ask what *distinguishes* H from K rather than what makes them formally similar.

9.2.1 Human vs. physical capital as investment vehicles

According to some authors, as Lucas (1988) and Barro and Sala-1-Martin (1994), one key difference between physical and human capital in the two-sector models of growth of Section 1.1 above is that human capital production is relatively intensive in human rather than physical capital, i.e., that it is possible to educate individuals by employing other already educated individuals rather than machines.

Another, maybe more significant difference between K and H has been pointed out by models that use "education" or "human capital accumulation" as a label for a productive factor which is not rewarded according to social returns because of factor-market imperfections: self-financing constraints prevent returns from being equalized across heterogeneous individuals or production units and, even when cross-sectional homogeneity is assumed, externalities may introduce a wedge between private and social returns to investment.

The dynamic role of human capital's public good characteristics has been pointed out by early endogenous growth theorist like Romer (1986) and Lucas (1988, who writes on p.97 that "*...human capital accumulation is a social activity,* *involving groups of people in a way that has no counterpart in the accumulation of physical capital".).* To the extent that much human-capital investment takes place in informal settings rather than in organized markets, the allocative role of prices is diminished, and externalities have paramount importance: this observation underlies the recent "social capital" approach of Bènabou (1994 a, b) and others who concentrate on the educational spillovers that characterize the human capital production function *G(., .)*, rather than aggregate activity externalities in the form *F (K, H, S(H))*.

Macroeconomic models of imperfect financial markets build a general equilibrium framework on the microeconomic analysis of Becker (1975) and Becker and Tomes (1986), based on the observation that *"access to capital markets to finance investments in children separates the transmission of earnings from the generosity and resources of parents"*. These authors had already recognized the difficulty faced by an individual to borrow funds to invest in human capital. The common justification for imperfections in human capital financial markets is based on moral hazard problems. Imperfections stem from the fact that the effort expended by an agent in looking for a job, or on the job itself, cannot be perfectly monitored or enforced. Thus, an individual who is insured against low earnings, or is financed by someone else, has incentives to "shirk," i.e., to provide less effort than he potentially could. Moreover, the fact that human capital is an illiquid asset -slavery being unlawful in most socioeconomic settings- lowers the monetary cost of shirking because it cannot act as a debt's collateral.[6] As to the consequences (rather than the sources) of financial market imperfections, Becker (1975, p.98) emphasizes the basic insight that human capital, being embodied in specific individuals, faces strongly decreasing returns to scale at the individual level: hence, misallocation of investment in human capital can have important efficiency and distribution effects.

We should note, however, that these market failures may well be relevant to many other factors of production. As to financial market imperfections, Becker (1975, p.80) writes that *"while it is undeniably difficult to use the capital market to finance investments in human capital, there is some reason to doubt whether otherwise equivalent investments in physical capital can be financed much more easily"*. As to externalities, they may result from physical (as opposed to human) capital accumulation or from other productive activities, as in the seminal work of Arrow (1962) on learning-by-doing. To ground a distinction between physical and human capital on financial markets imperfections and spillovers, we should argue that market failures are more relevant to human than to physical capital investments. Concentrating on financial markets' failures, three arguments can be advanced to support this conclusion. First, as already mentioned, human capital cannot act as a debt's collateral, while physical capital, like machines and plants,

---

[6] Galor and Zeira (1993) and Banerjee and Newman (1993) endogenously derive human capital financial markets imperfections assuming that the borrower could, with some probability of success, avoid debt repayment by moving to other places but that the lender would spend such an amount in keeping track of him to disincentive its escape.

can partially guarantee the lender against repayment's defaults. Second, as observed by Becker (1975, p.80), *"financing may prove a more formidable obstacle to investors in human capital because they cannot postpone their investment so readily"*, i.e. if a lender is unwilling to finance a young person's investment in human or physical capital because the investor is likely to be ignorant of his abilities and of the investment opportunities available, then, since the return on education is negatively related to the age of the investor, it would be more difficult to postpone the investment in human capital. Third, if a person does not invest at all in education he could still be able to find a job and be paid as an unskilled worker, while it is necessary to invest in order to obtain some return from physical capital: thus, human capital investments (unlike physical-capital ones) entail opportunity costs-whose importance can be better appreciated if we consider that human capital investments' decision are usually taken by parents when children are very young and that the children themselves are, from the parents' point of view, a potential source of income as they can be employed in some manual work. In this case, even if parents could leave negative bequests to their children, they could still have an incentive not to acquire education for them because this would entail the loss of an income source: the relevant financial market imperfection in this case is the fact that children cannot borrow, simply for age-related legal reasons, to subsidize the parents' investments in their education.

To the extent that "human capital" is peculiarly subject to such market failures, then it is a particularly suitable vehicle for contemporary macroeconomic theory's concern for distribution and market imperfections (and for the policymaking and political issues which derive from them).

9.2.2 The many roles of education

On the basis of the arguments reviewed so for, investment in human capital differs from physical investment in two crucial respects: it has many of a *public good*'s characteristics, as its returns spill over to individuals other than the one being educated, and it is a *distorted private good* with imperfect rewards to individual decisions.

The macroeconomic literature typically identifies (formal, or informal) education with investment in human capital; this view seems now largely accepted: the main function of education would be to increase "the stock of skills and productive knowledge embodied in people". In the 1970s and early 1980s, however, a different literature offered an intense and stimulating debate on various rates of education other than production of human capital. At least two competing views challenged the human capital paradigm, while going beyond earlier views of education as a consumption good.

On the one hand, Bowles and Gintis (1976) argued that (formal) education reproduces and promotes the same values like obedience, competition, etc., which are useful to and compensated by a capitalist economic system. Another view considered education as a means of overcoming imperfect information problems in the labor market when individuals are heterogeneous in terms of innate abilities (see, e.g. , Arrow, 1973; Spence, 1973; Stiglitz, 1974 and 1975). This literature, while recognizing the importance of primary education for skills acquisition, emphasized the role of higher education as a *screening device* rather than as an investment in cognitive skills.[7] Two assumptions are relevant to the *screening device* approach to education:

(i) innate abilities - an individual's comparative and absolute advantages - are the fundamental determinants of his or her on-the-job productivity;

(ii) firms are interested in hiring the most productive workers but, without incurring high selection costs, cannot distinguish them from others.

An individual's schooling record is then valuable, from the employer's point of view, in that it provides information on a worker's innate abilities.

This literature distinguishes *signaling* from *pure screening* models. In the former, an individual's productivity is his private (hence, asymmetric) information; in the former, information is symmetrically imperfect as neither firms non workers know the true productivity of heterogeneous individuals.

In the signaling model of Spence (1973), innate abilities take one of two levels, and firms assume that agents whose education level (as measured by, e.g., the number of school years) is greater than q^*_t belong to the group of abler workers. Conversely, educational levels smaller than q^*_t are associated to low productivity. Accordingly, firms offer greater wages to workers whose educational attainment q_t is such that $q_t > q^*_t$. If more productive individuals face a lower costs of attaining a given educational level, then they will indeed choose higher amounts of educational expenditure when maximizing their net income. At this stage, firms hire workers, observe their productivity, and update their estimate of the q^*_t critical level of education. In a signaling equilibrium, high-ability individuals have an incentive to *signal* their greater productivity to firms by acquiring an educational level ($q_t = q^*_t$) sufficiently high to be classified as relatively more productive, while low-ability agents prefer not acquire any level of education: thus, individuals self-select by providing information about their abilities through the choices they make.

When information about skills is symmetrically imperfect (as in Stiglitz, 1974 and 1975), education provides information about individuals' comparative and absolute abilities. In a pure screening model, education is informative about agents' innate skills not because of self-selection mechanisms, but through the direct and indirect information provided by school grades and tests results. This information is directly relevant to allocation of scarce educational resources to

---

[7] It is interesting to note, however, that the human capital and the screening paradigms need not represent alternative views of education. Layard and Psacharopoulos (1974) note that if education performs a mere screening role, then employers and governments should have tried to replace the educational system by a cheaper screening mechanism, based for example on psychological aptitude tests. In practice, however, the screening function of education may be "a natural by-product of its principal activity of providing knowledge" (Stiglitz, 1975, p.294): then, there cannot be any efficient screening without an investment in cognitive skills.

more skillful individual within the school system, but it is also (and perhaps more importantly) used by prospective employers to sort out more skillful workers and to efficiently match prospective workers with available jobs.

From an income distribution perspective, the human capital and the pure screening paradigms have very different implications. In the framework of Section 1.2.1 above, investments in human capital increase individuals'on-the-job productivity, so that the wage rate they can command increases too. Conversely, screening conveys information about individuals' productivities, but does not increase them: aggregate productivity can increase only as for as firms efficiently use these information to match workers with available jobs. Thus, from an individual's point of view, screening increases wage-income risk: in a no-screening equilibrium, income distribution must be completely egalitarian, as firms cannot disentangle a skillful from an unskillful agent's contribution to production; under perfect screening, conversely, aggregate production (net of screening costs) may well be higher, reflecting greater allocational efficiency, but individuals would be compensated according to their true productivities which (by assumption) are not affected by education.

9.3 Complex education and macroeconomic reality

The literature reviewed in the previous Section identifies important roles of education beyond those emphasized by macroeconomic growth models reviewed in Section 1 above. Education is not only aimed at providing "human capital" as a factor of production (which entails investment decisions not dissimilar from those relevant to physical capital accumulation, albeit more strongly affected by realistic market imperfections): it also aims at allocating heterogeneous individuals to the occupations which suit them best, with potentially important distributive implications. This Section briefly outlines how simple models may be used to explore the implications of "allocative" education for models of economic development and income distribution.

9.3.1 A simple model of matching

To model allocation of individuals to heterogeneous production opportunities, let production take place in n sectors indexed by $k = 0, 1/n, 2/n, ..., 1 - 1/n$, and let workers form a continuum of unit measure, indexed by $i \in [0, 1]$. Each k value represents a differentiated technology (or product) which offers suitable employment only to workers with certain characteristics. Let both sectors' and workers' one-dimensional characteristics be arranged around the unit circle (so that $k = 0$ coincides with $k = 1$, and $i = 0$ coincides with $i = 1$), and let worker i be suitable for employment in sector k if i is close to k.

For simplicity, we shall take the output y_{ik} of worker i employed in secret k to be positive only if i is closer to k than to any other sector

(12) $$y_{ik} = \begin{cases} f_i > 0, \text{ if } i \in [k - \frac{1}{2n}, k + \frac{1}{2n}]; \\ 0, \text{ otherwise} \end{cases}$$

where the quantity f_i produced by worker i if "well-matched" is independent of the particular sector where he is employed, and is otherwise left unspecified for now. In words, a worker is productive if employed in the sector which can best use his characteristics, and contributes nothing to production if employed by any other sector. Less extreme assumptions may be more realistic (for example, we might let productivity decline smoothly with the distance from the most suitable occupation) but would unnecessarily complicate the algebra, while yielding similar qualitative implications.

With i uniformly distributed on the unit circle, and equally spaced k values, this economy's aggregate production is readily computed in two extreme cases. If workers are perfectly assigned to their best occupation, they are all productive and the unit measure of workers produces

(13) $$\int_0^1 f_i \, di \equiv \bar{Y}(\cdot)$$

where we allow the perfect-matching level of output to be a function $Y(\cdot)$ of economic phenomena, discussed below, other than comparative-advantage matching of workers to sectors. At the other extreme, only $1/n$ units of labor are productive if workers are randomly assigned to occupations, and aggregate output is n times smaller at

(14) $$\frac{1}{n} \int_0^1 f_i \, di = \frac{\bar{Y}(\cdot)}{n}$$

To model intermediate cases between the extreme ones of equations (13) and (14), suppose workers' characteristics i are *ex-ante* unknown, but can be estimated through testing-so that education can contribute to labor's aggregate productivity by screening workers and allocating them to their most suitable occupations. We shall denote with $\hat{\imath}$ the resulting estimate of i_0, the worker's true characteristic, and think of $\hat{\imath}$ as the transcript of a vocational school. Performance in school provides an unbiased estimate of the worker's true characteristic, but schooling cannot provide full information: let

$$\hat{\imath} \sim U[i_0 - \sigma, i_0 + \sigma], \quad \sigma \leq \frac{1}{2}.$$

In words, testing preserves uniform distribution of workers' characteristics, but narrows the range of a given worker's characteristic from the no-information prior, [0, 1], to an interval of length 2σ on the unit circle.[8] If the transcript adds to the true θ an uncorrelated, uniformly distributed error with support of length 2σ, then estimated $\hat{\imath}$s are also distributed uniformly on [0, 1], and workers are allocated to

[8] Recalling that the characteristic is measured on a circle, $\hat{\imath} > 1$ has the some meaning as $\hat{\imath}-1$, and similarly $\hat{\imath} < 0$ coincides with $1 + \hat{\imath}$.

the occupation which appears to suit them best: secret k employs workers with "transcripts" $\hat{\imath}$ such trial

$$\hat{\imath} \in [k - \frac{1}{2n}, k + \frac{1}{2n}]$$

The precision of schooling information is indexed by $1/\sigma$. If $\sigma = 0.5$, schooling is completely uninformative as to workers' true characteristics; at the other extreme, $\sigma = 0$ makes it possible to match workers and sectors perfectly. The Appendix derives expressions for the economy's aggregate production in intermediate cases: the ratio of *ex-post* output to its perfect-matching upper bound introduced in equation (13) may be written as a function of n and σ:

$$(15) \qquad \frac{Y(\cdot;\sigma;n)}{\bar{Y}(\cdot)} = F(\sigma,n) \equiv \begin{cases} (1-\frac{\sigma n}{2}), & \text{if } 0 \le \sigma \le \frac{1}{2n}; \\ (\frac{1-\sigma n}{2\sigma n} + \frac{\sigma n}{2}), & \text{if } \frac{1}{2n} \le \sigma \le \frac{1}{n}; \\ \frac{1}{2\sigma n} & \text{if } \frac{1}{n} \le \sigma \le \frac{1}{2}. \end{cases}$$

The function $F(\sigma,n)$ is defined for all n and for $0 \le \sigma \le 1/2$, is continuous and, consistently with equations (13) and (14), takes values of unity and $1/n$ at $\sigma = 0$ and at $\sigma = 1/2$ (see Figure 1).

9.3.2 Specialization, human capital, and productivity

For any given value of the perfect-matching upper boundary $Y(\cdot)$ in equation (13), the output loss from imperfect matching of workers to sectors is larger for larger values of n, not only in the limit case of equation (14) but also in less extreme cases with $\sigma < 1/2$: we see in equation (15) and Figure 1 that output declines very sharply in σ for large values of n.

Figure 1. Aggregate production as a ratio of perfect-matching output

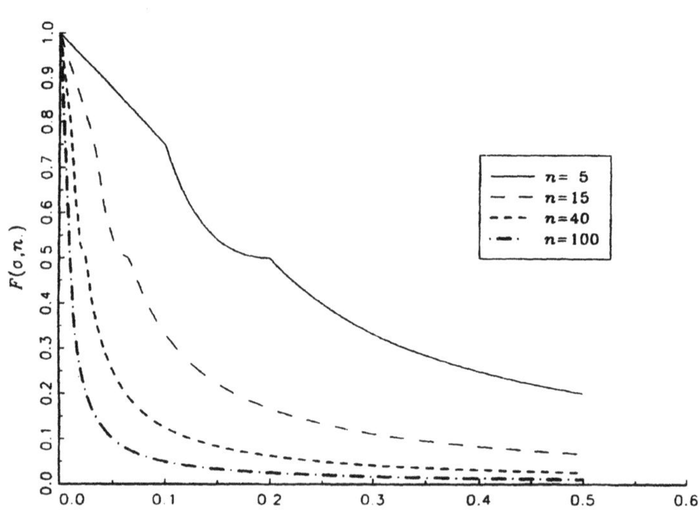

If we view the number of sectors n as an index of the degree of specialization in the economy under consideration, this simple model captures the realistic idea that allocation of workers to occupations is more important in a more sophisticated economy: a randomly chosen worker is more likely to be unsuitable to the very detailed tasks entailed by a highly specialized production structure than to the less well defined and more homogeneous tasks on which primitive technologies are based.

Of course, increasing sophistication and complexity of the economy's production process (indexed by n in the simple model above) can increase the productivity of labor and other non-accumulated factors through the channels considered by the models of Romer (1987), Grossman and Helpman (1991, Chapter 3), and others: a given amount of expenditure affords higher utility if a larger variety of differentiated products is available and preferences are convex, and more specialized intermediate inputs make it possible to produce standardized consumption goods more efficiently.[9] Following the literature, we may model this insight by writing aggregate output as a C.E.S. aggregate of the n sectors' production levels, with elasticity of substitution $1/(1 - \alpha)$: normalizing production of a well-matched worker to unity ($f_i \equiv 1$) and recognizing that only a $F(\sigma, n)$ fraction of the $1/n$ workers employed by each sector k is productive, we obtain the expression

$$(16) \qquad Y(n, \sigma) = \left[\sum_{k=1}^{n} \left(\frac{F(\sigma, n)}{n}\right)^{\alpha}\right]^{\frac{1}{\alpha}} = [n^{1-\alpha} (F(\sigma, n))^{\alpha}]^{\frac{1}{\alpha}} = n^{\frac{1-\alpha}{\alpha}} F(\sigma, n)$$

for aggregate output.

Thus, realistic "returns to specialization" for given precision of information can be readily introduced in out model by assuming that the amount of output produced by a worker when suitably employed by the closest sector k is an increasing function of n-or, in the notation introduced above, that

$$(17) \qquad f_i = f(n) \, \forall i \Rightarrow \bar{Y}(\cdot) = f(n)$$

with $f'(n) > 0$ representing returns to variety. By this assumption, aggregate production is an increasing function of n if all workers are "well matched" as in equation (13); but productivity of well matched workers should increase more than linearly with n for output under tandem assignment, in equation (14), to be an increasing function of n.

While out assumptions as to the structure of imperfect information on individual workers' comparative advantage (and the derivations in the Appendix below) are quite stylized, the resulting model brings out an insight of some generality. It is apparent in equation (15) that a more sophisticated economy (where n is larger)

---

[9] In other models, introduction of new goods is associated to "creative" destruction of existing productive processes (see, e,g., Grossman and Helpman, 1991, Chapter 4), possibly causing obsolescence of product-specific human capital or allocative information.

Figure 2: Aggregate production as a function of number of activities.

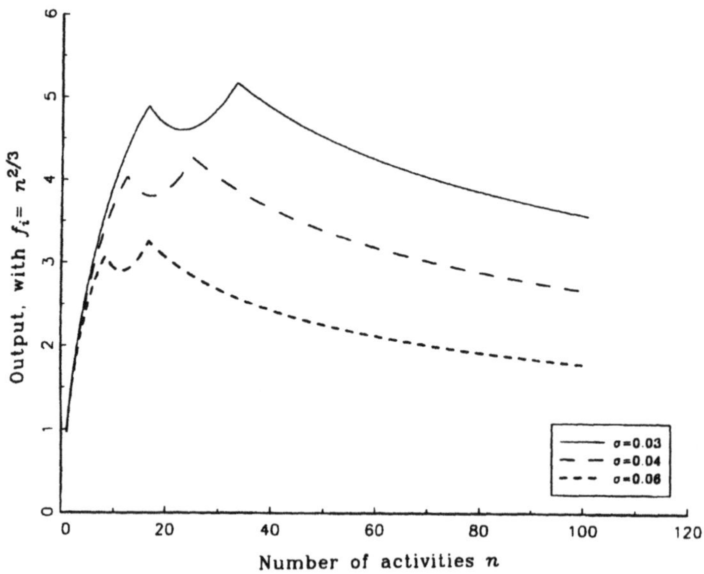

makes better use of information as to individual characteristics. Our simple formalization, in fact, implies that labor productivity might be decreased by introduction of more specialized progresses (a larger n): the measure of well-matched workers shrinks, for constant σ, as n increases, and this may more than offset their increased productivity reflected in a higher $f(n)$.

Figure 2 plots aggregate production for increasing specialization (larger n) and various values of σ. The Figure assumes that $f(n) = n^{2/3}$, or that $\alpha = 3/5$ in equation (16) (clearly, the qualitative message of the model would be similar for other values of α). For each value of σ, aggregate production is an increasing function of n for $n < 1/(2\sigma)$, as "variety" returns of a larger n dominate the output reduction due to imprecise allocation-which, as long as $n < (1/2\sigma)$, is linear in n by the expression in the first fine of equation (15). But output is not monotonic in n beyond that range: more specialization of production requires better and better matching of workers to sectors, and for $n > 1/\sigma$ aggregate output declines steadily in n in the Figure. The assumed 2/3 "returns to variety" elasticity of perfect-matching output to n is smaller than that of allocative losses - which have elasticity - 1 to the specialization index n in the first line of equation (15), and the quality of available information is so low as to more than offset by inappropriate matching the increasing productivity of well-matched workers.

Proper allocation according to comparative advantage is crucial in our simple setting. In Figure 2, we see that for each value of n aggregate production increases in the precision $1/\sigma$ of available information on workers' individual characteristics, and Figure 3 further illustrates the relationship between precision of allocative information (as indexed by $1/\sigma$) and labor productivity. Note, again,

Figure 3: Aggregate production as a function of schooling precision indicator.

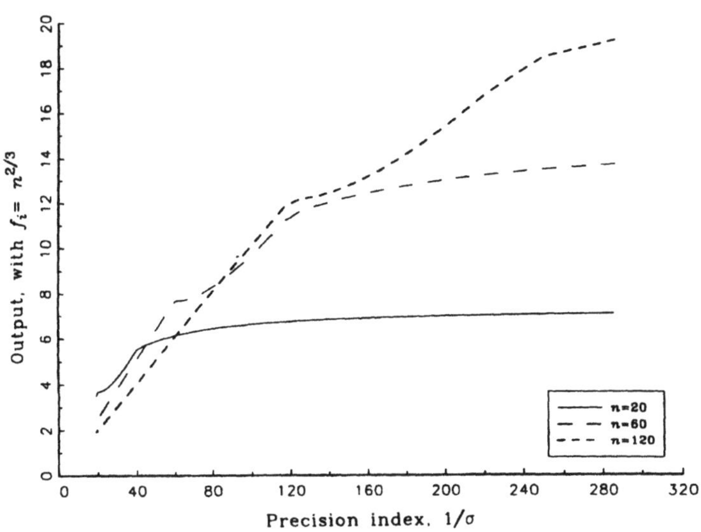

that as the elasticity of f_i to n is positive, but less than one, a larger n may or may not be associated to a larger output for given σ.

9.3.3 The supply of education and information

In essence, the model outlined above provides simple microfoundations for an aggregate (reduced-form) production function where information as to individuals' comparative advantage is a productive factor of sorts: in our notation, we may write

$$(18) \qquad Y(n,\sigma) = F(\sigma,n) f(n)$$

and note that a low value of a, by improving allocation of heterogeneous individuals, increases aggregate production just like traditional human capital or physical capital, which can be easily introduced in the model (see below).

Of course, we are not the first to point out that allocative information has an important role in determining individual (and aggregate) productivity: among others, Jovanovic (1991) proposes a model with assignment externalities across heterogeneous individuals, and Jovanovic and Nyarko (1994) study the productivity effects of Bayesian learning about different technologies. Existing models, however, do not focus on the possible allocational role of education, which is crucial for the present paper's purposes.

In the remainder of this Section, we informally (and very briefly) discuss the issues introduced by this wider view of education's economic role.

9.3.3.1 Social tradeoffs

To close the model, we may suppose that better information as to individuals' comparative advantage (a lower σ) may be obtained by costly education (or, in a dynamic environment, by other-and also costly-sources of information, such as observation of on-the-job performance and turnover), and proceed to evaluate the *social* tradeoffs involved by optimization of an appropriate (static, or dynamic) objective function. While the structure of a suitable model would in many respects be similar to those reviewed in Section 1 above, the informational role of education makes it necessary to address novel modeling choices. Space prevents a detailed specification of this conceptually straight-forward exercise in the present paper, but we briefly outline the relevant issues in what follows.

In a dynamic environment, aggregate resources might be devoted not only to aggregate consumption and accumulation of capital (or introduction of new technologies, or accumulation of human capital) along the lines of equations (1) and (5) above, but also to provision of the allocative information represented by σ in the model. The modeling blocks outlined so for let an individual worker's productivity depend on comparative advantage in the tasks allotted to him or her, indexed by the distance between the characteristic indices of workers (i) from those of their sector of employment (k); and on the degree of sophistication of the economy's production processes, represented by an increasing function $f(n)$.

To properly analyze the interaction of these features, and to study how the precision of matching information may be endogenously determined, we should of course recognize that a worker's productivity may be influenced by innate absolute advantage, technically advanced tools, health, and uncountably many other factors. Most relevantly for this paper's purposes, *education* may enhance productivity for *all* occupations as well as provide information on an individual's characteristics.

Such features may be readily introduced in out model by amending the specification in (17) of a well-matched worker's production, and writing

$$\text{(19)} \qquad f_i = h_i f(n) \ \forall i \qquad \Rightarrow \qquad \bar{Y}(\cdot) = f(n) \int_0^1 h_i di = f(n)\, H$$

where h_i is an *absolute advantage* index of worker i's (possibly education-enhanced) productivity in any occupation and H represents its aggregate counterpart as defined in equation (7) above.

A crucial modeling issue is that of whether the amount of screening provided by schooling may be chosen. independently of its other outputs, and especially of its role in the provision of standard human capital (represented by H): part of what is gauged by school grades is an individual's ability to absorb knowledge, which is impossible to gauge without actually teaching notions which (like Latin and Greek) may or may not be directly relevant to productive skills. This is essentially a technological issue, on which economics has very little to say. In Stiglitz's (1974, p.378) view, the educational system's contribution to aggregate productivity is *"to sort out individuals into their comparative advantages, and in doing so, it obtains a great deal of information about absolute advantage"* as a byproduct; and information on both innate comparative- and absolute-advantage may, in turn, be joint products of any teaching effort aimed at equipping workers with "human capital" skills: by teaching children to read, write, and count, one unavoidably finds out much information about their innate qualities.

Expenditure aimed at lowering σ may be thought as providing an accumulated factor of production if the information thus obtained remains useful beyond the current period; alternatively, one might recognize that information on any individual's skills and characteristics depreciates very quickly (because it cannot fast longer than the individual's lifetime and/or because its relevance to productivity is made obsolete by technological progress), and model the allocational role of education as a non-accumulated, essentially static choice along the lines of Grossman and Helpman's (1991, Ch.5.2) treatment of knowledge embodied in individuals.

9.3.3.2 Distribution, incentives, and financial markets

If there existed perfectly competitive and complete markets for all goods, individual decisions and real-life economic outcomes would be based on the social tradeoffs identified by the modeling issues just discussed above. Of course, however, such markets may hardly be assumed to exist or to function competitively for such complex goods as education, innovation, and heterogeneous labor. A careful and detailed treatment of interactions between the market imperfections emphasized by the work reviewed in Sections 1 and 2 above must be left to further research, but we can briefly outline here some modeling issues. Specifically, we discuss in what follows the possible *laissez-faire* determinants of heterogeneous workers' compensation, with particular attention to their implications for income inequality; the private incentives to obtain allocational information on one's own characteristics; and the interaction of financial-market imperfections with the screening role of education.

In an environment like that represented by this Section's model, where heterogeneous labor is endowed with sector- or firm-specific skills, individual workers may or may not be compensated according to their actual or expected productivity, depending on the structure of the labor market (see Thisse and Zenou, 1990, for a complete specification of a realistic market structure in a rather similar environment). At one extreme, we may suppose that a plurality of potential employers within each sector bids wages up to each worker's expected ($\hat{\imath}$-based) productivity: then, wages would be smallest for those workers whose estimated characteristics fall on the midpoint between a pair of sectors.

Alternatively, however, we might suppose that a single firm has perfect monopsony power within each sector-in fact, it would be essential to allow for such market power if profits finance the endogenous , costly creation of new and more specialized goods , along the lines of Romer (1987), Grossman and Helpman (1991), and others. Then, individual workers would be paid on the basis of their *outside option* rather than of their productivity on the job: somewhat paradoxically, wages would then be *lowest* for workers who appear best suited to one of the existing sectors (so that $\hat{\imath}= k$, in our notation), since they are also the least suitable to alternative occupations.

In both of these extreme cases, wages would be more sharply differentiated when σ is lower and expected productivity depends more strongly on the observable indicator $\hat{\imath}$ (see the Appendix). This result appears qualitative robust to many realistic extensions of our simple model: if schooling provides information on the characteristics of heterogeneous workers, and the market compensation of labor is to any extent based on such characteristics, then more precise information on individual characteristics is available increases wage inequality across *ex-ante* identical workers. Thus, the impact of progress on inequality may be peculiar in models of this type: at the some time as individual productivity is optimized, individual rewards become more unequal.

Turning next to the private incentives for individuals to provide such information, consider first the case of symmetric information about comparative advantage. As noted in Section 2, risk-averse workers should not willingly enroll in a schooling program that only promises to provide a more precise, *unbiased* estimator of his true ability-as this would increase the variance of their income without affecting its mean. Thus, for schoolbased screening to be viable in a decentralized economy it should be compulsory, and publicly financed. Alternatively, employers as a group might find it profitable to set up schooling programs of the type we consider; lump-sum payments could be used to entice workers to enroll, and firms'incentives to implement such screening programs should be stronger the greater are the returns to comparative advantage information (or, in our model, the larger is n).

When information about absolute-advantage productivity is asymmetric, i.e., individuals know their true ability but potential employers do not, then interesting interactions may arise between financial market imperfections and the informational role of schooling-as the self-financing requirement assumed by many macroeconomic models may be (constrained) Pareto-optimal: to help reveal private information, education should be expensive for the individual, hence it *must* be self-financed if uncertainty about individual ability cannot be resolved *ex-post*. Further research should study whether financial market imperfections may still make it impossible for schooling to deliver the constrained efficient outcome, along the lines of the models reviewed in Section 2: for example, because poor individuals cannot finance their own education even when they know they are capable.[10]

In reality, of course, schooling does increase a worker's overall ability, and individuals have private information as to their innate ability. Inasmuch as general human capital and allocative information are joint products of any schooling, there can be no presumption that *laissez-faire* interactions will provide the socially optimal amount of screening.

---

[10] Note, however, that if family background and financial resources are public information then the market can extract just as much information from a poor boy's limited education investment than from a rich boy's much target expenditure.

A realistic model should allow workers to invest in education in order to obtain general human capital (as in Becker, 1975) as well as to signal their ability to prospective employers (as in Stiglitz, 1974). In light of the above considerations, however, it is interesting to note that the informational/screening role of schooling records is much less important in the largely private U.S. schooling system than in the public (and compulsory) schooling institutions of European countries and Japan.

9.4 Concluding comments

In this paper, we have organized along simple but hopefully clear lines a brief review of the recent literature on the macroeconomic role of education, and we have examined its microeconomic foundations -arguing that the economic role of education, and the imperfections in its provision, are not necessarily as straightforward as reduced- form macroeconomic studies make them. While education may certainly increase a given amount of labor's productive potential (like machinery, know-how, and other ordinary factors of production would), it also gives information on the absolute and comparative advantage of each individual within a heterogeneous workforce.

No theory can do justice to all facets of reality, of course. The simple model outlined in Section 3 focuses on a specific role of education, namely, that of eliciting information on an individual's innate skills. This may not be as important as, e.g., task-specific training among the many functions of real-life educational systems and, except perhaps at the very early stages of primary education, it is likely to be performed by firms in the marketplace-as in Thisse and Zenou's (1997) model-rather than by formal educational institutions. As it stands, however, the model makes a simple point which may be of some relevance for theories of economic development: the allocative role of education should be more important in more advanced societies, where specialized tasks can put to good use a heterogeneous labor force's different skills, than in more primitive settings where "absolute" rather then "comparative" advantage is crucial both to an individual's innate productivity and to its enhancement through education.

In further work, this and similar insights may lead to a reevaluation of existing empirical evidence on linkages between aggregate measures of "human capital" (usually based on educational variables, such as literacy levels or years of schooling), growth performances, and inequality. Benhabib and Spiegel (1994) find that economywide educational attainment stocks, while empirically quite relevant to macroeconomic performance and growth in cross-sectional data (see also Barro and Sala-i-Martin, 1994, and references' therein), do not appear to enter the aggregate production function as "ordinary" inputs.

The quantity and type of education in specific economies may on the one band depend endogenously on technological achievements (rather than on capital-market imperfections) and, on the other band, it may be related to income distribution in more complex ways than those suggested by models where growth is driver by accumulation of human capital under self-financing constraints.

9.5 Appendix

What follows derives equation (15) in the main text (aggregate production as a function of the degree of specialization n and of available matching information as indexed by σ). Recall that a worker is *ex-post* productive only if his true characteristic i_0 lies within the interval of length $1/n$ centered around the index k of the sector which employs him, and that i_0 is distributed uniformly on an interval of length 2σ centered around the observable indicator $\hat{\imath}$. Accordingly, a worker whose indicator is $\hat{\imath}$ is certainly productive if all of his possible i_0 realizations lie within the sector's preferred interval, but is unproductive with positive probability in the other cases illustrated in Figure Al.

Clearly, $\hat{\imath}$ provides no useful informational content if $\sigma = 1/2$. In what follows, we derive aggregate-output expressions for $0 \leq \sigma \leq 1/2$.

It will be convenient to introduce the symbol

$$\xi \equiv 1/2n$$

for the half-width of each sector's range of closest skills, and to consider separately the three cases illustrated in Figure A2.

The probability that workers assigned to each sector are productive as a function of their observed indicator $\hat{\imath}$, plotted in the Figure, is also likely to be an important determinant of cross-sectional wage heterogeneity (see Section 3.3.2).

If $0 \leq \sigma \leq \xi$, then workers indexed by $\hat{\imath}$ such that $|\hat{\imath} - k| \leq \xi - \sigma$ are productive with probability one (as in the top panel of Figure Al); for other workers, the probability of positive output declines linearly in $|\hat{\imath} - k|$ with slope $-1/(2\sigma)$, as in

Figure A1. Probability that worker with indicator $\hat{\imath}$ is productive in sector k

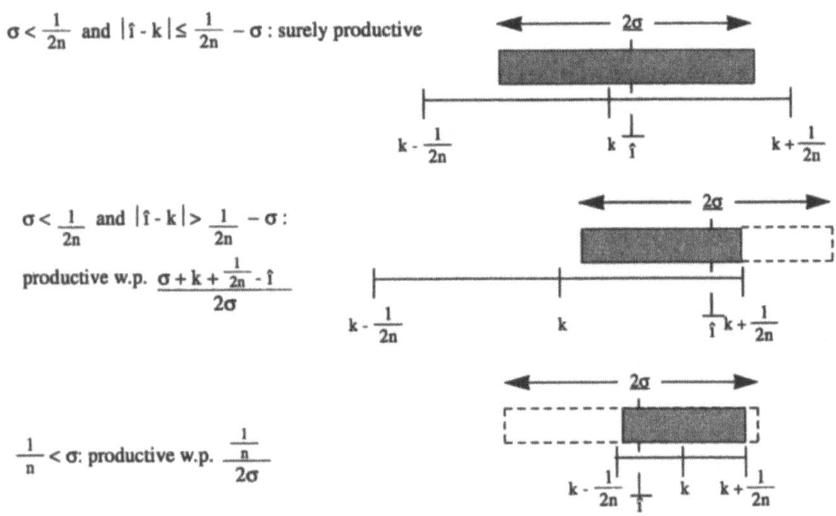

Figure A2. Production probability density whitin each sector's closest $\hat{\imath}$ neighborhood

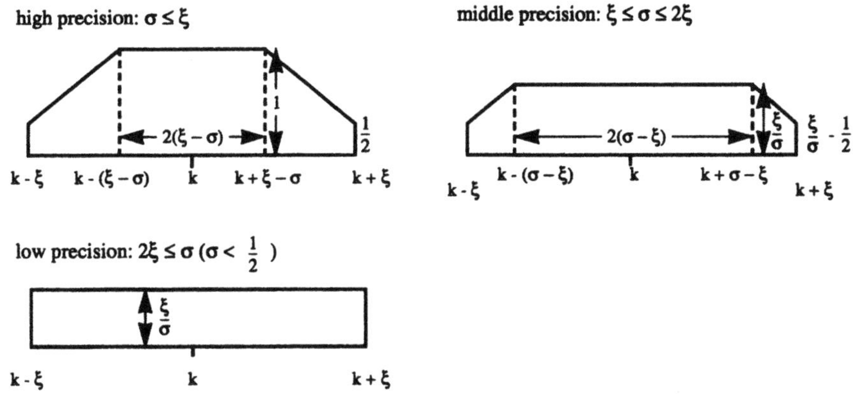

the middle panel of Figure Al, and reaches 1/2 for workers whose $\hat{\imath}$ is equidistant from two different values of k. The measure of productive workers in sector k is measured by the area of the geometric figure in the top panel of Figure A2, or

(A1) $$2(\xi-\sigma) + (\sigma-2\xi)(1+\frac{1}{2}) = 2\xi - \frac{1}{2}\sigma$$

Recognizing that $\xi = 1/2n$, the aggregate output of the n sectors is

(A2) $$Y = n(\frac{1}{n} - \frac{1}{2}\sigma)\int_0^1 f_i di = \bar{Y}(\cdot)(1\frac{\sigma n}{2})$$

If $\xi \leq \sigma \leq 2\xi$, then workers for whom $|\hat{\imath} - k| < \sigma - \xi$ are productive with probability ξ/σ (as in the bottom panel of Figure Al); the probability of positive output by others again declines linearly in $|\hat{\imath} - k|$, with slope $-1/(2\sigma)$, reaching $\xi/\theta - 1/2$ for workers at the boundary between two sectors. The measure of productive workers in each sector corresponds to the area of the geometric figure in the middle panel of Figure A2,

(A3) $$2(\sigma-\xi)\frac{\xi}{\sigma} + (2\xi-\sigma)(\frac{\xi}{\sigma}+\frac{\xi}{\sigma}-\frac{1}{2}) = 2\frac{\xi^2}{\sigma} - \xi + \frac{\sigma}{2}$$

With $\xi = 1/2n$, in the aggregate of the n sectors we have

(A4) $$Y = n(2\frac{1}{4n^2\sigma} - \frac{1}{2n} + \frac{\sigma}{2})\int_0^1 f_i di = \bar{Y}(\cdot)(\frac{1}{2n\sigma} - \frac{\sigma n}{2} - \frac{1}{2})$$

Finally, if $2\xi \leq \sigma \leq 1/2$ then each of the workers assigned to each sector k is productive with probability ξ/σ (as in the bottom parrei of Figure Al). In the aggregate,a proportion ξ/σ of the unit labor measure is productive and produces

(A5) $$Y = \int_0^1 \frac{\xi}{\sigma} f_i di = \frac{\bar{Y}(\cdot)}{2n\sigma}$$

References

Alesina A., Rodrik D. (1991), "Distributive Politics and Economic Growth", NBER Working Paper no.3668, March.

Arrow K. (1962), "The Economic Implications of Learning by Doing", *Review of Economic Studies* 29, pp.155-173.

Arrow K. (1973), "Higher Education as a Filter", *Journal of Public Economics 2*, pp.193-217.

Barro R., Sala-i-Martin X. (1994), *Economic Growth*, McGraw-HiLL.

Becker G. (1975) *Human Capital: A theoretical and Empirical Analysis, with Special Reference to Education*, New York: NBER/Columbia University Press.

Becker G., Tomes N. (1986), "Human Capital and the Rise and Fall of Families", *Journal of Labor Economics 4*, pp.Sl-539.

Bénabou R. (1994a), "Education, Income Distribution, and Growth: the Local Connection", CEPR Discussion Paper No.995.

Bénabou R. (1994b), "Human Capital, Inequality, and Growth: a local Perspective", *European Economic Review 38*, pp.817-826.

Bénabou R. (1995), "Unequal Societies", Mimeo.

Benhabib J., Spiegel M. (1994), "The Role of Human Capital in Economic Development: Evidence from aggregate cross-country data", *Journal of Monetary Economics 34*, pp.143-173.

Bertola G. (1993), "Factor Shares and Savings in Endogenous Growth", *American Economic Review 83*, pp.1184-1198.

Bertola G. (1995), "Uninsurable Shocks and International Income Convergence", *American Economic Review (Papers and proceedings)*, 1995.

Bowles S., Gratis H. (1976), Schooling in Capitalist America, New York: Basic Books.

Cass D. (1965), "Optimum growth in an Aggregative Model of Capital Accumulation", *Review of Economic Studies 32*, pp.233-240.

Checchi D., Ichino A. e Rustichini A. (1994), "Talis Pater Talis Filius ?", Mimeo.

de Bartolome C. (1990), "Equilibrium and Inefficiency in a Community Model with Peer Group Effects",' *Journal of Political Economy 98*, pp.110-133.

Durlauf S. (1992); "A Theory of Persistent Income Inequality", NBER Working Paper no.4056.

Durlauf S. (1994), "Spillovers, Stratification and Inequality", *European Economic Review 38*, pp.836-845.

Femandez R., Rogerson R. (1994), "On the Political Economy of Education Subsidies", Federal Reserve Bank of Minneapolis Research Department Staff Report 185.

Galor Q., Zeira J. (1993), "Income Distribution and Macroeconomics", *Review of Economic Studies 60*, pp.36-52.

Glomm G., Ravikumar B. (1992), "Public versus Private Investment in Human Capital: Endogenous Growth and Income Inequality", *Journal of Political Economy 100*, pp.818-834.

Grossman G., Helpman H. (1991), Innovation and Growth in the Global Economy, Cambridge, MA: MIT Press.

Jovanovic B. (1991), "Coordination and Spillovers", New York University Economic Research Report no.65.

Jovanovic B., Nyarko Y. (1994), "The Bayesian Foundations of Leaming by Doing", NBER Working Paper no.4739.

Layard R., Psacharopoulos G. (1974), "The Screening Hypothesis and the Returns to Education", *Journal of Political Economy 82*, pp.985-998.

Ljunqvist L. (1993), "Economic Underdeveloment: the Case of a Missing Market", *Journal of Development Economics 40*, pp.219-239.
Loury G. (1981), "Intergenerational Transfers and the Distribution of Earnings", *Econometrica 49*, pp. 843-867.
Lucas R. (1988), "On the Mechanics of Economic Development", *Journal of Monetary Economics 22*, pp.3-42.
Mincer J. (1958), "Investment in Human Capital and the Personal Income Distribution", *Journal of Political Economy 4*, pp.281-302.
Mulligan C., Sala-i-Martin X. (1993), "Transitional Dynamics in Two-Sector Models of Endogenous Growth", *Quarterly Journal of Economics 108*, pp.737-773.
Okun A. (1975), Equity and Efficiency: The Big Tradeoff Washington D.C.: The Brookings Institution.
Perotti R. (1993), "Political Equilibrium, Income Distribution, and Growth", *Review of Economic Studies 60*, pp.755-776.
Persson T., Tabellini G. (1994), "Is Inequality Harmful for Growth ?", *American Economic Review, 84*, pp.600-621.
Rebelo S. (1991), ""Long Run Policy Analysis and Long Run Growth", *Journal of Political Economy 99*, pp.500-521.
Romer P. (1986), "Increasing Returns and Long Run Growth", *Journal of Political Economy 94*, pp.1002-1037.
Romer P. (1987), "Growth Based on Increasing Returns Due to Specialization", *American Economic Review Papers and Proceedings 77*, pp.56-62.
Romer P. (1990), "Endogenous Technological Change", *Journal of Political Economy 98*, pp.S71-S102.
Saint-Paul G.~ Verdier T. (1993), "Education, Democracy and Growth", *Journal of Development Economics 42*, pp.399-407.
Schultz T. (1961), "Investment in Human Capital", *American Economic Review 51*, pp.1-17.
Solow R. (1956), "A Contribution to the Theory of Economic Growth", *Quarterly Journal of Economics 70*, pp.65-94.
Spence M. (1973), "Job Market Signaling", *Quarterly Journal of Economics 87*, pp.355-379.
Stiglitz J. (1974), ""The Demand For Education In Public and Private School Systems", *Journal of Public Economics 3*, pp.349-385.
Stiglitz J. (1975), "The Theory of 'Screening', Education, and the Distribution of Income", *American Economic Review 65*, pp.283-300.
Tamura R. (1991), "Income Convergence in an Endogenous Growth model", *Journal of Political Economy 99*, pp.522-54.
Thisse, Jacques-François and Yves Zenou (1995), "How to Finance Education when the Labor Force is Heterogeneous?", presented at F.E.E.M. / Beijer I.I. of E.E. Conference on Creation and Transfer of Knowledge: Institutions and Incentives
Uzawa H. (1965), "Optimum Technical Change in an Aggregative Model of Economic Growth", *International Economic Review 6*, pp.18-31.
West G. (1964), "The Role of Education in Nineteenth-Century Doctrines of Political Economy", *British Journal of Educational Studies 12*, pp.161-172.

10. How to Finance Education when the Labor Force is Heterogeneous?

Jacques-François Thisse, Yves Zenou*

One of the key-problems in modern economies is the splitting of the cost of education between workers, firms and the government. Ever since the pioneering work of Becker (1964), the demand for education is analyzed with investment models. The concept of human capital appears to be central for the study of several phenomena. In particular, Becker (1964) introduces a fundamental distinction between *general* human capital and *specific* human capital. The former refers to a general level of ability which has value to all firms in the economy. The latter refers to specific abilities which have values to only one firm. It is part of the conventional wisdom that the cost of general training is to be borne by workers only, whereas the cost of specific training is to be shared by firms and workers because both parties can share the corresponding returns (Hashimoto (1981) and Parsons (1986)). The potential mobility of workers across firms would then determine the way the cost and the return of education are shared.

Our purpose is to set up links between human capital and job matching theories, while recognizing explicitly that firms are heterogeneous and operate on an imperfectly competitive labor market. The aim of matching models is to explain mobility across firms rather than within firms. However, the concept of matching can also be applied to the match between heterogeneous workers and firms in terms of skills. The latter definition of job matching is adopted here. Hence *job matching constitutes here the process whereby heterogeneous workers are matched to different jobs.* The existence of specific human capital implies that both firms' requirements and workers' skills are heterogeneous. The heterogeneity on the workers' side is the cornerstone of the literature on job matching, as initiated by Jovanovic (1979a and b). Indeed, had workers been perfectly homogeneous, the possibility of a mismatch between a job and a worker would vanish. Furthermore, the hiring process is conducted by a representative competitive firm. Such an approach disregards the fact that firms are themselves heterogeneous in their job requirements.

---

* CORE, Université Catholique de Louvain and Ceras, Ecole Nationale des Ponts et Chaussées; CORE, Université Catholique de Louvain and ERMES, Université de Paris II-Assas

We thank Paul David, Jonathan Hamilton, Margaret Stevens, Xavier Wauthy and Stefano Zamagni for their useful comments. Financial help from the European Union "Human Capital and Mobility" program is gratefully acknowledged by the second author. Both authors also thank the PIRVILLE (CNRS) for its financial support.

Table 1: Public and private expenditure for education as a percentage of GDP in 1991

	USA	Japan	France	Canada	Germany	Sweden	UK	Netherlands
Public	5.5	3.7	5.4	6.7	4.0	6.5	5.3	5.6
Private	1.5	1.3	0.6	0.7	1.5	0.0	-	0.1

Source: Education at a glance, OECD 1993, p.66.

In our model, the process of job matching drives the formation of wages on the labor market instead of explaining quits and lay-offs. Specifically, workers have heterogeneous skills while firms have differentiated requirements. Indeed, as shown by Stevens (1994a), firms have an incentive to differentiate their skill requirements in order to obtain market power in the labor market. For a common level of general training, workers will be hired by firms for which the job matching is as good as possible because firms are able to observe the specific skills of their workers and because firms and workers share the cost of specific training. In equilibrium, the self-selection process of workers generates no quits or lay-offs since workers share with their employers the return of the investment in specific human capital. More precisely, each worker earns a net income higher that the best alternative offered by the other firms in the market, which is here endogeneous. Furthermore, firms pay wages lower than workers' productivity after training. This confirms Becker's argument about the sharing of cost and return. In particular, we are able to study the impact of the cost sharing between firms and workers on the structure of the labor market.

In addition, we assume that the general human capital is directly provided by the government through a nationwide education system which is financed by a tax paid by the workers. This assumption can be justified by the fact that general education is a public service in the vast majority of OECD countries. This is illustrated by Table 1 which gives public and private educational expenditure relative to GDP for the main OECD countries in 1991. It is seen that only Japan, Germany and United States have non-negligible private spendings on education while Nordic countries, Canada and France have very low private ones. On average, public funding covers almost 90% of education costs in advanced industrialized countries, thus making our assumption reasonable.

We model education as an *impure* public/private good in that the general training cost exhibits scale economies. In particular, this allows us to contrast the two polar cases in which education is a pure private or public good. To the best of our knowledge, this problem has not been formally tackled yet. Consequently, we combine here the two polar cases of training considered by Becker, i.e., "perfectly general" training, provided by the government and "completely specific" training, supplied by the firms.

In our model, there is an externality at work in that the global productivity depends on the division of labor: the larger the number of job requirements, the higher the average wage after training. However, the corresponding gain associated with the entry of a new firm is not fully captured by the entrant. Under free entry, the division of labor itself becomes endogeneous. Because of the externality mentioned above, it is not necessarily socially optimal. Moreover, we will see that the division of labor also depends on the firms' entry costs as well on the unit training cost.

Our setting can be described as follows. In the first stage, the government chooses the socially optimal level of general training while, in the second stage, wages are set up by oligopsonistic firms at a Nash free-entry equilibrium. When firms compete to hire workers, both firms and workers know the part of the specific training cost they have to pay for. When the government decides how much to invest in general training, it anticipates the market equilibrium arising in the second stage. It is shown that the optimal level of general human capital rises with the total population of workers. Furthermore, wages increase with the level of general human capital. Hence, the following trade-off arises: increasing the general human capital leads to higher wages but requires a higher tax. Wages also rise with the size of the labor market because the average matching between a firm and a worker improves. In general, we show that the above trade-off is such that workers are better off in a large labor market. Finally, at the social optimum, we find that firms must bear the totality of the specific training cost of the workers.

The approach developed in this paper bears some formal resemblance with Kim (1989) who also assumes that the matching between a firm and a worker is measured along a *skill characteristic*. This is in accordance with the spatial competition model of product differentiation à la Hotelling-Lancaster (see Anderson, de Palma and Thisse (1992, ch.8) for a survey), which captures the idea of an heterogeneous population of agents. Hence, in order to meet the needs of a particular firm, a worker must be specifically trained and the corresponding cost increases with the distance between the workers' skill and the firm's requirement. Furthermore, he also considers a free-entry equilibrium where firms earn zero profits, like in Salop (1979). However, the questions addressed here are very different from those investigated in Kim's paper.[1]

The remainder of the paper is organized as follows. The model is introduced in the next section. As usual, we solve the problem by backward induction. In section 2, we deal with the second stage which consists in determining the equilibrium market wage for a given level of investment in general human capital. The optimal level of general human capital chosen by the government, corresponding to the first stage, is characterized in section 3. Section 4 concludes the paper.

---

[1] Note that the solution to the Nash bargaining between firms and workers proposed by Kim is inaccurate because of a mispecification of the threat point. Instead of a flat wage, the bargaining wage can be shown to increase with distance from firms' job requirement. (see Haimilton, Thisse and Zenou, 1997) This casts serious doubt on the validity of his conclusion.

[2] The difference in workers' skills is analogous to the difference in consumers' tastes in models of horizontal product differentiation developed in industrial organization.

10.1 The Model

Consider n firms producing a homogeneous good sold on a competitive market (this good is taken as the *numéraire*) and a continuum of workers. The workers have the same general human capital (GHC hereafter) but heterogeneous skills. There is no *a priori* superiority or inferiority among workers' skills: there are just different (for example, they have a degree in engineering but in different fields).[2] Skills are supposed to be innate but the common level of skill is determined by the education policy of the government. Formally, it is supposed that workers have the same level of GHC, represented by g units of the numéraire, but have different skills x continuously distributed along the circumference C of length L; C represents the *skill space* and L expresses the degree of diversity in workers' skill. For simplicity, we assume that the skill characteristic x is uniformly distributed with density Δ on the circumference C and that the job requirement by firm i is unique and given by $x_i = (i\text{-}1)\, L/n$, for i=1,..,n. This is in accordance with Salop (1979) and many other papers which have used this framework.

It is supposed that the level of GHC, denoted g, is provided through a nationwide education system. This system is financed by the government by means of a lump-sum tax levied on the workers. Since we consider a free entry equilibrium, firms' profits are equal to zero. The objective of the government is then to maximize the sum of the workers' utilities. The social cost of producing a level g of GHC for a population of size ΔL is defined as follows:

$$\text{(1)} \qquad C(\Delta, g) = [\lambda + (1\text{-}\lambda)\,\Delta\, L]\; c\,(g)$$

where $c(g)$ is the cost of producing the output g for one individual; this function is increasing and convex in g. For a given value of g, (1) involves a fixed cost given by $\lambda c\,(g)$ and a proportional variable cost corresponding to $(1\text{-}\lambda)\, c\,(g)\, \Delta L$. In this expression, $\lambda \in [0, 1]$ is an index measuring the degree of publicness of the good g which determines the relative importance of the fixed cost. At one extreme, the educational system whose output is g supplies a pure public good, in which case $\lambda = 1$ and the cost $C(\Delta, g) = c(g)$ is independent of Δ. At the other one, $\lambda = 0$ and $C(\Delta, g) = \Delta\, L\, c(g)$ so that the good g is purely private and produced under constant returns to scale with respect to the population size. For $0 < \lambda < 1$, the good g is neither purely public nor purely private. This is a semi-public good whose production exhibits economies of scale with respect to the population size. Such a specification is likely to correspond to the production cost function of the education system.

The labor market is now described. A firm's market power increases when it chooses a technology requiring a specific training that protects its "location" against infringements by rival employers. To this effect, we suppose that firms have a technology such that workers can produce only when they perfectly match the firms' job requirement. Since workers are heterogeneous, they need to be trained within the firm. The aim of this training is to give the worker the skill corresponding to the firm's job requirement. Thus, after this training, the matching is perfect and, without loss of generality, the worker then produces g units of the numéraire. The more distant the skill of a worker is from the firm's job requirement, the larger the training cost. More precisely, once a worker is hired by firm i, his skill is observed by the firm and his labor contract stipulates that he has to bear a training cost given by a linear function $\alpha\, s\, |x-x_i|$ of the difference between the worker's skill x and the firm's job requirement x_i. In this formulation, $\alpha \in [0, 1]$ is the share of the specific training cost borne by the worker which is supposed to be given and chosen through some institutional mechanism not described here. When $\alpha = 1$, the worker pays his entire specific training cost whereas the firm bears this cost when $\alpha = 0$. It is standard in labor economics to consider a two period model in which workers, first, invest in specific human capital and, then, decide to stay or to quit. In order to avoid unnecessary complications, we have collapsed the two periods into a single one with the aim of focusing on the process of competition on the labor market.

In the present context, worker x earns in firm i a wage net of training cost higher than the opportunity net income he could get from the second nearest firm (hence he accurately anticipates that gross wages are equal across firms as shown below). Observe that here the opportunity wage is endogeneous unlike what can be found in existing contributions. Furthermore, firm i pays this worker a wage lower than his productivity in this firm. Therefore, both parties have an incentive to share the cost of training (though not formally modeled here, it is worth noting that training has the nature of an investment for both agents). This remains true for each worker x and each firm i. For simplicity, as in Hashimoto (1981), Parsons (1986) and Stevens (1994b), we assume that the training cost is shared between firms and workers according to some given ratio which lies between 0 and 1.[3]

Firms choose simultaneously their wage level, $(w_1,..,w_i,..,w_n)$. The net wage is therefore equal to $w_i - \alpha\, s\, |x-x_i|$. Each worker then chooses to be hired by the firm which gives him the highest net wage. Since firms anticipate the choice of workers, they hire all the workers who want to work at the prevailing wages, $(w_1,..,w_i,..,w_n)$, because they know that the workers hired have skills close to their job requirement. The wage w_i cannot be lower than the reservation wage which is assumed to be the same across workers because they have the same GHC and are differentiated by their skill which do not imply any superiority or inferiority; without loss of generality, this wage is set equal to zero. Furthermore, w_i cannot exceed the productivity g, for otherwise firm i would make a negative profit.

10.2 Labor Market Equilibrium and Specific Human Capital

Assume that the level of GHC is fixed and given by g. The second stage of our problem can then be solved as follows. Let i be the representative firm. Given two wages w_{i-1} and w_{i+1}, firm i's labor pool is composed of two sub-segments whose

[3] See Hashimoto (1981) for an endogeneous determination of this ratio in the special case of one firm and one worker.

outside boundaries are given by marginal workers $\bar{x}$ and $\bar{y}$ for whom the net wage is identical between firms i -1 and i, on the one hand, and firms i and i+1, on the other. In other words, $\bar{x}$ is the solution of the equation:

$$w_i - \alpha s\,(x_i - \bar{x}) = w_{i-1} - \alpha s\,(\bar{x} - x_{i-1})$$

so that

$$\text{(2)} \qquad \bar{x} = \frac{w_{i-1} - w_i + \alpha s\,(x_i + x_{i-1})}{2\alpha s}$$

In this case, firm i attracts workers whose skill characteristics belong to the interval $[\,\bar{x}, x_i\,]$ because the net wage they obtain from firm i is higher than the one they would obtain from firm i - 1. Clearly, workers belonging to the interval $]\,\bar{x}, x_i\,]$ are hired by firm i - 1. In a similar way, we show that:

$$\text{(3)} \qquad \bar{y} = \frac{w_i - w_{i+1} + \alpha s\,(x_i + x_{i+1})}{2\alpha s}$$

Consequently, firm i's labor pool is defined by the interval $[\,\bar{x}, \bar{y}]$, which is illustrated by figure 1. Its profit function can be written as:

$$\text{(4)} \qquad \Pi_i = \int_{\bar{x}}^{\bar{y}} \Delta\,[\,(g-w_i) - (1-\alpha)\,s\,|x-x_i|]\,dx$$

$$= \Delta\,\{(g-w_i)\,(\bar{y}-\bar{x}) - \frac{(1-\alpha)\,s}{2}\,[(\bar{y}-x_i)^2 + (\bar{x}-x_i)^2]\}$$

The profit function (4) is continuous in (w_{i-1}, w_i, w_{i+1}) and concave in w_i. Therefore, there exists a Nash equilibrium in wages. Given the symmetry of the problem, the Nash equilibrium is such that all firms set the same wage that can be obtained by applying the first order conditions to the profit function (4):

$$\text{(5)} \quad \frac{\partial \Pi_i}{\partial w_i} = \Delta \left\{ -(\bar{y}-\bar{x}) + (g-w_i)\left(\frac{\partial \bar{y}}{\partial w_i} - \frac{\partial \bar{x}}{\partial w_i}\right) - (1-\alpha)s\left[(\bar{y}-x_i)\frac{\partial \bar{y}}{\partial w_i}(\bar{x}-x_i)\frac{\partial \bar{x}}{\partial w_i}\right]\right\} = 0$$

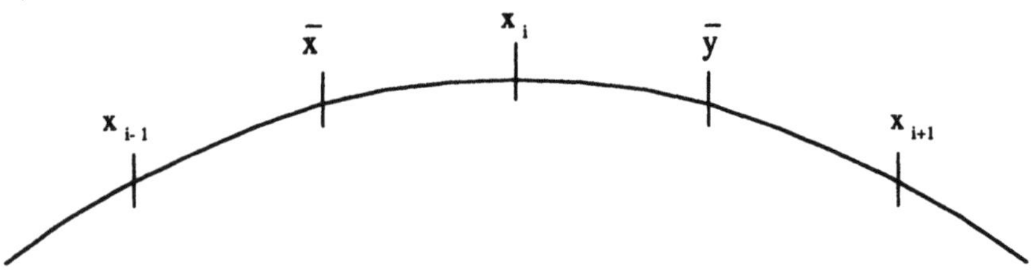

Figure 1: Firm i's labor pool

Combining (2), (3) and (5), and equalizing the equilibrium wages, we obtain

(6) $$w^*_i = g - \frac{(1+\alpha)\, sL}{2n}$$

The equilibrium given by (6) is unique because the first order conditions define a system of linear equations in w_{i-1}, w_i et w_{i+1}.

Proposition 2.1.a: *If* $g \geq (1+2\alpha)\, sL/2n$ *and if* $\alpha > 0$, *there exists a unique Nash equilibrium in wage with full employment and the equilibrium wage level is given by* (6).[4]

First, observe that for each worker the equilibrium net wage falls in between the opportunity wage offered by the second nearest firm and the worker's productivity (see also Hashimoto (1981) and Stevens (1994b) for a similar result in the case of one firm and one worker). Hence, in equilibrium, firms and workers share the returns of investment in specific human capital. Second, it follows from (6) that the common equilibrium wage is an increasing function of the level of GHC of the workers (g) and a decreasing function of the cost of mismatch (s), since a higher specific training cost s gives to each firm more monopsony power on its labor pool. Note also that the wage paid to workers is a decreasing function of the share that they bear in financing their specific training cost. Indeed, diminishing α is equivalent to a decrease in s which, other things being equal, rises the wage of the workers who are perceived as less differentiated by the firms. On the other hand, each firm must bear an increase in the specific training cost. Since, at the symmetric equilibrium, the number of hired workers is the same regardless of the value of α, only the first effect has an impact on the equilibrium wage. Last, by definition of a Nash equilibrium, competition among heterogeneous firms precludes the emergence of poaching effects. Furthermore, no worker can be better off by changing job since he should incur new training cost while receiving the same gross wage.

In the above, we have considered the equilibrium market configuration that prevails under full employment. However, another configuration with unemployment may emerge. In this case, some workers do not accept to work because their net wage is negative. Assuming for simplicty that unemployment benefits are zero, the profit function now writes as follows:

(7) $$\Pi_i = \frac{\Delta}{\alpha^2 s}\left[2\alpha g w_i - (1+\alpha) w_i^2\right]$$

and the monopsony wage is:

[4] Note that an equilibrium does not exist for very low value of α because the net contribution of remote workers to the firm's profit is negative for deviations above the candidate equilibrium. We obviate this technical problem by assuming that firms must hire all the workers who want to work for it at the offered wage.

(8) $w^*_i = (\frac{\alpha}{1+\alpha})g$

Proposition 2.1.b: *If $g < (1+\alpha)\, sL/2n$ and if $\alpha > 0$, there exists a unique Nash equilibrium in wage with unemployment and the equilibrium wage level is given by* (8).

The following comments are in order. First, contrary to the previous case, the monopsony wage does not depend on the number of firms. This reflects the total monopsony power of each firm. Second, the nature of unemployment is twofold. On the one hand, unemployment is *voluntary* since some workers refuse to work for a net negative wage. On the other hand, this unemployment can also be viewed as *involuntary* since the productivity of unemployed workers is higher than their reservation wage given by their training cost (Lindbeck, 1993, p.47). In a perfectly competitive market, these workers would be employed since their net wage would be positive. Here there is a direct link between the skill of workers and the existence of unemployment. *This is because their skills are too far from the ones required by firms that some workers are unemployed.* If they were a technological shock that would change firms' requirement, these workers could then be hired. Last, if the assumptions on the parameters in Propositions 1a and 1b are not satisfied (i.e., $(1+\alpha)\, sL/2n \leq g < (1+2\alpha)\, sL/2n$), there is still a Nash equilibrium: labor pools just touch and all workers are hired at the wage $w_i^* = \alpha\, sL/2n$.

Since we want to study the free entry equilibrium, we focus on the full employment configuration where the equilibrium wage (6) varies with the number of firms. It is readily verified that the equilibrium profits are given by:

(9) $\Pi^*_i(n) = \Delta\,(1+3\alpha)s\,\frac{L^2}{4n^2}$

When the number of firms rises, the equilibrium wage increases because two adjacent firms now compete for workers whose job matching is better. This wage tends to the competitive wage g, while the profits tend to 0, when the number of firms becomes arbitrarily large. *The competitive model of the labor market is thus the limit of the spatial competition model with job matching.* Moreover, the restriction $\alpha > 0$ is imposed in order to allow for the determination of each firm's labor pool. Indeed, when $\alpha = 0$, i.e., all the specific training cost is supported by the firm so that the self selection process of workers may no longer work. However, we assume that workers do the socially optimal thing and apply to the closest firm.

In fact, the number of firms does not become arbitrary large since the entry of a new firm implies a positive fixed cost F. At the long run equilibrium with free entry, the number of firms is obtained by equating profits to zero. Therefore, we have

(10) $\Pi^e = \Pi^*_i(n) - F = \Delta\,(1+3\alpha)s\,\frac{L^2}{4n^2} - F = 0$

the solution of which is given by:

$$\text{(11)} \quad n^e = \frac{L}{2}\sqrt{\frac{(1+3\alpha)s\Delta}{F}}$$

As expected, the number of firms increases with the size of the market and decreases with the level of fixed costs. In this model, the size of the market is expressed through two parameters, L and Δ. Nevertheless, the sensitivity of the number of firms with respect to these parameters is not the same: n^e varies linearly with the size of the skill characteristic space (L), whereas it varies according to the square root of the density of workers (Δ). The number of firms rises with the specific training cost because an increase in s allows the incumbents to set a lower wage and, thus, to earn higher profits. Moreover, at a given n, the equilibrium wage (6) rises when α falls. However, in the long run, the number of firms decreases with such a change in α, which favors a decrease in the equilibrium wage (6). Consequently, the impact of a reduction of α on the long run wage is *a priori* indeterminate. This wage is given by:

$$\text{(12)} \quad w^e = g - (1+\alpha)\sqrt{\frac{sF}{\Delta(1+3\alpha)}}$$

Contrary to the prediction of mainstream labor economics where workers are homogeneous, *the long run equilibrium wage rises with the number of workers* (Δ) because the average matching between workers and firms is better in a large market due to the greater number of firms.[5] Moreover, w^e still decreases with the training cost in specific human capital. Finally, it is readily checked that the RHS of (12) reaches its maximum when the workers bear one third of their training cost. However, the gross wage paid to the workers is not an appropriate measure of their welfare since they bear some of their training cost. On average, a worker receives a surplus given by his wage minus the average training cost equal to $\alpha sL/4n$. Using (11) and (12), the long run worker surplus $SW(n^e)$ can be shown to be given by

$$SW(n^e) = g - \left(1+\frac{3\alpha}{2}\right)\sqrt{\frac{sF}{\Delta(1+3\alpha)}}$$

Differentiating this expression with respect to α yields

$$-\frac{9\alpha}{4(1+3\alpha)}\sqrt{\frac{sF}{\Delta(1+3\alpha)}}$$

which is always negative. Consequently, we have:

---

[5] This result is related to those of Ethier (1982) and Romer (1987) who show that a market is more efficient when the number of specialized tasks (here the firms) increase. In particular, we may expect that wages are higher in urban areas than in rural areas. A similar result can be found in Zenou and Smith (1995) in a different framework.

Proposition 2.2: *Workers are on average better off when $\alpha = 0$.*

The social optimum is reached when the number of firms minimizes the sum of the fixed costs and of the total specific training costs. This amounts to minimizing the following function with respect to *n*

$$(13) \quad SC(n) = nF + 2n \int_0^{\frac{L}{2n}} \Delta s \, dx = nF + \frac{\Delta s L^2}{4n}$$

which is a convex function of *n*. Thus, the socially optimal number of firms is obtained by equating the first derivative of (13) with respect to *n* to zero:

$$(14) \quad n^0 = \frac{L}{2} \sqrt{\frac{s\Delta}{F}}$$

Observe that the socially optimal number of firms is in general smaller than the equilibrium number. In particular, the former corresponds to half the latter when the specific training cost is fully borne by the workers. The following question then suggests itself: does there exist a value of for which the socially optimal number of firms is the long run equilibrium number? Comparing (11) and (14) yields immediately $\alpha = 0$. Thus, we have:

Proposition 2.3: *If all workers are hired, it is socially desirable that the firms bear the totality of the workers' specific training cost.*

Hence, while the general training cost is financed by the government, it is socially optimal that the specific training cost is totally financed by the firms. This result suggests that countries where firms finance totally the specific training cost of workers (as, for example, Germany or Japan) would be more efficient than those where this cost is mainly borne by the workers (as in the United States). It is worth noting that the optimal sharing is the same as the one maximizing the worker surplus.

10.3 The Optimal Choice of General Human Capital

We now turn to the optimal determination of the level of GHC by the government. Because firms earn zero profits, we suppose that the objective of the government is to maximize the sum of the workers' utilities. For simplicity, a worker's utility is assumed to be linear in his income. Since the tax is lump-sum and since the government anticipates the symmetric equilibrium on the labor market, this amounts to maximizing the wage bill paid by the firms minus the total lump-sum transfers paid by the workers minus the training costs. Since these costs are independent of *g*, we do not include them in our social welfare function. The instruments of the government are the level of general human capital, *g*, and the value of the transfer, *t*. Finally, in choosing its optimal policy, the government must break-even. Formally, the problem of the government can be expressed as follows:

$$\max_{g,t} \Delta L \, (w_i\,(g) - t) \tag{15}$$

subject to

$$\Delta L \, t = C\,(\Delta, g) \tag{16}$$

where $w_i\,(g) \equiv w^e$ is the long run equilibrium wage given by (10), while $C\,(\Delta, g)$ is given by (1).

Solving the budget constraint with respect to t, introducing the corresponding expression in (15), and replacing $w_i\,(g)$ by its equilibrium value (12) yields to the social surplus equal to

$$S\,(g) \equiv \Delta L \left(g - \sqrt{\frac{\alpha s F}{\Delta}} - [\lambda + (1-\lambda)\,\Delta L]\, c\,(g)\right) \tag{17}$$

which is maximized by the government with respect to g. Applying the first order conditions leads to the equation

$$\Delta L = [\lambda + (1-\lambda)\,\Delta L]\, c'\,(g^*) \tag{18}$$

Since the marginal value of a unit of GHC is 1 for each worker (see (12)), (18) amounts to a Samuelsonian optimality condition in that the sum of the individual marginal utilities of g equals its marginal production cost when $\lambda > 0$. When $\lambda = 0$, the good is purely private and (16) implies that its marginal utility for each individual equals its marginal production cost.

Assuming that $c'(0) < (\Delta L)/[\lambda + (1-\lambda)\,\Delta L]$, i.e., the cost of producing the first unit of GHC is smaller than its marginal social value, the equation above has a single positive solution g^* since the RHS of (18) is strictly increasing in g. Taking the total differential of (18) with respect to Δ gives

$$\frac{dg^*}{d\Delta} = \frac{L\,[1-(1-\lambda)\, c'\,(g^*)]}{[\lambda + (1-\lambda)\,\Delta L]\, c''\,(g^*)} \tag{19}$$

which is strictly positive by (18) when $\lambda > 0$. On the other hand, when $\lambda = 0$, (18) implies that $dg^*/d\Delta = 0$. Furthermore, since

$$\frac{dg^*}{d\lambda} = \frac{(\Delta L - 1)\, c'\,(g^*)}{[\lambda + (1-\lambda)\,\Delta L]\, c''\,(g^*)} \tag{20}$$

g^* increases (decreases) with λ if the population size is larger (smaller) than 1. When λ increases, the fixed cost of producing g is augmented. Since the corresponding burden is spread over all the workers, it would seem socially desirable to increase the level of GHC when the labor force grows. However, this argument is valid only when the population of workers is large enough because the "fixed" cost generates too heavy a fiscal burden when this population is small. In this case, it turns out to be preferable to reduce the optimal provision of education with the aim of cutting down the increased overhead cost $\lambda c(g)$, despite the negative effect on the wage level. On the contrary, when the population is large, the wage effect always dominates.

Finally, the larger is the size L of the skill space, the higher is the optimal value of GHC (for $\lambda > 0$) since

(21) $$\frac{dg^*}{dL} = \frac{\Delta\,[1-(1-\lambda)\,c'\,(g^*)]}{[\lambda+(1-\lambda)\,\Delta L]\,c''\,(g^*)} \quad >0 \text{ if } \lambda >0$$

Hence, for a given population density, enlarging the skill spectrum is always beneficial to the workers. Accordingly, we have shown

Proposition 3: *When the good g is not a pure private good, the optimal supply of general human capital increases with the population density and with the size of the skill space. When g is a pure private good, the optimal supply of general human capital is independent of the population density and of the size of the skill space. Furthermore, g^* increases with the degree of publicness of GHC if and only if the population size exceeds one.*

Thus, the level of GHC is high in a large labor market because there are increasing returns in producing g. This is confirmed by the empirical study of Rauch (1993) who observes that big cities are likely to be associated with more general human capital.

It follows from the budget constraint (16) that the socially optimal level of taxation is given by

(22) $$t^* = \frac{[\lambda+(1-\lambda)\,\Delta L]\,c\,(g^*)}{\Delta L}$$

This value is feasible if and only if the corresponding wage is not lower than the reservation wage (normalized here to zero) so that t^* must belong to the interval $[0, w_i(g^*)\text{-}sL/2n]$. We assume throughout that this condition holds.

The impact of increasing the population density on t^* is ambiguous. Indeed, if increasing Δ leads to a higher level of GHC and, therefore, a larger budget, this additional expenditure is spread over a larger population so that the net effect is not *a priori* determined. Taking the total differential of (22) with respect to Δ and using the first order conditions shows after some manipulations

(23) $$\frac{dt^*}{d\Delta} = \frac{\lambda}{\Delta^2 L\,c''\,(g^*)} \quad [(c'\,(g^*))^2 - c\,(g^*)\,c''(g^*)]$$

Thus, the sign of $dt^*/d\Delta$ is the same as $[(c'\,(g^*)]^2 - c\,(g^*)\,c''(g^*)$. In words, when $c(g)$ is not "too" convex, g^* increases rapidly with Δ so that the tax paid by each worker rises with the population density. On the contrary, when $c(g)$ is "very" convex, we have the opposite effect: g^* increases only by a trifle and the individual payment to finance the education system decreases. Similarly, it can be shown that

(24) $$\frac{dt^*}{d\lambda} = \frac{(\Delta L-1)}{\Delta L\, c''(g^*)}\,[(c'(g^*))^2 - c(g^*)\, c''(g^*)]$$

so that the sign of this expression depends on the same condition as in the one above if and only if the total population exceeds one; and conversely if the population size is lower than one. Similarly, we have

(25) $$\frac{dt^*}{dL} = \frac{\lambda}{\Delta L^2\, c''(g^*)}\,[(c'(g^*))^2 - c(g^*)\, c''(g^*)]$$

whose sign is the same as that of $dt^*/d\Delta$.

Finally, observe that if we take the limiting case where the general education is a pure private good, i.e., $\lambda = 0$, each individual maximizes $w_i(g) - c(g)$ with respect to g since each worker totally finances his education whereas the government maximizes $\Delta\,[w_i(g) - c(g)]$ with respect to g. Thus, in this case, the individual maximization yields the socially optimal level of GHC.

In order to illustrate the results of this section, assume that $c(g)=\gamma g^2$. Then, the solution to (18) is

(26) $$g^* = \frac{\Delta L}{2\gamma\,[\lambda+(1-\lambda)\Delta L]}$$

while

(27) $$t^* = \frac{\Delta L}{4\gamma\,[\lambda+(1-\lambda)\Delta L]}$$

Since $t^*=g^*/2$, the comparative statics on t^* is the same as on g^*. In other words, a larger market (both in term of Δ and L) leads to more general human capital and a higher educational tax per worker. As observed in section 2, the parameters Δ and L have a similar impact on the social optimal number of firms, but increasing Δ reduces the average matching cost while these costs are independent of L.

Turning now to the net long run wage, we obtain from (12)

(28) $$w^* \equiv w_i(g^*) - t^* = \frac{\Delta L}{4\gamma\,[\lambda+(1-\lambda)\Delta L]} - \sqrt{\frac{\alpha s F}{\Delta}}$$

This expression is the outcome of three different effects: (i) the matching between workers and firms, (ii) the education policy of the government, and (iii) the cost of education. The wage rises with both the population density Δ and the size L of the skill space, regardless of the value of the sharing parameter α. Consequently, *under quadratic cost, workers are always strictly better off in a large labor market.* Indeed, a bigger market results in lower average matching costs and a higher level of GHC. These two effects push wages up high enough to dominate the increment in the educational tax.

10.4 Conclusion

Though the model used in this paper is specific, it is our belief that it captures some basic features of the interaction between a heterogenous labor market and the different forms of human capital. In particular, by using tools borrowed from spatial competition, we have been able to determine the equilibrium wage on a market where firms behave strategically and to study the optimal splitting of the cost of education between workers, firms and the government. Among other things, we have been able to confirm formally the argument of Becker regarding the sharing of the cost and the return to the specific investment in a more general setting with heterogeneous agents. Furthermore, we have seen that the degree of publicness of education is a key factor in explaining the labor market structure. It also follows from our analysis that a large and concentrated population favors the emergence of highly specialized labor markets on which workers have a high level of education and earn high wages. This is in accordance with the intuition of Lucas (1988) for whom the existence of cities can be explained by a finer division of labor (see also Berliant and Zenou (1995) for a model in which city structure crucially depends on labor specialization). We wish to emphasize that our result is derived from some fundamental microeconomic underpinnings and does not rely on the *ad hoc* specification of externalities.

Much work remains to be done. First, the assumption of perfect observability of the worker skills should be relaxed. One way to proceed is to suppose that the productivity of each worker is subjected to random shocks. This can be done by appealing to discrete choice theory which has proven to be very useful to study the joint influence of observable and unobservable characteristics in product differentiation (see Anderson et al. (1992, ch.9)). An alternative approach is to follow Jovanovic (1979a et b) and to assume that both the worker and the firm learn the quality of the job matching over time.

Second, our model could be embedded into an urban setting with the aim of determining the equilibrium and optimal population sizes. If firms are located at the Central Business District, workers will have to compete more for land and to commute over larger and larger distances as the population grows. The combination of these two effects makes workers worse off. A trade-off arises between the advantages of the division of labor and the costs of urban congestion.

Third, while we have assumed that general education is provided by a national government, it would be interesting to investigate the case of competing local governments. As a first approximation, a local government could be viewed as a coalition of firms acting cooperatively in the provision of GHC while still behaving noncooperatively on the labor market.

Fourth, and last, our model has exploited a formal analogy between the skill characteristic and a horizontally differentiated characteristic. Ever since Lancaster (1979), it is customary in industrial organization to consider a second type of characteristic, called vertically differentiated, that can be reinterpreted here as another skill characteristic along which workers can be ranked by their productivity. Such a kind of heterogeneity would allow us to determine the segmentation of the labor market between highly-educated and poorly-educated individuals and to analyze the implications for the economy.

References

Anderson S., de Palma A. and J.-F. Thisse (1992), *Discrete Choice Theory of Product Differentiation*, Cambridge (Mass.) : MIT Press.

Becker, G. (1964), *Human Capital*, New York: Columbia University Press.

Berliant, M. and Y. Zenou (1995), "Labor Specialization and City Formation", CORE Discussion Paper n. 9508.

Ethier, W. (1982), "National and International Returns to Scale in the Modern Theory of International Trade", *American Economic Review*, 72, 389-405.

Hashimoto, M. (1981), "Firm Specific Human Capital as a Shared Investment", *American Economic Review*, 71, 475-482.

Hamilton J., Thisse J.F. and Zenou Y. (1997), "Skill acquisition and wage competition with heterogeneous workers and firms", CORE Discussion Paper n. 9722

Jovanovic, B. (1979a), "Job Matching and the Theory of Turnover", *Journal of Political Economy*, 87, 972-990.

Jovanovic, B. (1979b), "Firm-Specific Capital and Turnover", *Journal of Political Economy*, 87, 1246-1260.

Kim, S. (1989), "Labor Specialization and the Extent of the Market", *Journal of Political Economy*, 97, 692-705.

Lancaster, K. (1979), *Variety, Equity and Efficiency*, New York: Columbia University Press.

Lindbeck, A. (1993), *Unemployment and Macroeconomics*, Cambridge (Mass.): MIT Press.

Lucas, R. (1988), "On the Mechanics of Economic Development", *Journal of Monetary Economics*, 22, 3-42.

Parsons, D. (1986), "The Employment Relationship", in *Handbook of Labour Economics*, Ashenfelter, O. and R. Layard, eds., Amsterdam: North-Holland.

Rauch, J. (1993), "Productivity Gains from Geographic Concentration of Human Capital: Evidence from the Cities", *Journal of Urban Economics*, 34, 380-400.

Romer, P. (1987), "Growth Based on Increasing Returns Due to Specialization", *American Economic Review (Papers and Proceedings)*, 77, 56-62.

Salop, S. (1979), "Monopolistic Competition with Outside Goods", *Bell Journal of Economics*, 10, 141-156.

Stevens, M. (1994a), "A Theoretical Model of On-The-Job Training with Imperfect Competition", *Oxford Economic Papers*, 46, 537-562.

Stevens, M. (1994b), "The Sharing Hypothesis for Specific Human Capital", London, CEPR Discussion Paper N1124.

Zenou, Y. and T. Smith (1995), "Efficiency Wages, Involuntary Unemployment and Urban Spatial Structure", *Regional Science and Urban Economics*, 25, 547-573.

11. The Role of Educational Institutions in the Art Sector: From Academies to the Grand Tour

Ilaria Bignamini and Enrico Castelnuovo*

The establishment of Academies of Art in Florence and Rome during the second half of the Sixteenth Century marked the beginning of a new era[1]. Early academies ennobled art education, contributed to the social and cultural rise of the artist and established new career patterns. The changes they introduced regarding the creation, selection and transfer of theoretical and visual knowledge were long lasting, but they will be better understood when compared with the context of art education during the Middle Ages and when due attention is paid to changes which took place before their establishment.

11.1 Starting-points

During the Middle Ages there was no difference, from a social point of view, between artists and craftsmen and their education was similar[2]. For centuries the production and consumption of works of art pivoted mostly upon major monastic and episcopal centres. Some education and training was offered by workshops and *scriptoria*, where manuscripts were copied and illuminated, and where gold and silver works as well as ivory sculptures, stained-glass windows, enemals and embroideries were produced. Schools attached to cathedral churches probably played an important role, but little is known about them as to the transmission of art knowledge. The life of St. Bernwardus, Bishop of Hildesheim early in the Eleventh Century, suggests that before becoming a great art patron the Saint had practised various arts and he might have been introduced to art techniques at the school of the Cathedral.

Presumably early in the Twelveth Century, an important treatise on various arts, *De Diversis Artibus*, was written in a monastery of southern Germany. Its author, probably a priest, is known as Theophilus[3]. The treatise is divided into three books; the first dealing with painting and its techniques (mural, table, parchment painting, illumination etc.), the second with glass (especially stained-glass) and the third with gold, silver, bronze and copper works. It is unlikely that

* Department of the History of Art, University of Oxford; Scuola Normale Superiore di Pisa

[1] Pevsner, 1940; *Academies of Art between Renaissance and Romanticism*, 1986-87; Castelnuovo and Bignamini, 1991.

[2] Castelnuovo, 1987.

[3] Theophilus, 1986.

Theophilus' book was addressed to individual artists and craftsmen; more probably its readers were instructors at major educational centres attached to monastic and episcopal workshops and schools. Information provided by this treatise ranges from ways of organizing a workshop, or a furnace, to descriptions of the tools, materials and techniques. The author, who shows some direct experience of monastic workshops, might have been a practising artist, possibly a goldsmith like Roger of Helmarshausen, an important maker of liturgical objects of the Twelveth Century, with whom Theophilus has been identified by some authors. In those days goldsmiths were at the top of art hierarchy.

Altough monasteries tended to monopolize art production, works were also made for municipalities and courts. In the Eighth Century a Longobard Duke of Spoleto presented the Abbey of Ferentillo, near Terni (Umbria), with a marble slab carved in relief. It shows the features and bears the signature of a certain "Ursus magester" (Master Ursus), who presumably was running a workshop producing sculptures. Similarly, the goldsmith Wolwinius portrayed himself in a gold altar commissioned for the Church of St. Ambrose by Angilbertus, Bishop of Milan in the Nineth Century. Wolwinius, who is wearing secular dress and is crowned by the patron saint while he presents him with the altar, signed his work and described himself as an "aurifaber".

Monastic workshops and schools of cathedral churches are not the only centres to take into account when considering the transmission of artistic knowledge. The renaissance of urban life during the Eleventh and Twelveth Centuries led to the rapid growth of the demand for art works, to the increase of workshops, to the creation of itinerant groups of stone-cutters, or sculptors, and to the rise of skilled artists describing themselves as masters (the word used was "magister"), which implied that they had completed their apprenticeship and were running a shop. Guilds were later to codify rules regulating workshops, their production and the relationship between masters and apprentices.

New attitudes towards artists and art production were soon to become manifest, especially in central Italy. A particularly interesting document is the *Libro dell'arte*, an art treatise written by the Florentine painter Cennino Cennini in Verona at the turn of the Fifteenth Century. A pupil of Agnolo Gaddi, Cennini illustrates how artists's workshops were run in Florence during the previous century, and gave details about art techniques and working methods. But, above all, Cennini's *Libro* contains an apology for painting, a technique that thanks to innovations introduced by Giotto and the wide-spread reputation of that artist had reached the pinnacle of the hierarchy of the arts. Giotto is regarded as the model-artist and also as the painter ennobling the activities of Cennini himself, who reminds the reader that he had been a pupil of Agnolo Gaddi who, in his turn, had been a pupil of his father, Taddeo Gaddi who had long been working in Giotto's workshop. In Cennini's book, passages on the figure of the painter and his training are related to his defense of painting and, as a consequence, they are rather idealized. In spite of this the book includes interesting information about the painter's education. Painters are obviously advised to stay as long as possible in the workshop of a good master, but it is also recommended that they assidously study works by other contemporary masters and concentrate on those by one artist only.

The idea of the superiority of painting over other arts can also be traced in the rules of guilds. In 1355, for instance, the members of the painters's guild of Siena proudly wrote: "noi siamo per la grazia di Dio manifestatori agli uomini grossi che non sanno lectera de le cose miracolose..." (We are, by the grace of God, illustrators for ignorant and illiterate men who cannot read of miraculous things.) In this case, the superiority of painting was clearly instrumental to the visualization of religious matters.

By the end of Fourteenth Century, new attitudes towards the artist (now a figure distinguished from mere craftsmen), towards the painter (ideally a pupil of Giotto and other famous masters), and towards painting (at the pinnacle of the hierarchy of the arts), had produced substancial and enduring changes in the social and cultural standing of artists, especially in Florence. This had the further effect of making a new institution necessary, which offered artists a more comprehensive educational curriculum than traditional workshop training. Fifteenth Century court patronage, humanism and theories about drawing, or *disegno* (the artist's prime idea), which developed mostly in Florence, did the rest and created the conditions for the establishment of academies of art.

11.2 Early Academies of Art

By the time Academies were established, drawing had long been regarded as a central feature of art education. Workshop training and apprenticeship progressed from copies after works by masters to simple drawings of individual parts of the human body, to studies of the whole human figure and to compositions with more than one figure. Young artists also copied three-dimensional models, drew after the nude model and studied drapery. Yet, these stages varied considerably from workshop to workshop, and not all the masters had the same professional skills. Not many of them could offer young artists a good selection of works to copy or could afford the cost of living models posing before the artists. (Usually male models, though the employment of undressed female models is documented in Venice from the late Fifteenth Century.) Moreover, the study of theory, geometry, perspective and human anatomy, though pursued by individual artists, was rarely taught to apprentices and young artists as part of their workshop training.

Early academies such as the Florentine Accademia del Disegno and the Roman Accademia di S. Luca were established for this purpose, but it took sometime before they actually offered all the educational services, which are described in their rules. Services they offered to patrons were initially more effective. The case of the Accademia del Disegno is telling.

The Accademia del Disegno was created in 1563, shortly before Cosimo I de' Medici (Duke of Florence from 1537) became Grand Duke of Tuscany in 1569[4].

[4] Barzman, 1986-87; Wazbinski, 1987.

The Academy was part of a wider institutional system for the arts pivoting upon the Court. There was, in fact, a close relationship between Cosimo's political ambitions, his patronage of the arts and institutional events such as the publication of the first edition of Giorgio Vasari's *Vite de' più eccellenti architetti, pittori, et scultori italiani* (1550), the project of the Gallery of the Uffizi (begun by Vasari in 1560), the creation of the Accademia del Disegno (1563), the establishment of public exhibitions of works of art executed by academy members (at the Ospedale degli Innocenti, the first was held in 1564) and the second edition of Vasari's *Vite* (1568). These institutions celebrated the supremacy of the Florentine School and the munificence of Cosimo himself. The seventy-five artists admitted to the Academy during the first year (1563-64) came from shops of artists working for Cosimo and his family, that is to say those of Vasari, Bronzino, Michele di Ridolfo Ghirlandaio, Pier Francesco Foschi, Giovan Angelo Montorsoli and Bartolomeo Ammannati. Members of noble familes such as the Medici, Strozzi, Ricasoli and Alammani families were also admitted to the Academy. The Academy liberated the artists from traditional guild control; the incorporation of it as a guild on its own right dates from 1570. It also promoted the arts of *disegno* as liberal pursuits, grounded in theory. In principle both artists and noble amateurs were given the possibility of attending classes on geometry, astronomy and natural philosophy as well as anatomy, life drawing and drapery studies. In practice, it was only in the Seventeenth Century that nude male models were regularly employed and that anatomy was studied, probably not so much through dissections performed for the artists at the hospital of Santa Maria Nuova, as stated in the Academy's rules, but through studies of three-dimensional *écorché* figures made by sculptors under the supervision of anatomists.

The Academy, of course, created new discriminations among the artists. It divided them into two groups; those granted membership and those excluded from it, namely artists working for lesser patrons and the market, and artists practising the so-called minor or applied arts, even including those who enjoyed court patronage.

The Accademia di S. Luca, established in Rome in 1593, was a less discriminating institution[5]. Federico Zuccari, its first Principe, designed it as an open educational body; young artists from all over Italy and Europe could apply to it. Although artists practising minor or applied arts were excluded, a parallel institution called Compagnia, or Confraternita, di S. Luca was created for them. The Academy also took over the old guild, or Università dei Pittori, and supervised the whole body, including *garzoni*. The educational curriculum designed by Zuccari for the Roman Academy was similar to that of the Florentine Academy except for anatomical dissections, which were not allowed in the Papal States. But, as in Florence, the real activity of the Academy was confined to debates about the rules and no regular teaching was offered until the mid-Seventeenth Century.

---

[5] Pietrangeli et al., 1974.

11.3 'Printed Academies'

More effective for the transmission of visual knowledge were drawing manuals, or 'printed academies', published from the beginning of the Seventeenth in Italy and elsewhere in Europe[6].

'Printed academies' were part of a larger production of art treatises and engravings, which made a vast body of visual knowledge widely available. These manuals supplemented workshop training and academy education and served as substitutes for actual institutions. They were addressed to professional and young artists, to amateurs, craftsmen and house-painters. The authors sometimes were well-known artists, but more often they were unsuccessful artists combining the activities of shop-keepers, print-dealers, pigment sellers, and drawing and painting instructors. *The number of buyers is remarkable; Dr. William Salmon's Polygraphice* -an English best seller addressed to both apprentices and professional artists- went through eight editions in nearly thirty years (1672 to 1701) and by 1701 it had sold 15,000 copies. Women were important buyers, and the market of 'printed academies' (like that of novels) soon became dominated by middle and upper-class female audience. Moreover, 'printed academies' had a wider geographical impact than art schools and they reached the provinces much earlier. The lessons they taught were more accessible than those of actual academies because they combined the functions of a preparatory school with that of an academy. Finally, they popularized the theoretical and visual lessons taught at academies, they popularized styles and stimulated a general taste for art. The modern art world would have never existed it were not for the dramatic growth of the art press in the Seventeenth and Eighteenth Centuries, especially the production of engravings after the Antique and the Old and Modern Masters, book illustrations, magazines and 'printed institutions' of all sorts (printed academies, collections, museums and tours of Italy and other countries). These printed educational institutions are central to any consideration of the transfer of visual knowledge, and many of them offered models as selective as those of actual academies.

11.4 The French Academy

No academy was as effective in selecting models for the artists and no academy was so powerful in imposing itself as a model-academy as the state-created and state-directed Académie de Peinture et de Sculpture. Established in Paris in 1648, it was reorganized by the King's Private Advisor, Jean-Baptiste Colbert, in 1663, and directed by the Premier Peintre du Roi, Charles Le Brun[7]. Like the Florentine Accademia del Disegno, but on a grander scale, Colbert's Academy was part of a wider institutional system which pivoted on the Court-State and its Government.

6 Bolten, 1985; Bignamini, 1996.

7 Duro, 1996; Locquin, 1912; Crow, 1985; Montagu, 1989.

Decisions were centralized to a degree never attempted before and never achieved later, not even by the dictatorships of our century. In addition to the Paris Academy of Art, Colbert created a French Academy in Rome (1666) and innumerable other institutions including the Académie de Danse (1661), Académie d'Inscriptions et Belle Lettres (1662), Académie des Sciences (1666), Académie de Musique (1669), Académie d'Architecture (1671), royal manufactures such as the Gobelins (1662) and Savonnerie (1663), and also a school of design and one for apprentices attached to the Gobelins (1666-67).

As soon as Colbert was appointed Private Advisor to Louis XIV in 1661, he also became Vice-Protector the Académie de Peinture et de Sculpture. (He later was Surintendent des Monuments, 1664, and Protecteuer de l'Académie de Peinture, 1672.) In such capacity, he ordered that all the *breviataires*, the licensed artists, should become members of the Academy. The monopoly of the Academy had begun and it extended to annual exhibitions, later called the *salons*, which started in 1664 (but were not held regularly until 1737). Only the members of the Academy were allowed to exhibit their works at the Salon and in the *livret*, or catalogue (the first was published in 1699), their names were listed according to their status within the Academy: first the Premier Peintre du Roi, then the Recteurs, Professeurs, Académiciens and Agrées. In 1661 the Grand Prix de Rome was also created; the best students were sent to Rome to complete their education. To insure that they worked there with the guidance of French masters an Académie de France à Rome was created in 1666. Subjects of religious and mythological history selected by the Academicians for the Prix de Rome left little freedom to young artists, who had to conform to the classical taste favoured by the political and artistic establishments. The publication of theoretical treatises such as those by Roland Fréart de Chambray (*Idée de la Perfection de la Peinture*, 1662) and Roger de Piles (translation of Du Fresnoy's *De Arte Graphica*, 1668), and the Academy's monthly *conférences* (started in 1667 and read by Academicians and amateurs such as André Félibien) were instrumental to the creation of an academic aesthetics. The principal aim of the *conférences* was to confirm that history painting was the pinnacle of the academic hierarchy. History painting was seen as encompassing all the genres (portraiture, landscape, genre, still-life) and therefore could be regarded as truly universal. The privilege for students sent to Rome was that they could study the best models of history painting and attain perfection in the study of the human body. But even in this students had no freedom in selecting their models; works to copy by Italian masters and classical sculptures were selected by the Academy Director. Young sculptors also had to make marble and plaster copies to send home for the use of the Paris Academy and to decorate Versailles and other royal gardens. Finally, while drawing after the living model, they were required to retain the perfected anatomical features they had studied while copying works by approved masters, especially Raphael and Annibale Carracci. Drawings produced by the French painter Louis Chéron at the academies in Paris and Rome, and drawings he later made at private academies in London are telling. They show how the anatomical features for drawings after the living male model were closely derived from copies of Carracci's ignudi (especially those painted on the ceiling of Palazzo Farnese, Rome). Those for drawings after the female model he exectuted in London in the early 1720s were still, after forty years he had left from Rome, closely derived from Raphael (especially those at the Farnesina, Rome).[8]

At the time Chéron was in Rome, in the late 1670s, the French Academy had virtually taken over the Accademia di S. Luca. Gian Pietro Bellori's influential book on the *Vite de' pittori, scultori ed architetti moderni* (1672) had been dedicated to Colbert and the Director of the Paris Academy, Charles Le Brun, had been elected Prince to the Roman Academy in 1675. This mirrored a more general political alliance between France and the Papal States, which lasted until the end of the Seven Years War in 1763.

11.5 The English Academy

The Edict of Nantes was signed in 1685. Chéron, like other Huguenot artists, was excluded from membership of the French Academy and, more generally, from the artist's profession. Like many Huguenots, he emigrated to England, where he established one of the first academies of art in London.[9] A few decades earlier, from the 1670s, many Dutch artists, auctioneers, dealers and print-sellers had also started to migrate to England, due to a serious recession in Holland creating dramatic unemployment. These two large emigrant artistic communities (along with artists who arrived from Italy, France, the Flanders, Germany, Switzerland and Sweden) imported new teaching methods and visual models into England. The Dutch imported also the female academy nude (the undressed female model was not employed at academies in Italy and in France until early in the Nineteenth Century) and they were responsible for the introduction of new marketing techniques including regulated auctions and auction catalogues.

The art maket in Seventeenth Century Holland was far more expanded than in other countries due to the absence of commissions from the Church, the Court and from wealthy aristocratic patrons, but it was a market inhibited by guild control, which was incapable to discriminate through prices between the different qualities of art products. The case of one of the gretest painters ever, Johannes Vermeer (who never made a living) is telling.[10]

When considering transfers of knowledge, one should never forget contributions from immigrants. Their chances to succeed have always depended on the political and social character of the country they migrated to and on the 'openess' of sectors they entered into. Artist-immigrants, who arrived in England in those days were particularly lucky because they met with a society far more liberal and open than any other in Europe (expect for Holland), with a booming economy, and with a national school of painting, sculpture, architecture and design in its infancy. In

---

[8] Bignamini and Postle, 1991, ills. 7-10 pp. 11-12.

[9] Bignamini, 1986-87; id., 1988.

[10] Montias, 1989.

fact, a modern British School was still to be created, the art market had not been organized yet, and the number of potential buyers of art works was so high as to make the lack of patronage from the Court and encouragement from the Government quite irrelevant. Immigrants played an active role in creating new outlets such as auction houses and print-shops, and they were founding members of early art clubs and private academies, whose multi-national and heterogeneous character was markedly different from that of any continental counterparts.

From its origins in 1673 until 1768, the English Academy was an unregulated and self-supporting private institution created and democratically directed by the artists themselves in response to growing market demands. It was open to all artists with no discrimination of nationality, religion or professional expertise. It contributed to the development of a mature system of interacting institutions. Advertisments on the press were widely used to publicize sales, auctions, exhibitions, illustrated books and other initiatives including subscriptions to academies. The whole artistic community was involved in the decoration of public leisure gardens, most of them worked as stage designers and book illustrators. Unlike in Italy or in France, annual public exhibitions of contemporary art were created before the Royal Academy, and public exhibitions were open to all the artists. Data about the number of the visitors are telling. The first exhibition of the Society of Arts was held in London in 1760 and it lasted for less than three weeks. There were 12,000 visitors (about 1,000 a day) and there was a total of 6,582 catalogues sold at 6 pences each. The following year there were 13,000 visitors and each of them was requested to buy a copy of the catalogue sold at 1 shelling.[11]

In half a century the unregulated English Academy had become a fully modern institution, whereas most other European academies only achieved similar results later, after the collapse of the Ancien Régime. The teaching offered by the market-oriented English Academy was inevitably different from that of other European academies. It was, first of all, a life-class meeting in the evening, when most subscribers were able to join after the working day. Experienced artists supervised the work of young artists but no title of professor was used, nor were instructors always the same individuals. Sculptors made plaster models for the artists to copy and those who wished to perfect in drawing after the Antique could attend the Duke of Richmond's Gallery of Casts, while young designers could attend William Shipley's Drawing School and compete for annual premiums offered by Society of Arts for designs of machines, textile materials etc. Finally, famous anatomists attended private academies established in London from 1720 onwards in order to learn how to draw from the human figure, which enabled them to supervise artists when making illustrations for their anatomical treatises. William Cheselden, one of the foremost anatomists of the Eighteenth Century, attended the Academy established by Louis Chéron in London in 1720, and he employed artist-members of the Academy to produce illustrations for his influential treatise, *Osteographia,* of 1733. Anatomists also helped sculptors to cast *ecorché* figures

[11] For further quantitatives data, see Bignamini, 1990.

and they performed dissections for the artists and anybody who was curious about them. (In England, dissections were frequently advertised in the press.)

Familiarity with the practice of drawing after the living model and anatomy, the use of both male and female models, the lack of pre-selected visual models, and the necessity to meet diversified market demands enabled the English Academy to become competitive with other European schools. But an unregulated Academy without official recognition could hardly be a major institution and help to promote the British School at home and abroad. This was a good enough reason as to convince many British artists to fight for the establishment of the Royal Academy of Arts. Artists particularly sensitive to this issue came from the ranks of traditional history painting and portraiture, and some of them, like Joshua Reynolds, the first President of the Royal Academy, had visited Italy in order to complete their education. Those artists thought that no real British School could ever exist without the backing of an official institution, without a programme of formal education offered to young artists, prestigious annual exhibitions organized by the Academy, and an aesthetics conforming to the hierarchy of pictorial genres. Academy lectures were initiated and, more generally, the artistic profession became institutionalized.

The Royal Academy of Arts, established in London in 1768, partly answered to the new requirements but, unlike its European counterparts, it was a private institution which supported itself with fees paid by subscribers, and with tickets and catalogues sold at annual exhibitions. These were markedly different from the French *salons;* all artists could submit their works for exhibition at the Royal Academy, not only academy members. The most important feature the Royal Academy shared with its continental counterparts was the institutionalization of the artist's career.

To be an Academician meant to be promoted to professional skill and perfection, independent of actual results. In 1735 Voltaire, who was proud of being a member of no academy, had pronounced severe words about French institutions:

> "Nous n'avons pas un grand peintre depuis que nous avons une académie de peinture, pas un grand philosophe formé par l'Académie des Sciences; je ne dirai rien de la française".
> (We had no great painter since the Académie de Peinture was established, no great philosopher educated at the Académie des Sciences; I will not even mention the Académie Française.)[12]

In 1761 William Hogarth recalled Voltaire's words in order to discourage his colleagues from following Reynolds and his friends in their attempts at establishing a regulated Royal Academy.[13] Hogarth was defeated, but the banner of rebellion and intellectual freedom was soon taken up by William Blake, who in c.1808 wrote fierce words against Reynolds and the Academy:

[12] Voltaire, 1954, letter no. 919, pp. 191-92.

[13] Hogarth, 1966-68, p. 92.

> "Reynolds's Opinion was that Genius May be Taught & that all Pretence to Inspiration is a Lie & a Deceit, to say the least of it...I do not believe that Rafael taught Mich. Angelo, or that Mich. Angelo taught Rafael, any more than I believe that the Rose teaches the Lilly how to grow, or the Apple tree teaches the Pear tree how to bear Fruit. I do not believe the tales of Anecdote writers when they militate against Individual Character".[14]

Individual freedom taken to such Romantic extremes was hardly compatible with an institution like the Academy, which perpetuated the traditional principle that the transmission of knowledge in art should be based on the practice of copying works by famous Old Masters selected for young artists by experienced Modern Masters -once workshop masters and now academy professors. Late in the Eighteenth Century and during the following century rebellious voices of all sort pronounced against the Academy. In countries that had long entered the modern and bourgeois era like England those voices gave expression to the rebellion of individuals or groups against an institution unable to answer the needs of modern art movements. In countries that had recently and dramatically entered the bourgeois era like France the voices of individuals were mixed with political ideologies, which contributed to transform the Academy yet again. Many French artists, including Jacques-Louis David, believed that the Revolution and new subject-matter would produce a novel Academy, but it did not take them long to realize that the Academy could only be transformed to a limited extent and that the Museum was far more promising an educational institution. In January 1794, David wrote:

> "...the Museum is not a mere collection of luxury goods, or trifles useful but for curiosity. It should become a great school. Masters will take their pupils there, fathers their children. The sight of works of genius will stimulate the growth of art and science seeds in those young people, who are naturally gifted for them. Time has come for you, the legislators, to stop ignorance half its way...save the Museum, save those works that can disappear in an instant and that avaricious nature might produce never again".[15]

11.6 Tours and Museums

Long before reformed institutions for the new bourgeois and industrial era were created, the Grand Tour and the 'grand museums' began to play a role far more crucial than that of any academy with respect to the creation, selection and transfer of visual and theoretical knowledge in the art sector.

The educational tour of Italy, the Grand Tour, made by the upper and educated classes of European society and by countless artists during the Eighteenth Century was the 'invisible academy' -or the virtual educational institution- of modern Europe.[16] One of the most relevant consequences of the Grand Tour were

[14] Blake, 1984, pp. 452-53.
[15] Quoted Castelnuovo, 1985, p. 145.
[16] Hibbert, 1987; Black 1992; De Seta, 1992; Wilton and Bignamini, 1996.

archaeological excavations carried out on an unprecedented scale and massive exports of art works from Italy, especially classical sculptures and Old Master paintings. This trade (and these activities) contributed to the enlargment of pre-existing collections and created the conditions for the establishment of new museums, where the aesthetic lessons of classical and Italian art were taught to wider sections of the educated public and to all the artists. The impact of 'museum towns' such as Rome, Naples and its environs, Florence and Venice and that of 'grand museums' such as the Vatican, the Uffizi and later the Louvre and British Museum on the creation and transfer of visual knowledge was unrivalled by any other institution. Their role was subsequently expanded by new tours to previously little explored cultural and geographical areas, which created the conditions for the establishment of new museums and collections of Medieval and non-European art during the Nineteenth-Century.

Tours and museums became more important than any academy or art school, and the training of the artist began to coincide with the education of the public until avant-garde mouvements reintroduced the notion that the upper crust of the art production should be accessible to the few only. But museums and the press helped again and commercial advertisments, for instance, have made wide use of incomprehensible formal solution originally elaborated by avant-garde artists.

References

"Academies of Art between Renaissance and Romanticism" (1986-87), ed. by A.W.A. Boschloo et al., in *Leids Kunshistorisch Jaarboek*, V-VI.

Barzman, K.-E. (1986-87), "The Florentine Accademia del Disegno: Liberal Education and the Renaissance Artist", in *Academies of Art between Renaissance and Romanticism*, (1986-87), pp. 14-32.

Bignamini, I. (1986-87), "The 'Academy of Art' in Britain before the foundation of the Royal Academy in 1768", in *Academies of Art between Renaissance and Romanticism*, (1986-87), pp. 434-50.

Bignamini, I., (1988), "Art Institutions in London, 1689-1768", in *The Walpole Society*, LIV, pp. 19-148.

Bignamini, I. (1990), "Osservazioni sulle istituzioni, il pubblico e il mercato delle arti in Inghilterra", in *Zeitschrift für Kunstgeschichte*, 54, 2, pp. 177-97.

Bignamini I. (1996), "The English Academy of Art. From the Renaissance to the Revolution", in *Europäische Sozietäts bewegung und demokratische Tradition*, ed. by K. Garber and H. Wismann, Max Niemeyer Verlag, II, pp. 1713-31.

Bignamini, I., and Postle, M. (1991), "The Artist's Model. Its Role in British Art from Lely to Etty", exh. cat., University Art Gallery, Nottingham, and The Iveagh Bequest, Kenwood.

Blake, W. (1984), "Annotations to Sir Joshua Reynolds's Discourses, c.1808", in Blake. *Complete Writings with variant readings*, ed. by G. Keynes, Oxford University Press.

Bolten, J. (1985), "Method and Practice. Dutch and Flemish Drawing Books 1600-1750", Pfalzische Verlangsanstalt; 1st ed. in Dutch, 1979.

Castelnuovo, E. (1985), "Arte, Industria, Rivoluzioni. Temi di storia sociale dell'arte", Einaudi.

Castelnuovo, E. (1987), "L'artista, in L'uomo medievale", ed. by J. Le Goff, Laterza, pp. 235-69.
Castelnuovo, E., and Bignamini. I. (1991), "Arte e società", in *Enciclopedia delle Scienze Sociali*, Istituto della Enciclopedia Italiana, I, pp. 319-39.
Crow, T.E. (1985). "Painters and Public Life in Eighteenth-Century Paris", Yale University Press.
De Seta, C. (1992). "L'Italia del Grand Tour da Montaigne a Goethe", Electa Napoli.
Duro, P. (1996), "Académie Royale de Peinture et de Sculpture", in *Dictionary of Art*, Macmillan.
Hibbert, C. (1987), "The Grand Tour", Thames Methuen.
Hogarth, W. (1966-68), "Hogarth's 'Apology for Painters', 1761", ed. and introduced by M. Kitson, in *The Walpole Society*, XLI, pp. 46-111.
Locquin, J. (1912), "La Peinture d'Histoire en France de 1747 à 1785", Henri Laurens; reprint Arthena 1978.
Montagu, J. (1989), "Roman Baroque Sculpture. The Industry of Art", Yale University Press.
Montias, J.M. (1989), "Vermeer and his Milieu. A Web of Social History". Princeton University Press.
Pevsner, N. (1940), "Academies of Art. Past and Present", Cambridge University Press; reprint Da Capo Press, 1973.
Pietrangeli, Carlo, et al. (1974), "L'Accademia Nazionale di San Luca (1974)", De Luca.
Theophilus (1986)," The Various Arts/ De Diversis Artibus", ed. by C. R. Dodwell, Clarendon Press.
Voltaire (1954), "Correspondance", Institut et Musée Voltaire, IV.
Wazbinski, Z. (1987), "L'Accademia Medicea del Disegno a Firenze nel Cinquecento. Idea e istituzione", 2 vols., Leo S. Olschki.
Wilton, A., and Bignamini, I. (1996), "Grand Tour. The Lure of Italy in the Eighteenth Century", exh. cat., Tate Gallery Publishing . In Italian: "Grand Tour. Il fascino dell'Italia nel XVIII secolo", exh. cat., Skira, 1997.

12. Oral Transmission in Indian Classical Music The Gharana System

Alaknanda Patel*

12.1 The Setting

Classical music, as it is practised in North India now, traces its origin from a treatise called the 'Natyashastra' of early Christian period, but music as part of ritual or aid to memorising texts has even earlier roots. The tradition of imparting knowledge was oral with a minimal use of the written word. Recorded, written material was necessary only as a key to reminder, a reminder to the storehouse of knowledge gathered through the master-disciple system of oral instruction. While lack of documentation makes it difficult to chronicle a precise historical development over about 2000 years, it is generally agreed that definitive references to the precursors of contemporary genres of art music are available from the 13th century onwards[1]. By this time, Muslim rule was well-established in North India with both Sultans and Sufi saints creating centres of musical performance. It is not surprising, therefore, that within a short period, Persian musical concepts started influencing the existing musical system, giving North Indian music, which had got rather structured and bound within a rigid frame[2], a new freedom. Under the patronage of Hindu-Muslim courts, new genres of performance practice in a composite style started evolving - a style that continues till today. In this over seven hundred years of changing political, economic and social milieu, music developed along two streams, one consisting of academics and the other, of performers. The scholars wrote commentaries, interpreted Sanskritic aesthetics and carried on research in an open, written academic tradition, not very different from that of medieval Europe. The performer, however, had to tread an entirely different path, a tight-rope walk between disclosure and secrecy, an optimisation act between present returns and future hold, with no legal guarantee anywhere.

---

* 12 Amee Society, Baroda, India
I thank the Centre of South Asian Studies, Cambridge and the Indira Gandhi Institute for Development Research, Bombay, for inviting me as a Visiting Scholar and giving me research facilities. I also thank Partha Dasgupta and Paul A. David for giving me an early version of their monograph on the "Socio-Economics of Science and Technology". Many of the issues raised by them helped me consolidate my own ideas.

[1] Widdess, 1981, p. 67.

[2] Sen, K.M.

Indian classical music is almost entirely improvised, improvisations based on certain inviolate ground rules. Creativity, innovation take place at a rehearsal, at a performance, at reflective moments; the artist is at once an epistemic scholar and a technical executor. The knowledge he has inherited or created is unwritten, uncoded - there is no notation for simple lines or rhythm, let alone for style or ornamentation. Even the song-text is not written down because the written word loses nuances and delicacy. In addition, much of what the singer presents or teaches is a part of his 'tacit' knowledge, derived from his own experience, sensitivity and insight. It is this complex system of knowledge that the classical artist tried to market as a commodity. Faced with a patronage system that had strong flavours of isolated monopoly situations, he developed a unique socio-economic institution called the 'Gharana'[3]. This paper is an attempt to describe the process of dissemination of knowledge under this system and to assess some of its impact.

Gharana comes from the word 'Ghar' meaning a house, a home or an abode. In some sense it may mean a family or even the family's place in the strata of society. In Khyāl, the predominant genre of classical singing since the early 1700's, Gharana denotes a particular musical style unbroken in a particular family for several (at least five) generations. It represents a family[4] of male musicians, a well-knit unit evolving, guarding and disseminating the distinctive style through its male members, some of whom are well-known performers and some who are not. The wealth of the Gharana is its knowledge - knowledge of song-texts, ragas (melodic modes), rhythm, ornamentation, which coupled with distinctive voice production, phrasing and sequence, produces a unified whole, an aesthetic form special to each family[5]. Gharana training, despite individual creativity, must reveal the artist's musical pedigree. A protective film spread around the knowledge continues this purity of style.

The Gharana system was based on two essential features, élite patronage requiring a certain amount of disclosure and the distinctive musical style requiring family secrecy. The music described here is essentially solo, with two accompanists on drums and sarangi[6] (or harmonium) and occasionally one, at best two, junior support singers at the back. With a complete absence of classical choral singing either ecclesiastic or operatic, it is a sharp contrast to the western

---

[3] During the Sultanate and most of the Mughal period, the predominant classical vocal form was the dhrupad, majestic and contemplative, sung in four different styles called 'Banis'. When the highly puritanic rule of Aurangzeb ended in 1707, patrons expressed a desire for a form more mobile and the introspective dhrupad was replaced by the colourful Khyāl, with Gharanas as the pivot of its evolution. Present Gharana members can generally trace their lineage back to early 19th century as Khyāl singers and even earlier when they are connected with dhrupad Banis. The golden age of Gharanas, as described here, started around 1840 and by the latter part of the 20th century, their importance started declining.

[4] Families discussed here are Muslim.

[5] Contemporary ethnomusicologists have given several definitions of the Gharana. Here the term is used in the way Gharana members themselves describe it.

[6] a bowed instrument.

forms. The congregational singing at the temples (Kirtan) or Dargahs (Qawali), though based on classical motifs, was not considered part of the pure genre.

12.2 Employment under the Patronage System

Since music conferences with tickets and a large audience did not start till well into the 20th century, patronage remained restricted to the élite. Records show that from the 15th century onwards, not only the imperial court at Delhi or Agra but kingdoms like Gwalior, Bijapur, Jaunpur, all took a lead role in the propagation of the arts. With the glory of the Mughal empire waning, by the mid-1700's artists started moving away from the Centre and slowly settled down at smaller courts like Baroda, Rampur, Vishnupur, Awadh, Jaipur, creating little islands for themselves. But these principalities went through changes of fortune themselves and as the new ruling class turned its back[7] on the prevalent art forms of India, the landed gentry, the new moneyed class in urban centres like Calcutta, turned patrons. As the 19th century wore on an assertion of national pride expressed itself through interest in the indigenous arts and classical music acquired a new 'halo'. Yet, with all that, patrons were few, jobs were limited and competition was strong. Members of different Gharanas made a large tribe with separate kin groups and each vied with the other for position and precedence. Mobility was not easy and, in any case, artists preferred a long, stable patron-client relationship. In some sense, the patronage system itself was a determining factor in the continuation of the Gharanas. In the absence of a broad base for performance, the performance source had to be kept restricted.

Service with a patron was subject to individual vagaries. Instances of artists being put on a pedestal and then ordered to step down are not rare either in India or in Europe. Awadh or Rampur, Dresden or Parma, a game of seesaw vis-a-vis the arts has been common. In centres where the patron had no particular interest in the arts and employed artists mainly for image, singers lived like the 'independent musician' of Europe. Where, however, the patron himself had musical interest or aptitude, music flourished according to the ruler's taste. In Vishnupur, Khyāl never made headway; Jaipur encouraged creativity through song-texts whereas the Banaras Maharajas gave a special boost to the art of the percussion. And in Rampur, all artists had to pay deference to the Nawab's teacher[8]. This in itself was not inimical to creative efforts but artists had fewer degrees of freedom. Europe followed a similar pattern for whether under Isabelle d'Este at Mantua or

---

[7] Orientalists like William Jones were interested in the ancient tradition of Indian music but there is no evidence of interest in the prevailing performance forms.

[8] Baroda is one instance where talented artists were employed more for prestige than actual interest. Court concerts were only on special occasions in front of the Maharaja and in his absence, to the unoccupied throne. This relative freedom allowed artists to concentrate on concerts outside. In contrast, the serious involvement of Rampur rulers in music restricted the lifestyle of artists, though the artistic atmosphere in the court itself was brilliant.

Frederick the Great in Prussia, court music had to follow the personal taste and accomplishment of the ruler[9]. In England, in addition to the ruler's taste, contemporary religious code influenced the extent and style of music[10].

The musician in Europe, although uncertain of his future and poorly paid, could at least try for a church benefice or some fees from tuition, but these doors were not open to his Indian counterpart. Those who did not secure a regular appointment visited different courts on festive occasions but such festivities were few and far between and local artists always tried to keep outsiders away. The only other road was to teach female singers, but that was socially taboo[11]. Private concerts were not common and generally not financially attractive. The most remunerative concerts were in the homes of women artists but again that had its social complications. Unlike in the West, there was no institution of impresario or agent which a retired musician or a minor one could turn to.

Pay was generally low; in the 1920's, while the one top musician in Baroda Court received Rs.150/-p.m., the salary for accompanists, junior artists, and young maestros of the future ranged from Rs.20/- to Rs.60/-p.m. There were no perquisites, only a gratuity on retirement. This may be comparable to the pay of present day young academies but the lack of job security, restricted opportunity for work and the responsibility of a large number of dependents kept living standards generally low. For example, court musicians could rarely afford a house or a carriage of their own. After late night concerts at the Palace most had to trudge home on foot. In Rampur, the pay was similar and although there was no pension or gratuity, extra benefits like housing, gifts on special occasions or sumptuous meals from the palace kitchen made up for it. Informants regret, however, that while court attachment gave artists a singular aura, the day the service ended, all connections with the court ended. Artists stayed on as long as possible and tried to ensure the job for the son when the father eventually retired.

Irrespective of where they lived, in urban centres or villages, Gharana families expressed much of the sentiments and codes of village society. As in the case of a rural patron-client relationship[12], an artist received psychological bonus and compensation from working for the same employer. He was aware that in case of an urgent need of money for an illness, a wedding or a funeral, a long and stable

---

9 (a) Fenlon, 1980; (b) Yorke-Long, 1954, ch.IV; (c) Haydn was a happy exception. Recalling his years at Esterhaza, he says, "as conductor ... I was free to alter, improve, add or omit and be bold as I pleased. Cut off from the world I had no-one to bother me and I was forced to become original". Robbins Landon, 1972, p.38.

10 (a) e.g., Disapproval of chromaticism, Mellors, p.89; (b) For an evaluation of the social and economic dependence of the musician on the attitude of the church, see Price, 1981, ch.IV. The relationship between the church as an employer and the composer as an employee was not always one of dominance. Before the 15th century the situation was more of laissez-faire than patron-client. Bowers in Fenlon, 1981, p. 15.

11 Female professional musicians were considered courtesans/entertainers, to be kept at strict social distance. Women of the gharana families were, therefore, forbidden to learn music even when they showed extraordinary talent.

12 Rudra, p. 254-255.

association with a patron would be of advantage and rarely tried to move away. For the great masters, changing centres was a political matter for court etiquette demanded formal acquiescence from the current employer before any move. Propriety has a kind of universality because this is reminiscent of the Este-Sforza quiet diplomacy and underlying competitive spirit over recruitment of artists in the late 15th century[13].

12.3 Training and Opportunities

Considering the restricted demand, the musician in India had to keep his wares continually polished; there was always a search for a new spark, always an attempt to bring a new dimension to his art. It was not peer group evaluation so much but peer group amazement that one was after. Families are full of tales about musical feats and competitions where, needless to say, one's own master won. In fact, myth-making was a hallmark of this era. In the absence of a standardised system of music and a reasonable yardstick of performance quality, attributed qualities like range, speed, triumph at soirées, length of practice, sheer personality, ability to start a fire or rain through the depth of a raga[14], were all a part of family legends.

Patronage of the arts by the privileged of the society is not special to India, it is a part of evolution of art almost anywhere. What made the Gharana unusual was its specificity to the family, its insistence on imparting the knowledge only to trusted kith and kin. As the reigning vocalist would have received training from a family elder, so he would transmit his knowledge only to a family junior, a son, a prospective son-in-law or a cousin. The enclosed Gharana tree of the Rampur Gharana gives an idea of the closed nature of the training[15]. Since the early Khyāl Gharanas were all Muslim where marriage within the family was the preferred custom, a family marital alliance as prerequisite for teaching created no special problems.

Once someone was accepted as a pupil, it was made very clear to him that a tradition of generations was being revealed; a tradition of voice culture, texts, raga analysis and all other inputs necessary for a finished performance. The young disciple knew he was entering an arena of responsibility and becoming a part of a rich legacy. To a hereditary singer, this legacy was sacred. The training was holistic, not just confined to the voice. There was guidance on diet, exercise, rest, service, loyalty and in the absence of an Indian Castiglone, on manners, courtly or otherwise. Stress was laid on 'adab' (manners) and 'dastoor' (propriety).

---

[13] (a) Lockwood, 1984,ch.XVII; (b) On losing a singer during her residence in Italy, Queen Christina of Sweden declared, "if he does not sing for me, he will not sing for long for anybody". Rosselli, 1992, p. 13.

[14] Ragas are supposed to create moods and even invoke the Gods. Legend, therefore, has it that the proper rendering of certain ragas can reach the elements and influence the atmosphere.

[15] See Appendix.

Loyalty and trust were of paramount importance; a lifelong commitment on the part of the student to the teacher, to the family, to the musical style was an integral part of the master-disciple relationship. This was ensured through ritualised ceremonies of discipleship such as the tying of the red thread as a symbolic blood-oath. The student was warned that a breach of this trust could lead to attracting the evil-eye with eventual loss of voice and other misfortune. In a cultural context where shamanistic practices, even now, are not unknown, the threat was real.

No monetary payment was made to the teacher during the student days but the student had to accept the responsibility of looking after the teacher and his family whenever needed and at whatever sacrifice. Apart from being a general help around the house, he had also to make a solemn promise of deferred payment on reaching the stage of an independent earner. Again, this is somewhat similar to the contractual system in 19th century Italy where talented pupils often undertook to pay a specified percentage of the take once they started earning. A well-known case is that of Enrico Caruso who in 1891 signed a contract of future payment with his teacher and literally bought 'himself out of it after his initial success'[16].

The training was oral: the master sang, the disciple followed; the master spoke, the disciple absorbed - no books, no notes. Young students woke up to the sound of music and practised the basics before dawn. As the day progressed, there were more sessions, either of active participation or of silent absorption. In between, the teacher explained certain rules or maybe certain intricacies of a raga or legends about the family. Visitors dropped in, sat on the floor and joined in, recalling old days and bygone masters. Through all this the strings of the tanpura[17] never stopped.

The intensive training, the need to be part of a lifestyle, necessitated that the student live with the teacher. In traditional Indian families with 'purdah' and other restrictions, it was difficult to admit non-family male members to the family fold. This was an added reason for accepting only relatives as students.

12.4 Property Rights and Innovations in an Oral Tradition

The primary reason for the restrictive teaching system was economic. In a world devoid of written records, there was no patent, no priority, no copyright, no royalty. The artist's knowledge was his intellectual property and without a formal coverage of rights, he had to guard it jealously. The ephemeral part of a performance would naturally disappear with each recital but the durable part had to be kept clear of trespassers. Insistence on secrecy and teaching only to trusted family was the only way, therefore, of avoiding misappropriation and ensuring rent. In a context where family wealth was passed down through the male line, the choice of a son as the preferred candidate for apprenticeship is not surprising. In a non-copyright situation one alternative may have been a state guarantee of regular

[16] Rosselli, 1991, p.31.

[17] A four-stringed drone instrument giving basic guidance on the pitch as also creating an atmosphere of serious music.

payment for creative work and open teaching[18]. This was tried in the state of Baroda in the 1920's but while the court had some of the greatest musicians of the country, the successful singers they nurtured were all sons, cousins or nephews, *not* outsiders. Trust possibly does not develop with public policy, nor custom change overnight.

The artist was always in a quandary, for disclosure was needed to impress a patron and secrecy was imperative to build an edge over the rivals. He, therefore, devised ways of expressing his vast knowledge while withholding information without seeming to do so. This he was able to accomplish precisely because of the 'oral' nature of his art. A basic tenet of raga music is that 'it is an art that must please' and there are many ways of pleasing a patron. One method was to avoid full disclosure of one's knowledge but give only a flavour of the possibilities. Complex virtuosity, difficult to copy, was an armour, as was slow, imaginative exposition, imagination being so personal. Misappropriation of the basic style or raga interpretation was uncommon because Gharanas had to keep their identity; the scramble was generally over texts. Since there was no accepted 'standard text' or a standard way of 'rendering a text', this was uncertain territory. Full compositions, therefore, would be presented only before the cognoscenti or on special occasions. Otherwise, improvisations would revolve around the first stanza and the second would either be suppressed or substituted[19]. Rare ragas, special to the artist's family, would be rendered only after he had established himself well at the court.

In many ways, the oral tradition resulted in greater musicality. Ragas changed over time or new ragas were developed as artists added their personal imagination to musical structures. Different Gharanas looked at the music from different perspectives and with no rigid restrictions on use of microtones or dominant/subdominant notes, raga images came almost like a marbled screen. While students were taught in strict Gharana discipline, on reaching solo status, within the given framework, they had the latitude to emphasise their comparative advantage and pursue their own conception of what was pleasing.

What led to enrichment also led to complications. There has always been a premium on the 'old' - an old composition, an uncommon raga, a rare ornamentation - something a family could claim to have been with them for centuries. Tansen (16th century) and Sadarang (18th century) were renowned names as composers and many of their songs have come down through generations. It is, however, not uncommon for later artists to set a poem to a raga, add one of these names to it[20] and then claim it as part of family legacy. Where no name appeared in the second stanza, it was possible to add a Gharana name and render it as one's own. To please a Hindu patron, a 'Krishna' or a 'Hari' would sometimes be substituted for the original 'Allah' and vice versa in case the patron was Muslim. The same song would at times have different claims of authorship

[18] Dasgupta and David, 1988, p.10-12.

[19] The song text has two sections, 'sthaii' - the first stanza and 'antara', the second.

[20] The second stanza often contains the name of the poet as a kind of signature.

leading to conflicts and disputes. With no certificate of authenticity, an undetected mix of the genuine and the imputed was bound to be; carbon dating was not possible. One socio-politically significant and unfortunate specific example is an old Agra Gharana song containing the signature of one Ibrahim, a Muslim poet. Some contemporary singers have substituted this Islamic part with two plain Hindi words, 'yé param'. In an unwritten oral tradition, modification for bias is not difficult.

A similar problem arose also with ornamentation and interweaving of the words with tune and rhythm. Khyāl is like an arabesque pattern based on some simple lines and credit for the intricate details was often claimed by many. While there was definite delineation of family styles, each would try to locate a bit of his own speciality in a rival's recital and often spread stories about stealthy borrowing by the competing family[21]. With no Royal Society acknowledging originality or allocating priority, Gharanas themselves tried to work out internal codes of behaviour. For example, a member of another family would not be accepted as a student, in the rare event that one wished to change sides, without formal permission of family elders, and inter-Gharana marriages would always be discouraged so as to prevent material crossing the lines. Texts would not be borrowed openly as that would imply a paucity of one's own material, and songs or raga made popular by specific families over time would be attempted only with adequate recognition. Artists at soirées would use extreme courtesy and decorum with one another even when their protégés argued outside about orders of precedence or tales of superiority and grandeur. Masters, despite the competitive spirit, displayed a sense of camaraderie, much like present-day academics; disputes generally arose through the excessive enthusiasm of disciples and admirers.

Unenforceable agreements present risk and require constant watch. The story of a 20th century diva highlights the problem. She refused to teach anyone throughout her life only to discover suddenly that her long-standing accompanist had learnt it all and initiated a Gharana himself!

It is interesting to see how Gharana singers coped with the changing attitude of the 20th century with its attempts to systematise the music. The pure scholar and the performer have generally not seen eye to eye, leading to the former's superior feeling of intellect and the latter's disdain for the one who stayed by the book with nothing to show through the voice. The Hindu-Muslim divide of this period helped widen the rift because the Brahminical scholar could not accept Islamic intrusions to the ancient Indian system and Muslim performers limited their horizon to the contribution of only the Muslim artists and scholars towards the

[21] One particular story has not changed in 700 years; Amir Khusrau in the 13th century is supposed to have hidden behind the Sultan's throne, listened to Gopal Nayak of the South and only then performed the musical feats he was famous for; in the 19th century, two young brothers repeated the trick, surreptitiously listened to the master of another gharana and mastered true virtuosity; and again in 1910's Fayaaz Khan of Agra Gharana learnt some ornamentation from Moijuddin Khan, an outsider, by hiding behind a screen in a Calcutta home.

development of Indian music[22]. By early 20th century there was a strong move by scholars to make the performance tradition a written one and bring the art out of its limited sphere by opening it for all and to all. The material now was mainly with Gharana artists who, as informants say, gave full courtesy and cooperation to the knowledgeable compilers, but divulged nothing of essence. To part with the family style was impossible from traditional, economic and emotional points of view; there was such fear of being dispossessed. The treasure may die for want of an heir but none but the authentic can reap benefit[23].

12.5 Families and Caste

Other than the interplay of supply and demand, there was a deep sociological justification for the Gharana. Despite the equality Islam professed, Islam in India was caste-ridden. Immigrant Muslims were of higher status and local converts, lower in the hierarchy, carried forward the Hindu caste system among themselves. One well-known singer of the early 20th century proudly claimed his Brahminical origin and regretted that the family forbade him from learning from a particular great master because of the latter's low caste. Both families were Muslim[24]. Besides, music itself had a questionable status. North Indian society enjoyed the performance, even respected artists but avoided social intercourse with them. Boys and girls of 'respectable' families were discouraged from listening to music, let alone learning.

The debate in Islam about the permissibility of music is unending. Gharana members, mainly devout Muslims, were often outcasts in their own community, specially in Rohilkhand where most artists came from. Singers, drum-beaters, were all necessary part of village rituals and a part of the Hindu jajmani system[25] but they belonged to the ritually low castes. Classical singing was not a part of this web of interconnected services but its practitioners could be descendants of such people or of itinerant musicians/mendicants and hence condemned to the lower rungs of society. Besides, the association of many artists with female professional singers made them doubly suspect, for the art of the female singer could be appreciated but the singers themselves had to be kept outside the margins of 'decent' society'[26]. Even between Gharanas, casteism operated. The number of

[22] Interview of Amir Khan, Jairazbhoi Collection.

[23] The 11th century traveller Alberuni, writing about the "peculiarities of (India's) national character" wrote that Hindus "are by nature niggardly in communicating that which they know, and they take the greatest possible care to withhold it from men of another caste". Alberuni, tr. Sachau, 1910, p.22.

[24] (a) Deshpande, 1979; (b) When radio broadcasts started in India in 1929, one well-known artist refused to perform because untouchable and lower castes would then hear the sacred music.

[25] Wiser, 1958.

[26] Peripheral to the serious platform of classical music, female singers formed a world of their own, passing on the art through female protégés, often daughters, actual or adopted. Patel, 1992.

generations a family had practised the profession marked its aptitude, legitimation and place in the cadre. Families kept elaborate genealogical tables to establish credibility. With the general social attitude towards professional musicians and caste-class conflicts among the Gharanas, it is doubtful that even if singers had left their doors open, many outsiders would have stepped in.

Considering the socio-economic context, the Gharana system was an intelligent model. Patrons and performers developed a kind of symbiotic relationship, each encouraging and reinforcing the other's vanity and fulfilling each other's needs. Certainly, the fear of peer group rivalry, uncertainty and limited scope of employment, kept artists constantly in search of fresh ideas, trying to make the next generation better and in general, trying to keep one step ahead of the competing families. To the extent rivalry produces better results, the Gharana system achieved it. It also developed a rare sense of loyalty and unity where the whole family was a mode of production and worked towards the making of an artist.

But there were negative fallouts on which not much has been written, fallouts from the isolation that the system imposed. While being a part of a specialised family gave certain advantages to the son or the cousin, talent is not family or gender specific; the closed circuit obviously kept many talents out and prevented market forces from bringing about a matching of talent and training. Besides, restrictive practices and conditional dissemination of knowledge led to an under-utilisation and erosion of the knowledge itself. In many cases, styles of embellishment and texts disappeared over the years with no scripts to enable a revival. Unlike in Western music, research into even 19th century forms can at best be a fair guess.

Generations of rivalry over music kept musicians apart even on matters of common interest. There has never been a musicians' union in North India and the one attempt in independent India to strike against the state broadcasting monopoly, the All India Radio, failed because of internal mistrust. The joint -stock companies floated by artists in Italy or the Accademia Filarmonica of Bologna would be unthinkable in the Indian situation. In an art form which is so personal and a relationship with a patron which is so individual, it is difficult for a Gharana artist to accept that a joint platform can lead to far greater welfare.

Gharana families tried to remain aloof from a changing society, the reality outside, with the consequence that members often lost a sense of reality about themselves - a true assessment of their ability and prospects. This was particularly true of those who did not succeed. Social history of music has researched on the well-known, the sought-after master, the Gharana wizard; surely, here too subalterns existed and one wonders what kind of economic organisation they formed. The subaltern, as far as possible, stayed on in the family profession. The better ones taught younger pupils, played the accompanying tanpuras and sang behind the teacher, often not graduating to solo status. Others looked after the family, the instruments, did household chores, sat around during practice sessions and generally lived with the pretence of music rather than actually making it. This large coterie gave rise to a considerable degree of disguised unemployment resulting in a far greater economic burden on the real earning members.

Socially this has left a deep scar. The non-productive group of people lived on a legendary past and unreal present, rarely searching for an alternative occupation. The somewhat non-conformist lifestyle of artists kept younger members away from prolonged formal education so that job opportunities were few. From the society's or even their own point of view the tragedy was that they got the wrong perspective on their place in society and when the umbrella over them was removed by heaven's prince or the earthly prince, they had nowhere to turn to. The great need for economic security and the resultant dependence on the patron gave the Gharana artists a kind of parasitic attitude that would not come to grips with reality even when sharp changes occurred in the society. In fact after 1947 when the musical climate of North India underwent dramatic transformation, Gharana members felt bewildered - they find it difficult even now to cope with a competitive talent market and patronage that comes from the public.

12.6 Contrasting with the West

It is often argued that since Western music has had systems of notation for over 3000 years[27], its evolution has been a total contrast to the Indian one where memory has been the primary method. This is possibly somewhat of an exaggerated generalisation, for cultural contexts within the West vary. Besides, the difference between Western and Indian art music is not just one of written/unwritten; they are divergent both in terms of musical concepts and socio-political backdrops.

In the West, patronage of classical music was not confined only to nobility but included the church and independent theatres. Choral singing, religious or secular, led to a demand for a large number of musicians who could be trained at institutions or by individual teachers. The composer, generally, has been a different entity from the performer, so while the former often had to adapt to the patron's taste and needs, the latter was more free. If hand-copied music restricted wider reach, music printing and publishing, possible by the 16th century, gave to a large public access to the great variety of music composed[28]. In Western music improvisation was a small part of a major notated whole; in India, it was the major part of a long unwritten whole[29]. While both traditions needed the guidance of a

[27] Rowell, 1992, p. 7.

[28] Musical texts were often printed in limited numbers but not published so that performances could be controlled. More interesting still, "In the 19th century full scores and instrumental parts for operas and large symphonic works were often printed but not published, so the owner could more effectively demand royalties or specific conditions of performance". Sadie, 1980, p. 260.

[29] Charles Burney on extemporisation, "vocal ornament which was never transmitted to paper, in the delivery of a few notes which lasted but a second; the genius squandered in the most evanescent performance, the memory of which died with those who had heard it". qtd. Lee, p.186.

master, Indian music with its non-tempered scale and microtones, narrow regions of different melodic structures[30], use of complex syncopated rhythmic variations, required intense and long association with one particular teacher, the Guru.

It is, therefore, not surprising that the propagation of this branch of knowledge has generally taken a different path in the West. Yet, there are areas where it is possible to find shades of similarity. Successful artists did not necessarily come from institutional training. Quantz, Frederick the Great's long-term musician, mentions in his journal his search for the right teacher and as late as the 1890's Caruso got his training in a master-disciple method.

The nearest approximation to an inbred organization has been the cori of the ospedali in Venice. These four institutions were exactly what they were meant to be, state sponsored orphanages or foundling homes. They flourished from 16th century onwards till the Napoleonic victory changed it all in 1797. Over the years, for whatever reason, economic, political, social or religious, music became central to their curriculum and in time, singers of the ospedali became part of the magic of Venice. Although eventually the inmates were no longer all orphans, nor teachers only from within, the basic rule was that senior girls taught the younger ones and the chorus worked under the best pupils of the ospedali. The inmates who did not get married stayed on, performed and taught till old age disabled them. Rules regarding training and performance were extremely strict and no one was allowed to become professional. Each institution specialised in different areas and was proud of its long association with teachers of the status of a Hasse or a Jombelli. It was a closed and controlled system with its own individual aesthetics; noting the similarity of surnames of musicians of the cori, one historian has wondered if the cori was not a training ground for the children of musicians or retirement homes for old artists[31].

Family tie has been a strong factor in the culture of Italy and as in India, for Italian singers too, music often came by way of inheritance[32]. Children of musicians were expected to take up music as a profession and were often taught by family elders. Where training was taken from outside, a father-child relationship was not uncommon. The long lineage of many artists or the complex interrelationship between families of artists is quite staggering. Puccini had four generations of musicians to inspire him, Rossini's father had been a trumpeter and a singer while Bellini had composer-organists as father and grandfather. "The family of Allesandro Scarlatti (1660-1725) - brothers, sisters, children, grandchildren -included opera singers and composers; they intermarried with a similar, less well-known family, the Uttinis, and a woman descended from this union gave birth in 1813 to Giuseppe Verdi"[33]. One could go on with such famous connections. Even among those not so well-known, "...family groups, typically made up of father, daughter, and son-in-law, or husband, wife, and wife's brother,

---

30 The same group of notes with subtle differences can denote several different ragas.

31 Baldauf-Bordes, 1993, p. 252.

32 Rosselli, 1992, p.92.

33 Ibid, p.92.

took the comic operas... around Italian towns..."[34]. Here, however, the similarity ends, for despite the importance of family heritage, it was possible for a carpenter (Francesco Marconi), a blacksmith (Titto Ruffo) and a horse trainer (Lorenzo Gibelli) to pursue a successful career in music in Italy.[35]

Whichever way an artist came to his profession, through family or otherwise, and whichever way he learnt his art, through an institution or an individual teacher, the 19th century musician in Italy operated under elaborate laws of contract safeguarding interests of the artists and organisers. Popular appreciation of the art gave a certain amount of market efficiency to the business of music creation. Despite what is called a period of cultural renaissance, the 19th century Indian artist preformed under very different conditions that made Gharana a necessity.

12.7 The Present Scene

With the start of broadcasting and now easy recording facilities, classical music has left the narrow domain of family art. Patronage is no longer confined to the privileged few; multinationals and Government institutions make it possible for organisations to arrange large music festivals. Institutions for the teaching of classical music are everywhere and talent permitting, anyone aspiring to be a professional singer can do so. Finding a teacher is not so difficult - and classical singing has ceased to follow strict Gharana styles. Jetsetting musicians are as common in India as anywhere in the West with successful musicians commanding adulation as well as high fees. The so-called 'Indian diaspora' to the West with its cross-cultural impact has created a fusion of Indian and Western music whether in the compositions of a Peter Maxwell Davies or in the symphonic attempts of Ravi Shankar.

The music, however, has changed. Precisely because it is still mainly unwritten and tacit knowledge forms such a large part of expertise and training that an open market leads to a certain dilution of the product. A wealth scattered is a wealth spread thin. Mass response is not necessarily an informed one or at best, is a mixed one and artists often need to make compromises to suit a varied taste. To the extent art forms reflect a changing society this is neither unforeseen nor unwelcome but it does lead to a confusion of genre. Khyāl singers today find it difficult to keep to its pristine quality and tend to incorporate idioms of popular culture. If the state or business houses decide to take a lead in the promotion of cultural forms, they have to take a decision about their ultimate aim. If preserving a tradition is at all a part of their objective, then the funding authorities have to create conditions to encourage the epistemic part of the genre. Otherwise technologists will become predators[36] and the back up of introspective creativity, an essential feature of classical music, will fade away.

---

[34] Ibid, p.93.

[35] Ibid, p. 100-102.

[36] Dasgupta and David, 1988, p. 51.

12.8 Appendix

Rampur Gharana (master-disciple relationship)

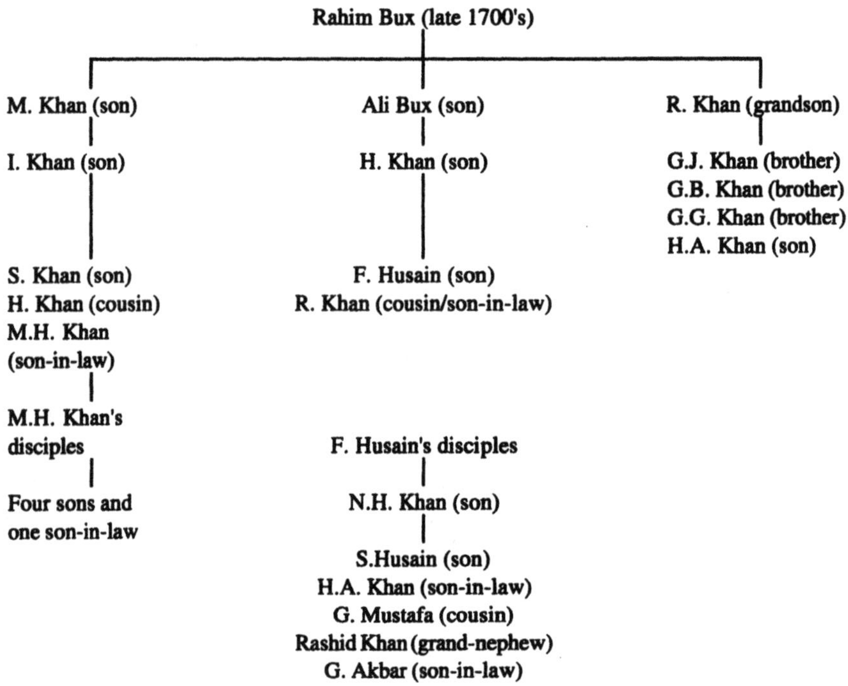

References

Al-Beruni's India, tr. E. Sachau (1910), Kegan, Paul, Tsench, Tribnur and Co. Ltd.

Baldauf-Bordes, J. (1993), "Women Musicians of Venice", Clarendon Press.

Bowers, R. (1981), "Obligation, Agency and Laissez-Faire: The Promotion of Polyphonic Composition for the Church in Fifteenth-Century England" in I. Fenlon (ed.), Music in Medieval and Early Modern Europe, Cambridge University Press.

Dasgupta, P. and David, P.A. (1988), "Priority, Secrecy, Patent and the Socio-Economics of Science and Technology", Centre for Economic Policy Research, Stanford University.

Deshpande, V. (1979), "Jaipur Gharana", Sajan Milap, Bombay.

Fenlon, I. (1980), "Music and Patronage in 16th Century Mantua", Cambridge University Press.

N.A. Jairazbhoi "Collection of tapes", Dartington College of Music, U.K.

Lee, V. (1970), "Studies of the Eighteenth Century in Italy", T. Fisher Unwin.

Lockwood, L. (1984), "Music in Renaissance Ferrara, 1400-1505", Clarendon Press.

Mellors, W. (MCML), "Music and Society in England and the European Tradition", Dennis Dobson Ltd.

O.H.I. of various Gharana singers and the late Satyendra Trivedi, by the author.

Patel, A. (1992), "Politics of Gender: Indian Classical Music", *Journal of Contemporary Thought*, Baroda.

Price, D.C. (1981), "Patrons and Musicians of the English Renaissance", Cambridge.

Robbins-Landon, H.C. (1972), "'Haydn", Faber and Faber.

Rosselli, J. (1991), "Music and Musicians in Nineteenth Century Italy", B.T. Batsford Ltd.

Rosselli, J. (1992), "Singers of the Italian Opera", Cambridge University Press.

Rowell, L. (1992), "Music and Musical Thought in Early India", University of Chicago Press.

Rudra, A. (1984), "Local Power and Farm-Level Decision-Making", in M. Desai, S.H. Rudolph and A. Rudra (ed.), *Agrarian Power and Agricultural Productivity in South Asia*, University of California Press.

Sadie, S. (1980) ed. "The New Grove Dictionary of Music and Musicians", MacMillan Publishers Ltd.

Sen, K.M. Bedottar Sangeet, Calcutta.

Widdess, D.R. (1981), "Aspects of Form in North Indian Alap and Dhrupads" in D.R. Widdess and R.P. Wolpert (eds.), Music and Tradition, Cambridge University Press.

Wiser, W. (1958), "Hindu Jajmani System", Lucknow.

Yorke-Long, A. (1954), "Music at Court; Four XVIIIth Century Studies", Widenfeld and Nicolson.

13. The International Transmission of Knowledge by Multinational Firms: Impacts on Source and Host Country Skilled Labor

James R. Markusen, Anthony J. Venables*

The effects of multinational firms are often assessed with casual arguments, sometimes tainted with political prejudices and nationalistic biases. In the 1960's and 1970's for example, there seemed to be a vague view among many developing countries that multinationals are exploitative in nature, and that they may undermine both economic development and political independence. The developed countries tended to take the opposite point of view, and naturally resisted developing countries' attempts at expropriation and excessive taxation. Beginning in the mid-1980's, these views seemed to reverse themselves. Many developing countries radically liberalized their foreign investment and foreign ownership policies, actively seeking to attract foreign capital. Direct foreign investment is now seem as a means not only to attract capital and provide jobs, but as a major source of technology transfer and learning-by-doing. Curiously, many developed countries (or at least political constituencies within those countries) have also reversed their view, now seeing multinationals as "exporting jobs" and thus a threat to domestic labor and engaging in "environmental dumping" to avoid high standards in their home countries.

The purpose of this paper is to take a limited step toward understanding these issues and evaluating the validity of the various positions. In order to do so, our first task is to summarize some of the empirical evidence about multinationals, so that we can begin from a model that fits the stylized facts.

Various studies have repeatedly linked multinational firms with the existence of "knowledge capital".[1] Empirical analyses find that multinationals are found in industries with (1) high levels of research and development relative to sales, (2) large numbers of scientific and technical workers, (3) new and/or technically complex products, (4) high levels of product differentiation and advertising. An indirect measure of knowledge capital or intangible assets is a residual value computed as the market value of a firm minus the value of tangible assets such as plant and equipment. Multinationals tend to be firms with high values for this variable, which is generally interpreted as knowledge capital or other firm specific assets. These assets range from fairly tangible ones such as blueprints, patents, and chemical formulae to intangible one such as organizational know-how and

* University of Colorado, NBER and CEPR; London School of Economics and CEPR

[1] Surveys of both theoretical and empirical research are found in Markusen (1995) and in Caves (1996). See also Brainard (1993a,b,c) and Ekholm (1995).

reputations for product quality. Conversely, these studies do not associate multinationals with high levels of physical capital intensity.

This association of multinationals with knowledge capital rather than physical capital makes sense for a couple of reasons. First, the managers and engineers are fairly mobile in that they are able to visit multiple, geographically separated plants at relatively low cost and even at short notice. Second, knowledge capital often has a joint-input or public-goods property in that it can be applied to additional production facilities at very low marginal cost. Thus blueprints or chemical formulae may require substantial initial investments, but can then be provided to additional plants at very low cost. Neither property is possessed by physical capital. Finally, the public-good property just mentioned also helps explain why firms transfer the services of these assets internally by acquiring or creating foreign subsidiaries rather than through arm's-length arrangements such as licenses. The same property that makes it easy to transfer a blueprint to a foreign plant makes its value easy dissipated to others outside the firm. Internal transfers are viewed as important for maintaining security for the firm's knowledge capital, including intangible forms of this capital such as reputations. This theory is supported by empirical evidence which, for example, shows that internal transfers are preferred to arm's-length transfers for products or processes which are new and/or technically complex.

What are the general implications of this view for understanding the consequences of direct investment? The first answer is that the knowledge-capital view of multinationals is helpful for understanding the nature of what is traded. Multinationals can be thought of as exporters of knowledge-intensive producer services to foreign subsidiaries. Multinationals export engineering, managerial, and marketing services through various channels ranging from blueprints to branch-plant visits by engineering and managerial personnel. In many cases, these service exports substitute for trade in goods, those goods being produced by the subsidiary instead of being imported from the multinational's home country. These investments are generally referred to as "horizonal" investments and often occur between countries of similar income levels (e.g., Ford and General Motors producing cars in Europe). In other cases, the service exports are complementary to trade in goods, a situation which often occurs with respect to investments in developing countries. Examples include electronics assembly plants located in low-wage countries, most of whose output is exported back to the multinational's home country with intermediate inputs shipped in the opposite direction.

This conceptual model then sets the stage for discussing the question of what are the effects of multinationals? This question requires however a well-defined counterfactual. We will assume for the purposes of this paper that the counterfactual is a situation where multinationals would exist in equilibrium, but they are simply excluded by some sort of prohibited investment ban. The effects of multinationals are then defined as what occurs with and without such a ban. An alternative would be to consider parameter changes (e.g., trade barriers) such that initially multinationals do not exist in equilibrium, but enter production following the parameter changes. We will not consider this latter possibility since it is somewhat difficult to interpret the meaning of such an exercise: what is the effect of multinationals versus the effect of the parameter changes per se?

The next section of the paper adapts our two-country, two-sector, two-factor static model developed in Markusen and Venables (1995 a,b, 1996). The factors are skilled and unskilled labor. One sector of the economy (X) is composed of three distinct activities: (1) creating firm-specific capital, (2) creating plant-specific capital, and (3) final production. Skilled labor is required in the production of firm-specific knowledge capital ("blueprints", etc.) and is also combined with unskilled labor in the creation of plant-specific fixed costs (producing capital equipment). Production of final output of X requires only unskilled labor. Our other work focusses almost entirely on the determinants of what types of firms are active in equilibrium and trade-volume issues, assuming that all three of these activities use factors in the same proportion, or indeed use only one composite factor.[2]

There are two countries, and the nationality of a firm corresponds to the location of its firm-specific capital. There are four firm types. There are two types (nationalities) of single-plant firms, corresponding to locations in the two countries. Each single-plant firm may or may not export to the other country. There are similarly two types (nationalities) of multinationals, each maintaining plants in both countries. The term "regime" denotes the set of firm types active in equilibrium. We review, with appropriate modifications, our earlier results as to how differences between countries in relative endowments of skilled labor and in country size (absolute endowments), along with trade barriers, determine the equilibrium regime.[3]

Multinationals fragment production geographically, locating skilled-labor-intensive activities and one production plant in their home country and an additional production plant in the host country. To put it differently, these firms change the nature of what is traded, from commodity exports to skilled-labor-intensive producer services.

Knowledge capital is exported to subsidiaries in exchange for repatriated profits, which in general-equilibrium take the form of exports of other goods and services back to the multinational's home country.

---

[2] Technologies and skill levels will be taken as exogenous in this paper. We are thus ignoring important questions concerning skill and knowledge "spillovers" from multinationals to local firms and workers. See Blomstrom and Kokko's article from this volume for an excellent review of existing work in this area.

[3] The model thus focusses on horizontal multinationals, and is not general enough to treat firms which, for example, export intermediate goods (e.g., computer chips) to branch plants for assembly into final products for shipment back to the source country. This latter type of vertical direct investment is not nearly so quantitatively important overall as horizontal direct investment. But it is of course of great importance from the point of view of some individual developing countries with major assembly operations (Singapore, Mexico, Malaysia).

Two questions arise. First, under what conditions do multinationals dominate in equilibrium and second, where are multinationals headquartered? Results show that multinationals are more dominate in equilibrium when the two countries are larger overall, and more similar in absolute economic size and in relative factor endowments. Thus the model predicts that multinationals are likely to be more important in the economic activity among the high-income developed countries than between those countries and lower-income developing countries. This implication is strongly supported by the evidence, including studies by Brainard (1993a,b,c) and Ekholm (1995).

The reason that direct investment is more important relative to trade for larger economies is that direct investment is a high fixed-cost option in the model whereas serving a foreign market by exports is a high variable-cost option. The high fixed-cost option is preferred the larger the markets. The reason that direct investment is more likely relative to trade as the two markets are more similar in size is a bit more subtle. Multinationals must invest the same plant-specific fixed cost in both markets, whereas national firms need only make this investment in one market. If the markets are of very different sizes, then multinationals are making a costly investment in a small market thereby saving only a small amount in transport costs. National firms located in the large country have a competitive advantage. Most of their sales are in their low-cost (no transport cost) market with only small additional costs incurred in exporting to the small market.

The reason that multinationals are more likely to dominate between countries with similar relative factor endowments is conceptually related. X-sector production is skilled-labor intensive relative to Y-sector production in the model, and multinational firms are skilled-labor intensive relative to national firms. If one country is very scarce in skilled-labor, then a situation with multinational firms only would make skilled labor very costly in that country. That would put multinational firms at a disadvantage relative to national firms located in the skilled-labor-abundant country, the latter locating their skilled-labor demands only in the country where skilled labor is abundant. Thus multinationals tend to disappear in the model when the countries are very different in their relative endowments. This is also consistent with the empirical evidence noted above that direct investment is much more important relative to trade among the high-income (skilled-labor-abundant) countries than it is between these countries and developing (unskilled-labor-abundant) countries.

With respect to the second question posed above, multinationals tend to be headquartered in the skilled-labor-abundant country, but ownership is distributed between the countries in proportion to their size when they different only in size, not in relative endowments. Again, this seems to fit well with casual empiricism, including the observation that some small but skilled-labor-abundant countries such as The Netherlands, Sweden, and Switzerland are home to many multinationals.

A question of major interest in the paper is the effects of multinationals on skilled labor in both source and host countries. This question is important in its own right, and of great important to subsequent dynamic work as to how multinationals encourage or discourage the accumulation of labor skills. We consider first two countries differing in relative factor endowments, but similar in size. We find that the entry of multinationals due to the removal of an investment barrier generally raises the real wage of skilled labor and also the skilled-unskilled wage gap in the skilled-labor-abundant country.

The effects of multinational entry on the real wage of skilled labor and the skilled-unskilled wage gap in the skilled-labor scarce country are ambiguous. On the one hand, MNEs export skilled-labor-intensive producer services (knowledge capital) that indirectly substitute for host-country skilled labor, thereby lowering the latter's relative return. On the other hand, national firms in the disadvantaged country may be severely handicapped by the scarcity of skilled labor so that few or possibly zero of them can enter when investment is banned. In such cases, the entry of MNEs headquartered in the advantaged country increases the demand for skilled labor in the disadvantaged country and raises the skilled-labor wage. The latter possibility is more likely when the difference in the relative endowments of the two countries is large and when skilled labor requirements in branch-plant fixed costs are significant. The entry of multinationals then raises the real wage of skilled labor in both countries; intuitively, branch plants requiring significant skilled-labor inputs replace transport costs which use only unskilled labor.[4]

The effects of multinational entry when the countries differ in absolute size is somewhat different. The effects on the wage of skilled labor are that the small country's skilled-labor wage rises and the large country's wage falls. Both results can be understood in terms of effects identified in the trade-industrial-organization literature. In the absence of multinationals, national firms headquartered in the advantaged country have a significant home-market advantage that allows them to dominate in equilibrium out of proportion to country-size differences. The entry of multinationals shifts ownership and therefore the demand for skilled labor more towards the smaller country. Thus the ability of all countries to engage in direct investment is particularly important for the smaller but skilled-labor-abundant countries such as The Netherlands, Sweden, and Switzerland.

Overall, the results are encouraging for smaller, skilled-labor-scarce countries, at least if the objective is to raise the return to skilled labor, thereby encouraging skill upgrading in the long run. Provided that host-country skilled labor is important in branch-plant fixed costs, the entry of multinationals raises the demand for skilled

---

[4] Rising skilled-unskilled wage gaps in both north and south have been discussed, modelled, and hypotheses tested in a number of recent papers, including Feenstra and Hanson (1995a,b), Cragg and Epelbaum (1995), and Berman, Machlin, and Bound (1995). Feenstra and Hanson model a continuum of intermediate inputs for an industry, ranked in terms of skilled-unskilled labor intensity. Investment liberalization moves capital to the south, and shifts the dividing line in the continuum in favor of more inputs produced in the south. The shifted inputs are skilled-labor intensive in the south, but unskilled labor intensive from the north's point of view, hence the relative demand for skilled labor rises in both countries. Cragg and Epelbaum focus on capital mobility and skill-biased technical change while Berman, Machlin, and Bound also find evidence of skill-biased technical change.

labor in skilled-labor scarce countries. The increase in multinationals' demand for engineers, technicians, and managers dominates the possible "substitution effect", in which imports of skilled-labor-intensive producer services substitute for host-country skilled labor. A related effect occurs when the countries differ in size. The removal of investment barriers leads to the loss of the home-market advantage effect for the large countries, leading to a transfer of production and firm ownership to the smaller countries, raising the demand for skilled labor in the latter.

13.1 Model Structure

The model has two countries (h and f) producing two homogeneous goods, Y and X. There are two factors of production, L (unskilled labor), and S (skilled labor). L and S are mobile between industries but internationally immobile. Y will be used as numeraire throughout the paper. Skilled labor is used for the firm-specific fixed cost of producing X, and plant-specific fixed costs use a combination of the two labor types. Unskilled labor is used in variable costs, and in addition there are transport costs between countries, specified as units of unskilled labor per unit of X exported.

Subscripts (i, j) will be used to denote the countries (f, h). The output of Y in country i is a CES function, identical in both countries. The production function for Y is

$$(1) \qquad Y_i = (aL^{\varepsilon}_{iy} + (1 - a)\, S^{\varepsilon}_{iy})^{1/\varepsilon} \quad i = h, f.$$

where L_{iy} and S_{iy} are the unskilled and skilled labor used in the Y sector in country i. The elasticity of substitution $(1/(1-\epsilon))$ is set at 5.0 in the simulation runs reported later in the paper.

Superscripts (n,m) will be used to designate a variable as referring to national firms and multinational firms respectively. (m_i, n_i) will also be used to indicate the number of active m firms and n firms based in country i. Hopefully, it will always be clear from the context what is being represented (e.g., n_i as a variable in an equation always refers to the number of national firms in country i).

In order to enter X production with one plant, a firm must incur a fixed cost in units of skilled labor, denoted F, and a fixed cost in units of unskilled labor G: national-firm fixed costs are thus $L_i = G$, $S_i = F$.

A two-plant multinational headquartered in country i incurs additional fixed costs in both countries. These include both skilled and unskilled labor costs in the branch-plant in country j, and possibly additional skilled-labor costs in the source country i. Total fixed costs for a two-plant multinational headquartered in country i are:

$$L_i = G \qquad L_j = \beta G, \qquad S_i = F + \gamma F, \qquad Sj = \delta F, \qquad 1 \geq \beta > (\gamma + \delta)$$

The inequality on the right expresses the assumptions that the second plant is more unskilled-labor intensive than the first, and that there are multi-plant economies of scale arising from the joint-input nature of knowledge capital. γ can be thought of as a technology-transfer cost. Later in the paper we report results for two simulations, ($\beta = 1.0, \gamma = \delta = 0.1$) and ($\beta = 0.75, \gamma = 0.1, \delta = 0.4$).[5]

Marginal factor requirements are constant in units of unskilled labor.[6] X^n_{ij} denotes the sales in country j of a national firm based in country i. Let w_i, and v_i, denote the prices of unskilled labor and skilled labor respectively in country i. A national firm undertakes all its production in its base country, so the cost function of one national firm in country i is given by

$$\text{(2)} \qquad w_i L^n_i + v_i S^n_i = w_i\,[c X^n_{ii} + (c + \tau) X^n_{ij} + G] + v_i F, \qquad i, j = h, f, \quad i \neq j$$

where c is the constant marginal production cost. c, F, and G are identical across countries. τ is the amount of unskilled labor needed to transport one unit of X from country i to country j, which we assume to be the same in both directions. In our calibration, national firms in the X sector are moderately more skilled-labor intensive that Y-sector firms. However, this is not particularly important in our model since we are focussing on the switch between national and multinational X-sector firms.

A multinational based in country i has sales in country j, X^m_{ij}. It operates one plant in each country incurring fixed costs, (G_i, $(1+\gamma)F_i$) in its base country, and fixed costs (βG_j, δF_j) in country j. Sales are met entirely from local production not trade. L^m_{ij} (S^m_{ij}) denotes a country i multinational firm's demand for unskilled (skilled) labor in country j. A firm type m_i thus has a cost function

$$\text{(3)} \qquad w_i L^m_{ii} + w_j L^m_{ij} + v_i S^m_{ii} + v_j S^m_{ij} = w_i\,[c X^m_{ii} + G] + w_j\,[c X^m_{ij} + \beta G] + v_i (1 + \gamma) F + v_j \delta F$$

In our calibration, multinational firms are generally more skilled-labor intensive than national firms, using more skilled labor for branch-plant fixed costs versus

---

[5] Reiterating what we noted above, our earlier papers either assume that all X-sector activities use factors in the same proportion (Markusen and Venables, 1995b) or use just a single factor (Markusen and Venables, 1996, 1995 a). Labor-market effects are only of second-order importance in those papers and not addressed. The technology-transfer cost (γ>0) is motivated by empirical results, especially those of Teece (1977, 1986), that direct investments require significant further investments in skilled-labor-intensive activities for multinational firms.

[6] Skilled workers could be used in X production but they enjoy no productively advantage over unskilled labor: engineers are no better on the assembly line than unskilled workers. Since skilled workers are differentiated from unskilled workers in the composite Y sector, all "excess" skilled workers not needed in fixed costs will be allocated to the Y sector in equilibrium.

the additional unskilled labor for transport costs used by national firms. This depends, however, on firm scale.[7]

Let $\bar{L}_i$ and $\bar{S}_i$ denote the total labor endowments of country i. Adding labor demand from n_i national firms, m_i multinationals based in country i, and m_j multinationals based in country j, gives country i factor market clearing:

$$\text{(4)} \qquad \bar{L}_i = L_{iy} + n_i L^n{}_i + m_i L^m{}_{ii} ++ m_j L^m{}_{ji}$$
$$\bar{S}_i = S_{iy} + n_i S^n{}_i + m_i S^m{}_{ii} ++ m_j S^m{}_{ji}$$

In equilibrium, the X sector makes no profits so country i income, denoted M_i, is

$$\text{(5)} \qquad M_i \bar{L}_i + v_i \bar{S}_i \qquad i = h, f$$

p_i denotes the price of X in country i, and X_{ic} and Y_{ic} denote the consumption of X and Y. Utility of the representative consumer in each country is Cobb-Douglas,

$$\text{(6)} \qquad U_i = X^{\alpha}{}_{ic} Y^{1-\alpha}{}_{ic}, \quad X_{ic} \equiv n_i X^n{}_{ii} + n_j X^n{}_{ji} + m_i X^m{}_{ii} + m_j X^m{}_{ji}$$

giving demands

$$\text{(7)} \qquad X_{ic} = \alpha M_i / p_i, \quad Y_{ic} = (1-\alpha) M_i$$

Equilibrium in the X sector is determined by pricing equations (marginal revenue equals marginal cost) and free-entry conditions. We denote proportional markups of price over marginal cost by $e^k{}_{ij}$, (k = n, m), so, for example, $e^m{}_{ji}$ is the markup of a country j multinational in market i. Pricing equations of national and multinational firms in each market are (written in complementary-slackness form with associated variables in brackets):

$$\text{(8)} \qquad p_i\,(1-e^n{}_{ii}) \le w_i c \qquad (X^n{}_{ii})$$
$$\text{(9)} \qquad p_j\,(1-e^n{}_{ij}) \le w_i\,(c + \tau) \qquad (X^n{}_{ij})$$
$$\text{(10)} \qquad p_i\,(1-e^m{}_{ii}) \le w_i c \qquad (X^m{}_{ii})$$
$$\text{(11)} \qquad p_j\,(1-e^m{}_{ij}) \le w_j\, c \qquad (X^m{}_{ij})$$

In a Cournot model with homogeneous products, the optimal markup formula is given by the firm's market share divided by the Marshallian price elasticity of

---

[7] An exception occurs when the two countries are *very similar* in our "Case 1" where skilled labor is of minor importance in branch-plant fixed costs. The removal of an investment barrier leads, in equilibrium, to multinational firms with significantly higher output per firm than the national firms they displace, a type of pro-competitive effect. The difference in equilibrium firm scale is enough to make the multinational firms less skilled-labor intensive than the displace national firms. In "Case 2" (more S used in branch-plant fixed costs) this result is reversed. In any case, when the countries become moderately different, these effects on factor prices are dominated by the shifting of production versus R&D activities between countries, which is more the focus of this paper than firm scale.

demand in that market. In our model, the price elasticity is one (see equation (7)), reducing the firm's markup to its market share.

This gives, (also using demand equations (7)),

$$(12) \qquad e^k_{ij} = \frac{X^k_{ij}}{X_{jc}} = \frac{p_j X^k_{ij}}{\alpha M_j} \qquad k = n, m \qquad i, j = h, f$$

There are four zero-profit conditions corresponding to the numbers of the four firm types. Given equations (8)-(11), zero profits can be written as the requirement that markup revenues equal fixed costs.

$$(13) \qquad p_h e^n_{hh} X^n_{hh} + p_f e^n_{hf} X^n_{hf} \leq w_h G + v_h F \qquad (n_h)$$

$$(14) \qquad p_f e^n_{ff} X^n_{ff} + p_h e^n_{fh} X^n_{fh} \leq w_f G + v_f F \qquad (n_f)$$

$$(15) \qquad p_h e^m_{hh} X^m_{hh} + p_f e^m_{hf} X^m_{hf} \leq w_h G + v_h (1+\gamma) F + w_f \beta G + v_f \delta F \qquad (m_h)$$

$$(16) \qquad p_f e^m_{ff} X^m_{ff} + p_h e^m_{fh} X^m_{fh} \leq w_f G + v_f (1+\gamma) F + w_h \beta G + v_h \delta F \qquad (m_f)$$

To summarize the X sector in the model, the eight inequalities (8)-(11) are associated with the eight output levels (two each for four firm types), the eight equations in (12) are associated with the eight markups, and the four inequalities in (13)-(16) are associated with the number of firms of each type. Additionally goods prices are given by (7), income levels from (5) and factor prices from factor market clearing equation (4) together with labor demand from the Y sector.

The model is quite complex and inherently involves inequalities making traditional analytical, comparative-statics methods of limited value. The problems introduced by inequalities are compounded by the fact that we have four different production activities (Y, X-sector output, national-firm fixed costs, multinational-firm fixed costs), all using factors in different proportions. In the next section, we will try to gain some intuition by reviewing analytical results from our earlier papers, and later in the paper simulate the full model which involves 51 non-linear inequalities.[8]

13.2 Forces at Work

In our related papers (Markusen and Venables, 1995 a,b, 1996) we were able to make considerable analytical progress, although partly at the expense of assuming the same factor intensity in all of the X-sector activities. Rather than re-derive some of the previous results here, we will simply explain the intuition behind them, and conjecture on some extensions to the case of differing factor intensities

[8] Reviewing points made separately earlier, multinational firms are moderately more skilled-labor intensive than national firms which are moderately more skilled-labor-intensive than Y production in our calibrations. But many of the interesting factor-market effects that we identify are driven by geographical rearrangement of activities when multinationals are allowed to enter rather than by the skilled-labor intensities of type-n firms versus type-m firms versus Y production per se.

considered in this paper. The focus of the section will be on the equilibrium regime as a function of relative and absolute factor endowments, and trade costs.

Suppose that the two countries are absolutely identical. Suppose also that we calibrate fixed costs and trade costs so that all four firm types could exist together in equilibrium. The higher total fixed costs of national firms are exactly balance by the higher trade costs of the national firms. Our previous work then establishes the following results.

(a) An increase in both countries' size ($\Delta M_h = \Delta M_f > 0$) will imply that only multinational firms exist in equilibrium. Large markets favor high fixed-cost multinational firms over high variable-cost national firms.

(b) An increase $v\Delta F = -w\Delta G > 0$ will imply that only multinational firms exist in equilibrium. A shift in fixed costs from plant-specific to firm-specific costs favors two-plant firms.

(c) An increase in trade costs will imply that only multinational firms will exist in equilibrium. High trade costs favor two-plant firms which pay no trade costs.

We thus have the result that multinationals are more likely to exist in equilibrium when incomes are high, firm-specific scale economies (fixed costs) are more important relative to plant level scale economies, and trade costs are relatively high.

Now consider the two countries differing in relative endowments, and assume that the countries are large, F is large relative to G, and trade costs sufficiently high that multinationals dominate in equilibrium when the two countries are initially identical. As the countries begin to diverge somewhat in relative endowments, with h becoming more skilled-labor abundant, the regime should stay fixed at (m_h, m_f). Firm numbers should adjust with more firms headquartered in country h, but there is no reason to expect the regime to shift for relatively small differences. But eventually relative factor prices in the two countries will begin to diverge along with the divergence in relative endowments. Holding the regime fixed, skilled labor will become more expensive in the skilled-labor-scarce country. This in turn disadvantages multinational firms, which must purchase skilled labor in both countries, relative to (potential) national firms headquartered in the skilled-labor-abundant country. At some point, type-n_h firms will enter and type-m_f firms will exit (not necessarily at the same parameter values) leaving a regime of type-m_h and/or type-n_h firms in equilibrium. While production in country f may continue, we would expect all firms to be headquartered in country h for sufficiently large differences in relative endowments.

Next, consider the effects of country size, making country h larger and country f smaller. Assume again that initially multinational firm only exist in equilibrium. As the country sizes begin to diverge, there is no inherent disadvantage created for type-m_f multinationals, although they should be fewer in number. If factor prices are the same in the two countries (consistent with $0 < m_f < m_h$), then type-m_f and type-m_h firms are in fact in a perfectly symmetric position. Each incurs the same total fixed costs, and produces the same amount in a given country's plant, with simply a larger output in the country-h plant than in the country-f plant.

As the country sizes diverge further, type n_h firms gain a potential advantage. Their (potential) sales are concentrate in their low-cost (zero transport cost) market, whereas multinationals must incur fixed costs in country f to serve an ever shrinking market there. Eventually, some type-n_h firms should be able to enter in competition with the multinationals.

Eventually, there will come a point where only type-n_h firms can exist in equilibrium. As country f's size and therefore demand becomes very small, it is simply not profitable for a firm to build a plant in f, regardless of whether or not skilled labor is used in plant fixed costs. Markup revenue cannot cover fixed costs.

Taken as a whole, these arguments suggest that country h firms will dominate when the countries are very different in size or in relative factor endowments, there will be a mixed regime (both in types and nationalities) as the countries converge, and eventually only multinationals of both countries will exist provided transport costs and/or income levels are sufficiently high.

It seems rather obvious that skilled-labor-intensive activities, which translate here into firm ownership, should be concentrated in the skilled-labor-abundant country, although the role of country size is perhaps less obvious. But the concentration of firm ownership in the large and/or skilled-labor-abundant country does not lead to any simple predictions about the effects of multinational entry due to the removal of an investment ban. For example, the entry of multinationals may decrease the X-sector demand for skilled labor in the large or skilled-labor abundant country h as branch plants are built in country f and/or type-m_f firms are able to enter. Or alternatively country h X-sector skilled-labor demand could increase as type-m_h firms replace type-n_f firms. Similarly, intuition suggests contradictory effects of multinational entry on skilled-labor demand (and hence wages) in host countries as we noted earlier. Exports of skilled-labor-intensive producer services by multinationals to branch plants may substitute for host-country skilled-labor demanded by local firms. But if there are few or zero local firms in the absence of multinationals, then multinational entry is likely to correspond to an increased demand for local skilled labor. With strong analytical results unavailable, we thus turn to simulation analysis in the next two sections.

13.3 Entry of Multinationals due to Investment Liberalization: Countries Differ in Relative Endowments

In this section, we assume that countries differ in relative endowments, but that they are of the same size, meaning that their GDP levels are about the same. A numerical model using Rutherford's (1994) non-linear complementarity algorithm solves a model of 51 non-linear equations and inequalities. Results for two simulations are presented. In case 1, $\beta = 1.0$, $\gamma = \delta = 0.1$: a second (branch) plant requires the same unskilled labor as the first plant, and 20% as much skilled labor as the first plant, 10% from each country. In case 2, $\beta = 0.75$, $\gamma = 0.1$, $\delta = 0.4$: a second plant requires 75% as much unskilled labor as the fist plant and 50% as much skilled labor, 10% from the country of headquarters and 40% in the country in which the branch plant is located. If the S/L ratio in Y is normalized to 1.0, the calibration implies an overall S/L ratio in the X sector between about 2.1 and 2.6 in case 1 depending on what firm types are active in equilibrium.[9]

Figure 1 presents qualitative results on the equilibrium regime for the relative endowment experiment, case 1. The countries are identical in the top row and their difference is maximized in the bottom row. The units of measure on the vertical axis is country f's endowment of skilled labor as a proportion of its initial endowment. Moving down a column, S is transferred from country f to country h, and L is transferred in the opposite direction in amounts needed to maintain approximate parity in the two countries GDP levels.[10] The horizontal axis measures the transport cost variable t as a proportion of marginal production costs (both use unskilled labor only). Results would be similar using a tariff variable except for a relatively small income effect. When the two countries are relatively similar and trade costs are low, the equilibrium regime involves intra-industry trade competition between national firm types n_h and n_f (region b). When trade costs are higher, the regime involves intra-industry investment competition between type-m_h and m_f multinational firms (region i).

As the differences in the relative endowments become larger (e.g., $\bar{S}_f \leq 0.6$ its original level) type-m_h firms can enter even at very low trade costs (region c), because low plant-specific costs in f (requiring cheap L_f) balance the low shipping costs enjoyed by type n_h firms (requiring expensive L_h). m_h firms are competitive with n_f firms because the former use more (inexpensive) S_h while the latter use more (expensive) S_f. At higher transport costs, only type m_h firms are supported in equilibrium (region h). As the difference becomes extreme, n_h firms can enter even at high trade costs because they use no S_f. Moving down a column for moderate transport costs, such as $\tau = .15$, we see a sequence about the same as that suggested in the previous section.

$$(m_h, m_f) \rightarrow (m_h) \rightarrow (m_h, n_h)$$

In order to understand exactly how multinationals affect variables of interest, we re-ran the model suppressing multinational firms (e.g., a prohibitive investment barrier), comparing the resulting values with those in which multinationals are endogenous. Figure 2 plots ratios of variables with and without multinationals for case 1, where the subscript n indicates "no multinationals". v_i now denotes the real wage of skilled labor in country i, the wage divided by the consumer price index (the unit expenditure function). r_i denotes the skilled-unskilled wage ratio v_i/w_i in

[9] If multinationals are banned, this can result in inefficiently many firms in protected markets producing at small scale, resulting in a significantly higher S/L ratio in the X sector as we shall note later.

[10] Note that the vertical axis of Figure 1 gives S_f as a proportion of its initial value. Skilled labor is transferred to country h holding the world endowment constant. Thus the row 0.4 in Figure 1 has a ratio of $S_f/S_h = 0.4/1.6$, or .25. The vertical axes in later regime diagrams (Figures 3, 5 and 7) are similarly defined.

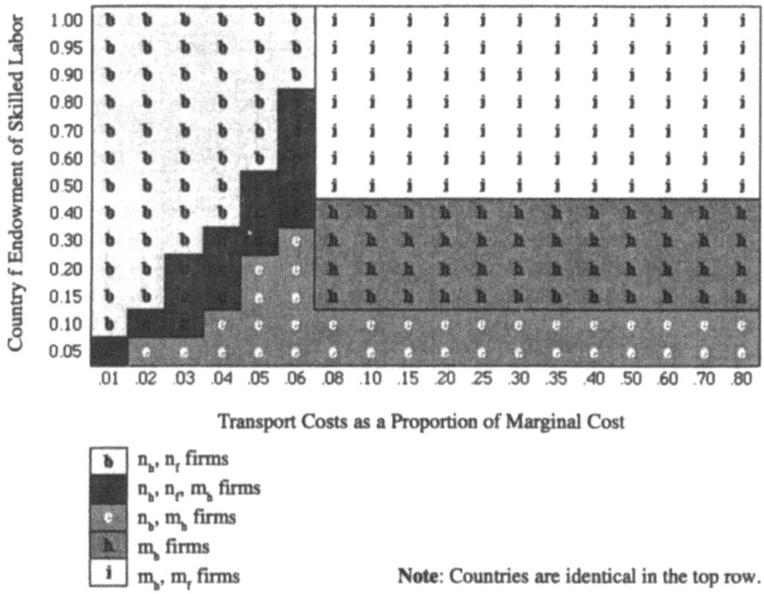

Figure 1. Relative Endowment Experiment.
Case 1: Low Skilled-Labor Requirement in Second Plant.

country i. The values run down the column for $\tau = .15$ in Figure 1, so countries are identical on the left-hand side of each panel, and their relative endowment differences maximized on the right-hand side.[11]

Looking at the left-hand side of each panel (countries identical) of Figure 2, we see that the removal of an investment barrier has virtually no effect on outputs in the two countries, but drops the real wage of skilled labor and skilled-unskilled wage ratio about 10% in both countries. This is an effect alluded to earlier in the paper: even though type-m firms are skilled-labor intensive relative to type-n firms at common scale, the removal of an investment barrier has a pro-competitive effect in which the multinational firms have a significantly higher output per firm than the national firms they displace. This reduces the skilled-labor intensity of the X sector and causes v_h and v_f to fall.[12]

[11] Apologies for the different scales for the two ratios in each of the panels of Figure 2. The fact that one of the ratios generally changes much more than the other makes the change in the smaller one difficult to evaluate if the two ratios are plotted on the same scale. We calibrated the two scales such that the ratios have the same position when the countries are initially identical and the ratios are equal.

[12] At the trade cost $\tau = .15$, the output per multinational firm is 48% higher than the national firms they displace. Normalizing the initial S/L ratio of the Y sector at 1.0 in Figure 1 when multinationals are present at $\tau = .15$, the S/L ratio in the X sector (and therefore for an individual X firm) is 2.12 with multinationals present, and 4.67 with multinationals suppressed. The change in firm scale overwhelms the fact that multinationals are more skilled-labor intensive at a common output level.

Figure 2

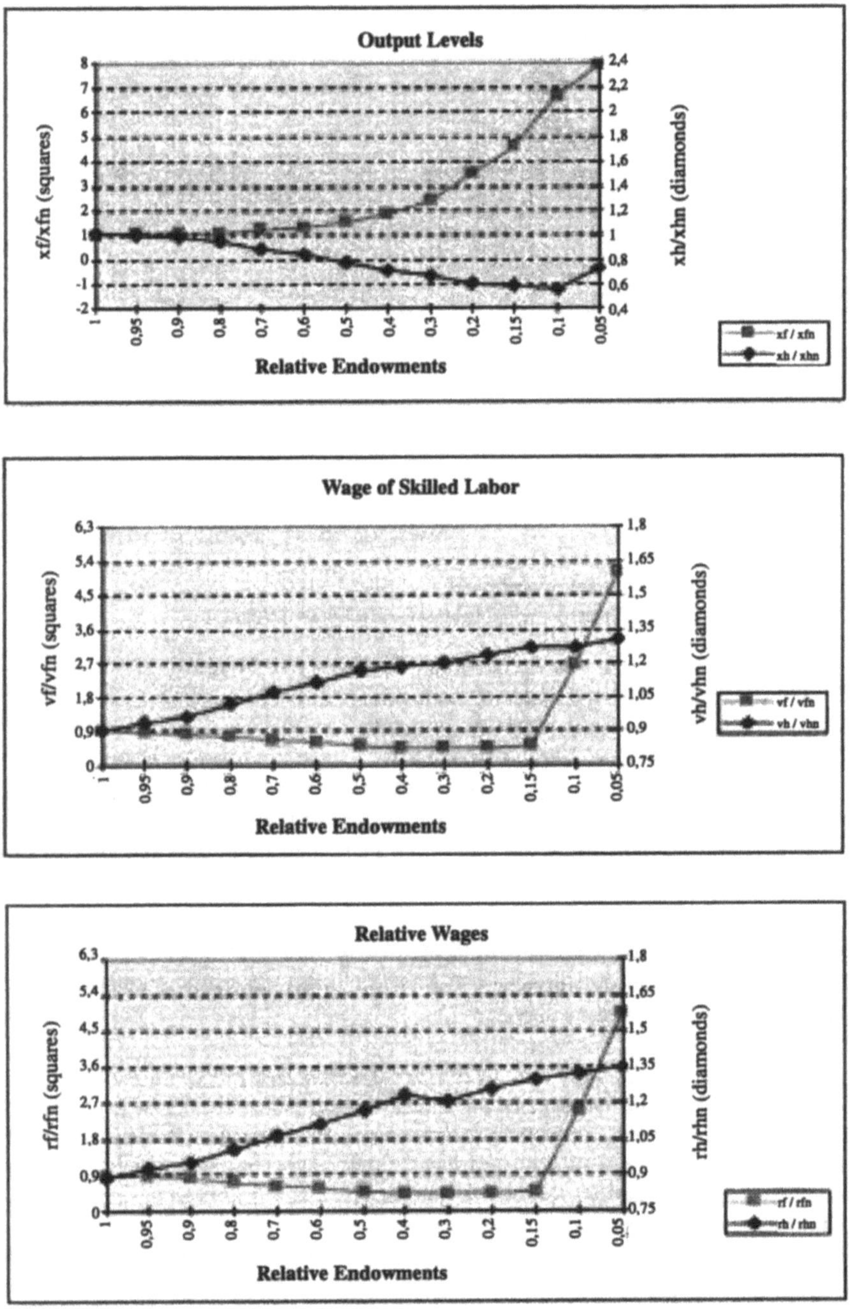
Output Levels
xf/xfn (squares)
xh/xhn (diamonds)
Relative Endowments
xf / xfn
xh / xhn
Wage of Skilled Labor
vf/vfn (squares)
vh/vhn (diamonds)
Relative Endowments
vf / vfn
vh / vhn
Relative Wages
rf/rfn (squares)
rh/rhn (diamonds)
Relative Endowments
rf / rfn
rh / rhn

Now consider relative endowments values in the range .90 to .15 in Figure 2. Entry of multinationals due to the removal of a trade barrier raises X production in country f and lowers it in country h (the top panel). Exports of knowledge capital by type-m_h firms compensate for the shortage of skilled labor in country f, and production (which is unskilled-labor intensive) is transferred from country h to country f. In some sense, the entry of multinationals does result in the "export" of unskilled jobs from the high-income country.

The middle and bottom panels of Figure 2 consider the real wage of skilled labor and the skilled-unskilled wage ratio respectively. Over the range .90 to .15 we see that the entry of multinationals due to the removal of an investment barrier raises the real wage of skilled labor and the skilled-unskilled wage ratio in country h, and lowers these same ratios in country f. The removal of the investment barrier permits the export of skilled-labor-intensive producer services (knowledge capital) from the skilled-labor-abundant to the skilled-labor-scarce country. Thus the result is fairly intuitive but of considerable importance for public policy. For example, multinationals do contribute to raising the skilled-unskilled wage gap in the high-income country, something that has been of great concern in these countries.

These results are reversed for country f (but not for country h) when the difference in relative endowments becomes extreme, values .10 and .05 in Figure 2. In such a situation, country f is so short of skilled labor that its national firm are uncompetitive in the absence of multinationals and production in country f is close to zero. The removal of the investment barrier removes this constraint on production, and the number of plants (type-m_h multinationals) increases about 10 fold as does production in country f. Hence the aggregate demand for S_f increases substantially even though each plant of a type-m_h firm demands much less S_f than did a type-n_f firm in the absence of direct investment. In this region, the removal of an investment barrier increases the real wage of skilled labor and the wage ratio in *both* countries, and home-country exports of skilled-labor-intensive producer services become a complement rather than a substitute for foreign-country skilled labor.

Figure 3 plots the equilibrium regime for case 2, in which β is lower from 1.0 to .75 and δ is raised from .1 to .4, thus making the second plant more intensive in host-country skilled labor, yet not as skilled-labor intensive as the first plant. All other parameter values are the same in the two cases. Figure 3 is qualitatively similar to Figure 1, and thus we shall not comment at length. The principle differences between Figures 1 and 3 is that (a) type m_f firms exist over a smaller range of parameter space and (b) the region in which type-m_h firms alone are supported (area h) is smaller in the case of Figure 3. Both results are intuitive due to the higher demands for skilled labor in country f by type-m firms.

Figure 4 conducts the same experiments for case 2 as were done for case 1 in Figure 2. Results are qualitatively very similar. Quantitatively, there is no drop in v_i and r_i when multinationals enter and the countries are identical (left-hand edge of the middle and lower panels). The higher skilled-labor intensity of type-m firms relative to type-n firms now about exactly balances the increase in firms scale in affecting the return to skilled labor when the investment barrier is removed. For

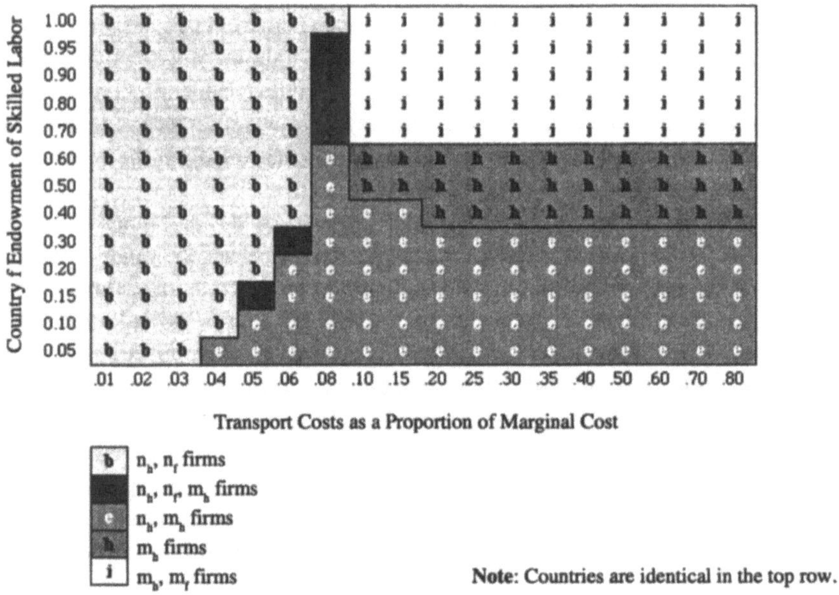

Figure 3. Relative Endowment Experiment.
Case 2: High Skilled-Labor Requirement in Second Plant

moderate differences in endowments we see the same effects that we did in Figure 2: the entry of multinationals shifts production from h to f, (v_h, r_h) rises and (v_f, r_f) falls. The principal difference between Figure 2 and 4 is quantitative. Now the range of endowment differences for which the entry of multinationals raises v_f (and therefore raises both v_h and v_f) is now significantly larger.

To summarize this section in which countries are of roughly equal GDP but differ significantly in relative factor endowments, we see several results. The entry of multinationals due to the removal of a prohibitive investment barriers transfers production from the high-income to the low-income country, and raises the real wage of skilled labor and the skilled-unskilled wage ratio in the high income country. The one ambiguous result relates to the real wage of skilled labor and the skilled-unskilled wage ratio in the low income country. If the difference in the relative endowments between the countries is not too large and branch plants require little skilled labor in the host country, then the entry of multinationals harms skilled labor in the low-income, skilled-labor-scarce country. The result is reversed if the relative endowment differences are large and/or host-country skilled labor is important in branch-plant fixed costs. In the latter case, exports of skilled-labor-intensive producer services by the high income country are complementary to skilled labor in the low-income country.

Figure 4

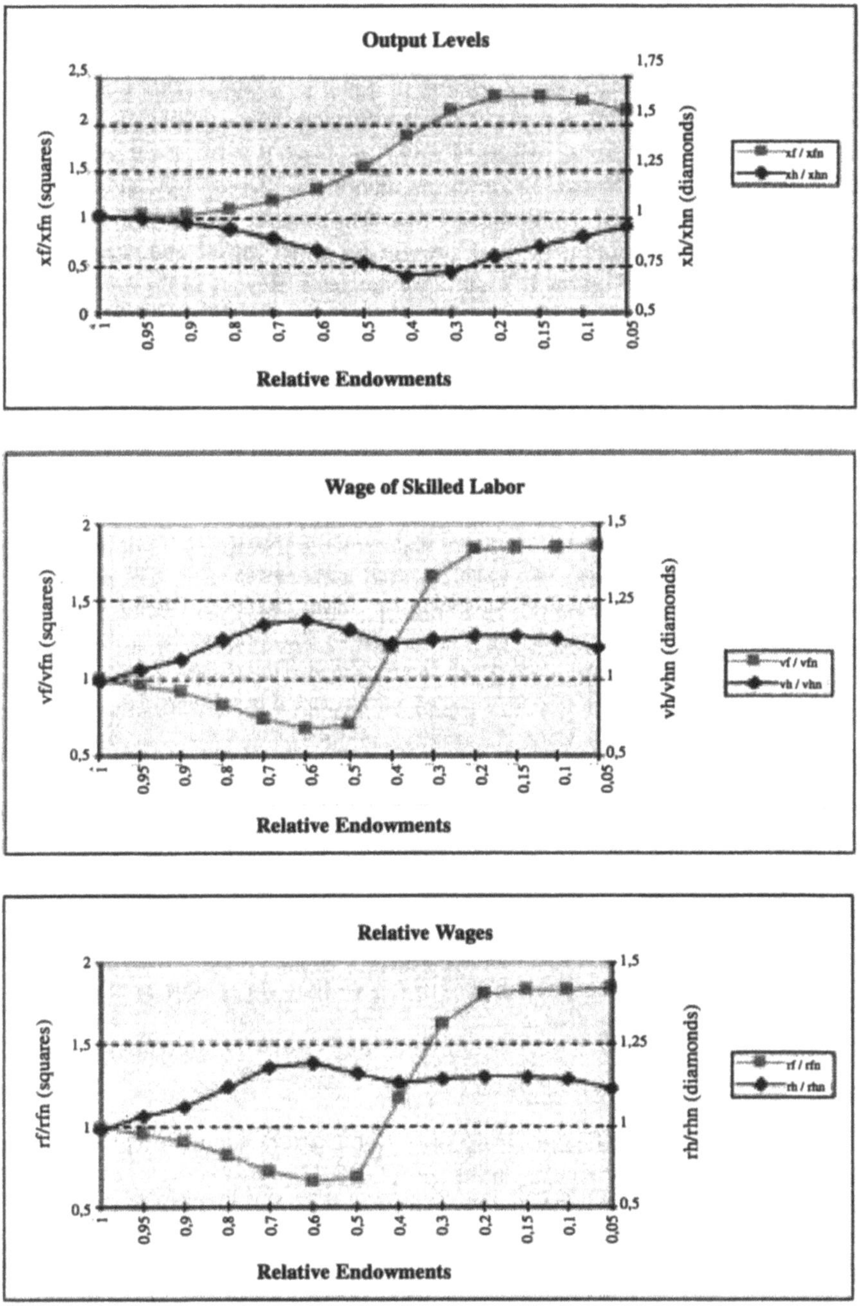
Output Levels
xf/xfn (squares)
xh/xhn (diamonds)
xf / xfn
xh / xhn
Relative Endowments
Wage of Skilled Labor
vf/vfn (squares)
vh/vhn (diamonds)
vf / vfn
vh / vhn
Relative Endowments
Relative Wages
rf/rfn (squares)
rh/rhn (diamonds)
rf / rfn
rh / rhn
Relative Endowments

13.4 Entry of Multinationals due to Investment Liberalization: Countries Differ in Size (Absolute Endowments)

Now consider pure country size effects, in which we hold the total world endowment constant but transfer factors in equal proportion from country f to country h, beginning with the two countries identical. The equilibrium regime is shown in Figure 5, where the values of case 1 are used: $\beta = 1.0, \gamma = \delta = 0.1$. As in the case of Figure 1, the two countries are identical in the top row and their size difference is maximized in the bottom row. The scaling on the vertical axis is country f's factor endowment as a proportion of its initial endowment. The horizontal axes of the Figures 1, 3 and 5 are the same. Results in the top halves of Figures 1 and 5 are similar. When the countries are not too different, intra-industry trade competition between national firm types occurs when trade costs are small, and intra-industry investment competition between multinational firm types occurs when trade costs are higher.

Figures 1 and 5 differ substantially in their bottom halves, when relative endowment and size differences are large. Consider low trade costs, and compare points in row-column (0.20 0.03) in the two diagrams. When the difference between countries h and f is in relative endowments (Figure 1), some country f-owned firms (type n_f) firms can exist. A small endowment of S_f will restrict the number of firms, but they are not hampered by a small domestic market. When the difference between countries h and f is in relative size (Figure 5), no country f-owned firms can exist. The small domestic market size in country f disadvantages country f firms sufficiently that they cannot be supported in equilibrium.

Now consider high trade costs and compare points in row-column (0.20, 0.50) in Figures 1 and 5. Here it is in the case of relative endowment differences (Figure 1) that no country f-owned firms can exist in equilibrium. Once endowment differences become sufficiently large that factor-price equalization cannot hold, type m_f and m_h firms cannot coexist with $v_f > v_h$. But when countries differ only in size (Figure 5), the two multinational firm types are in a symmetric position and the real wage of skilled labor can remain equalized across the countries despite the large difference in country size. Moving down a column of Figure 5 such as τ=.15, we see a sequence of regime shifts similar to that suggested by the partial equilibrium analysis of section 2.

$$(m_h, m_f) \rightarrow (m_h, m_f, n_h) \rightarrow (m_f, n_h) \rightarrow (n_h).$$

Figure 6 plots the same ratios as Figures 2 and 4, for the column τ=.15 of Figure 5. The two countries are identical at the left-hand edge of each panel of Figure 6 as before with their size differences being maximized at the right-hand edge. The top panel give production ratios with and without multinationals banned as in earlier figures. The ratio X_f / X_{fn} quickly goes off to infinity as the denominator goes to zero with a moderate difference in country size: with multinationals banned, the smaller country specializes in good Y. As in the case of the countries differing in

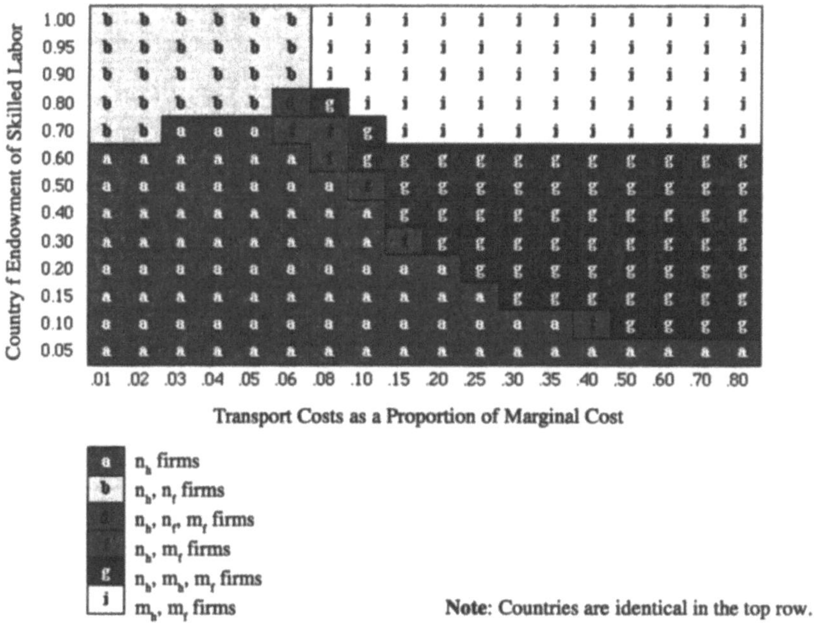

Figure 5. Relative Endowment Experiment.
Low Skilled-Labor Requirement in Second Plant

relative endowments, the removal of an investment barrier shifts production from country h to country f, in this case from the larger to the smaller country.[13]

The country-size experiment, however, yields rather different results from the relative-endowment experiment in Figure 2 with respect to the real wage of skilled labor and the skilled-unskilled wage ratio. In the present case, the removal of an investment ban raises v_f and r_f and lowers v_h and r_h.[14] The reason for this is familiar from the industrial-organization approach to trade. With the countries different in size and multinationals banned, type-n_h firms have a significant advantage over type-n_f firms in that the former have their sales concentrated in their low-cost (no transport costs) domestic market, whereas type-n_f firms have a small domestic market. Thus in equilibrium, country f generally has fewer firms than its relative size would predict. Even after accounting for size, X-sector demand for skilled labor is greater in country h than in country f, and the real wage of skilled labor is higher in the former. The introduction of multinationals removes country size as a "determinant of comparative advantage", shifts production and firm ownership to country f, and thus increases v_f and decreases

[13] Of course, we see from Figure 5 that, when size differences are very large, multinationals do not enter even in the absence of an investment barrier.

[14] As we shall see shortly, this does not imply that $v_f > v_h$! We are only referring here to the change in these prices when an investment barrier is removed.

Figure 6

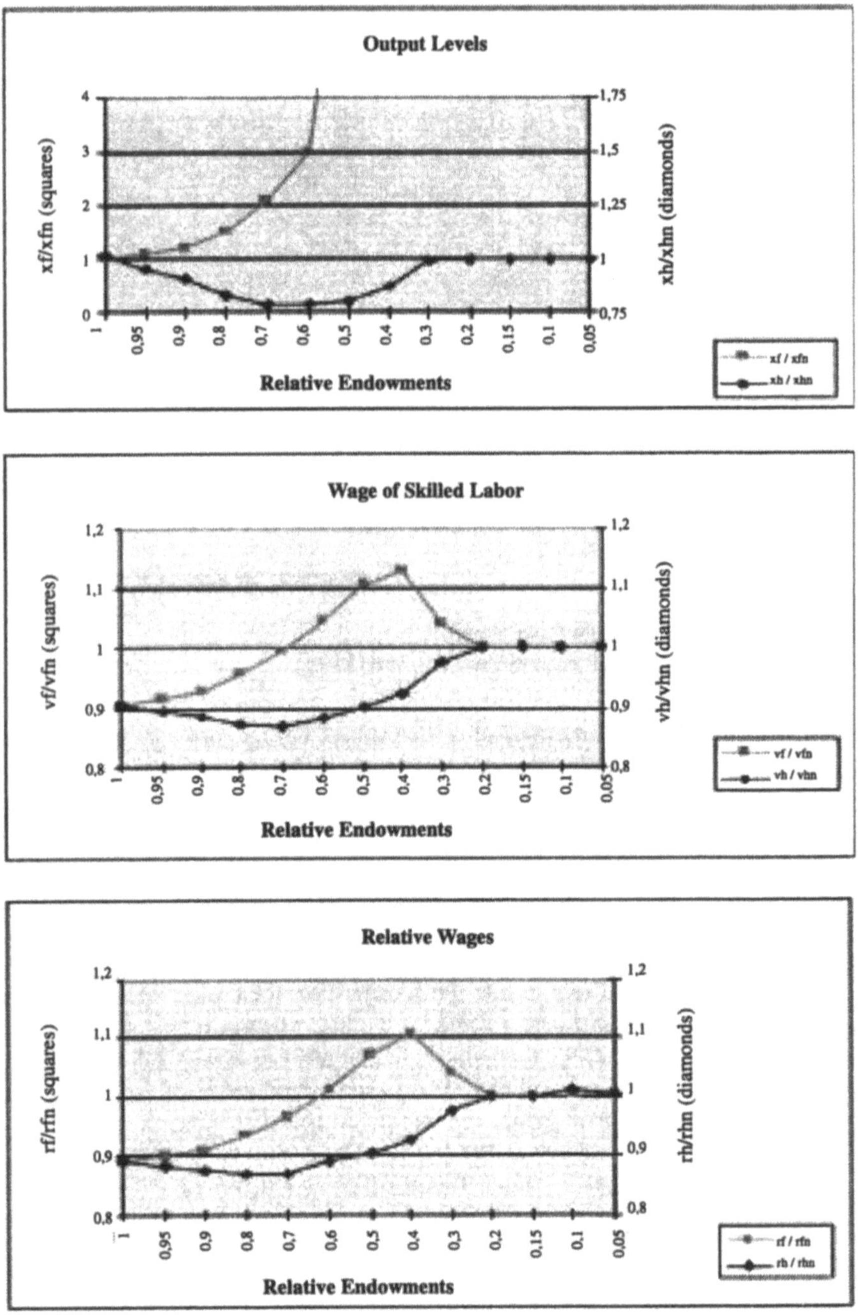
Output Levels
xf/xfn (squares)
xh/xhn (diamonds)
4
3
2
1
0
1,75
1,5
1,25
1
0,75
1
0,95
0,9
0,8
0,7
0,6
0,5
0,4
0,3
0,2
0,15
0,1
0,05
Relative Endowments
xf / xfn
xh / xhn
Wage of Skilled Labor
vf/vfn (squares)
vh/vhn (diamonds)
1,2
1,1
1
0,9
0,8
1
0,95
0,9
0,8
0,7
0,6
0,5
0,4
0,3
0,2
0,15
0,1
0,05
Relative Endowments
vf / vfn
vh / vhn
Relative Wages
rf/rfn (squares)
rh/rhn (diamonds)
1,2
1,1
1
0,9
0,8
1
0,95
0,9
0,8
0,7
0,6
0,5
0,4
0,3
0,2
0,15
0,1
0,05
Relative Endowments
rf / rfn
rh / rhn

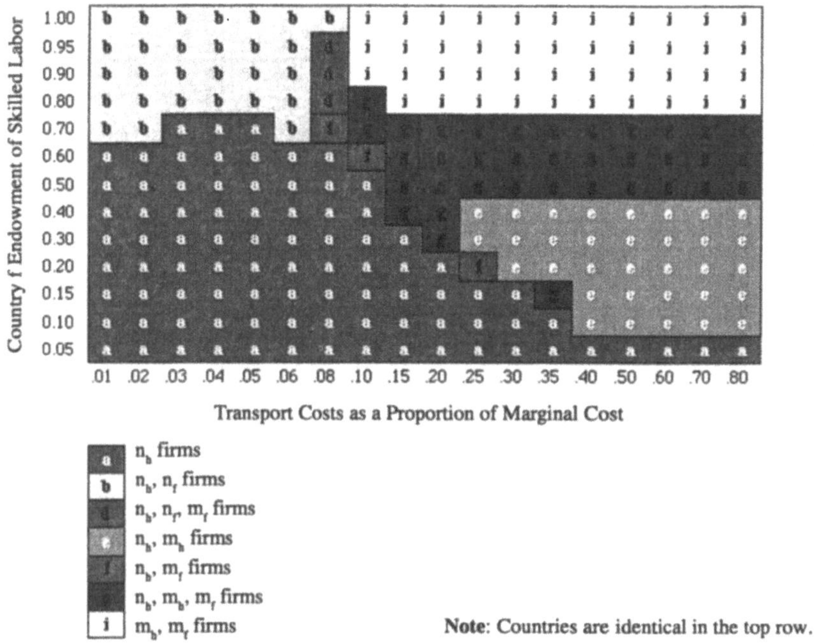

Figure 7. Absolute Endowment Experiment.
High Skilled-Labor Requirement in Second Plant

v_h.[15] Conversely, one could say that the removal of the investment barriers causes country h to "lose competitiveness", not only as the favored site for production (also true in case 1, Figures 1 and 2), but as the favored site for firm headquarters.

Figures 7 and 8 present the equilibrium regime and data series (again for τ=.15) for case 2, where branch plants are more skilled-labor intensive. Since the results are qualitatively very similar to those of case 1 in Figures 5 and 6, we simply present these results without comment.

13.5 Summary and Conclusions

The purpose of this paper is to make a conceptual contribution toward understanding how multinational firms are modifying international economic activity. This is quantitatively important, insofar as multinational firms are increasing dominating this activity, with direct investment growing much faster relative to trade over the last two decades.

We construct a model which we believe reflects the extensive empirical evidence that multinationals tend to be found in knowledge-intensive industries.

[15] When the different in size becomes very large, we see from Figure 5 that multinationals do not produce in country f even though they could. Thus all of the ratios in Figure 6 become equal to one when the absolute endowment number is less than or equal to .30.

Figure 8

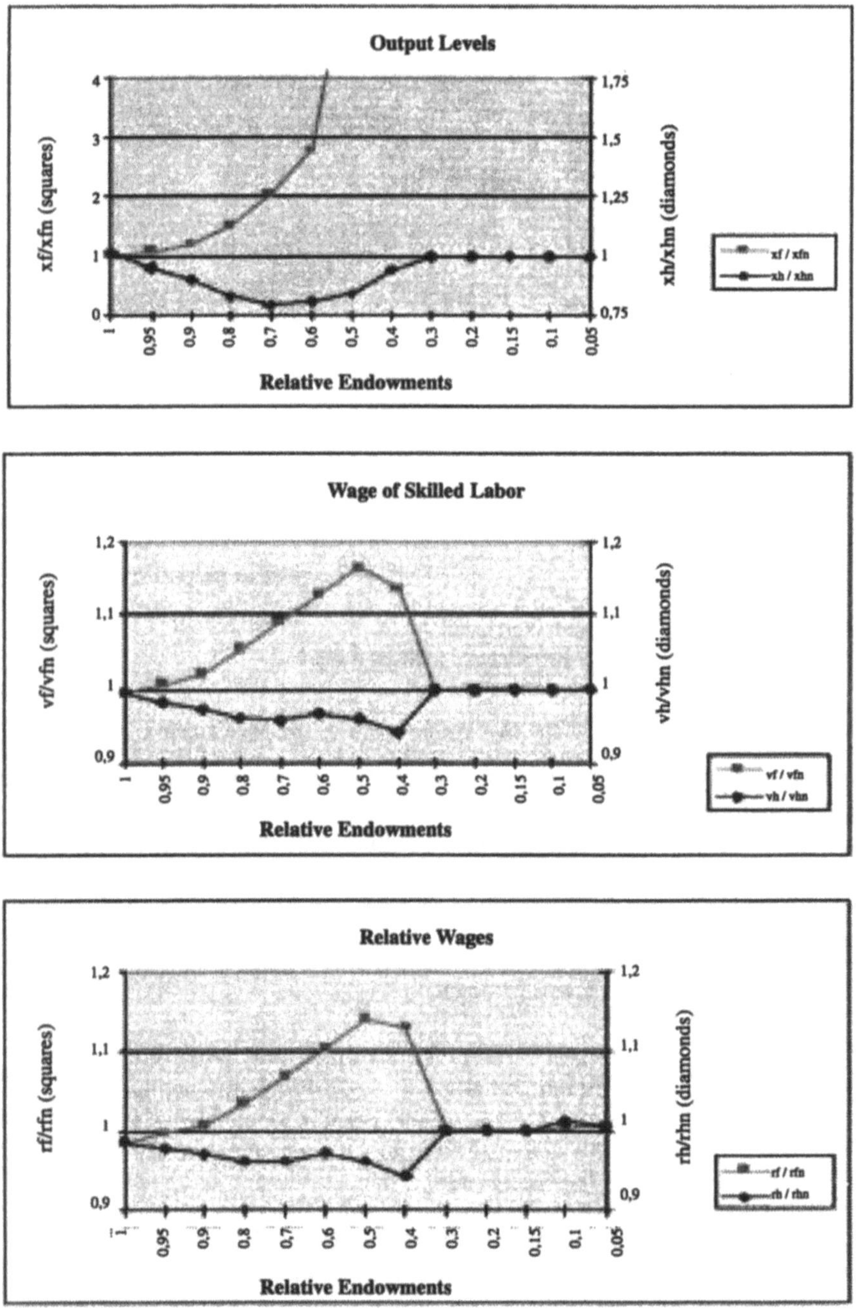
Output Levels
xf/xfn (squares)
xh/xhn (diamonds)
4
3
2
1
0
1,75
1,5
1,25
1
0,75
1
0,95
0,9
0,8
0,7
0,6
0,5
0,4
0,3
0,2
0,15
0,1
0,05
Relative Endowments
xf / xfn
xh / xhn
Wage of Skilled Labor
vf/vfn (squares)
vh/vhn (diamonds)
1,2
1,1
1
0,9
vf / vfn
vh / vhn
Relative Endowments
Relative Wages
rf/rfn (squares)
rh/rhn (diamonds)
1,2
1,1
1
0,9
rf / rfn
rh / rhn
Relative Endowments

This makes sense in light of the transportability of knowledge capital, and the partial joint-input nature of knowledge-based assets which allows them to be supplied to additional production facilities at very low cost.

Our model has scale economies at the level of the firm (e.g., research and development) and also plant-level scale economies. The fixed costs for creating a branch plant abroad are significantly less than the fixed costs for a single-plant national firm, reflecting the joint-input nature of knowledge capital. Single-plant national firms must incur a shipping cost between markets, making exporting a high variable-cost option for serving a foreign market, while multinational production is a high fixed cost option.

Multinationals dominate in equilibrium when the markets are large, and branch plant fixed costs are modest relative to shipping costs. With respect to difference between countries, multinationals dominate when the countries are relatively similar in absolute and in relative factor endowments. The intuition is that, when the countries are very different, national firms headquartered in the "advantaged" (large and/or skilled-labor abundant) country have a competitive edge over multinational firms which must locate capacity in the disadvantaged country. These results are consistent with the empirical evidence which indicates that direct investment is much more important relative to trade among the developed countries than it is between the developed and less developed countries.

Turning to issues of particular interest for this conference, an important point is that the displacement of national firms by multinational firms changes the nature of what is traded. National firms serve foreign markets (if at all) by exports of goods which embody a range of factor content, ranging from skilled labor in R&D to physical capital and unskilled labor in routine final production and handling. Multinational firms geographically fragment the production process, often exporting only skilled-labor-intensive producer services (services of knowledge capital). These include engineering, managerial, marketing, and financial services. More routine and less skilled-labor-intensive activities are transferred to the host country.

The focus of the last two sections of the paper is to try to understand the consequence of this geographic fragmentation of production for key variables, especially the demand for and wages of skilled workers in different countries. Intuition is not a particularly good guide, because it suggests two contradictory forces at work. First, the exports of knowledge capital to a host economy may naturally substitute for the services of skilled labor in that country, thus lowering the wage of skilled labor and discouraging the long-run accumulation of skills. This is the intuition suggested by a traditional Heckscher-Ohlin model. However, the intuition relies on the assumption that the multinationals displace local firms which were demanding more skilled labor prior to the arrival of multinationals. In the present type of model with increasing returns to scale in the X sector, it is entirely possible that a small and/or skilled-labor-scarce country cannot support much production in the absence of multinationals, and thus the entry of multinationals actually raises the demand for local skilled labor and exports of knowledge capital are a complement for host-country skilled labor.

Simulation results confirm that this tension does exist when two countries differ in relative factor endowments. The wage of skilled labor in the skilled-labor-scarce host country is more likely to benefit from the removal of an investment ban when local skilled labor is important in branch-plant fixed costs and when the two countries differ significantly in relative endowments. In these circumstances, the removal of an investment ban increases the wages of skilled labor in both countries, as more skilled-labor-intensive multinationals displace less skilled-labor-intensive national firms.

The results are rather different when the two countries differ in absolute endowments (size). In this case, a ban on multinationals gives the large country a significant home-market advantage, and firms are headquartered in the large country out of proportion to the difference in country size. There is correspondingly a high demand and wage for skilled labor in the large country. The removal of an investment ban shifts some plants and headquarters to the smaller country, thus shifting aggregate demand for skilled labor from the larger to the smaller country. The removal of the investment ban thus raises the real wage of skilled labor in the small country and lowers it in the larger country.

Overall, the results are cautiously optimistic if the objective is to raise the real wage of skilled labor in host countries in the short run and encourage its accumulation in the long run. In many circumstances, the intuition of the Heckscher-Ohlin paradigm (imported knowledge capital and local skilled labor are substitutes) does not hold up in this world of increasing returns and imperfect competition.

References

Berman, Eli, Stephen Machin, and John Bound (1995), "Implications of Skill Biased Technological Change: International Evidence", Working Paper.

Blomstrom, Magnus and Ari Kokko (1995), "Multinational Corporations and Spillovers to the Host Country", working paper, Stockholm School of Economics.

Brainard, S. Lael (1993a), "A Simple Theory of Multinational Corporations and Trade with a Trade-off between Proximity and Concentration", NBER Working Paper No. 4269.

Brainard, S. Lael (1993b), "An Empirical Assessment of the Factor Proportions Explanation of Multinationals Sales", NBER Working Paper No. 4583.

Brainard, S. Lael (1993c), "An Empirical Assessment of the Proximity-Concentration Tradeoff between Multinational Sales and Trade", NBER Working Paper No. 4580.

Caves, Richard E. (1996), *Multinational Enterprise and Economic Analysis*. London: Cambridge University Press. (second addition)

Cragg, Michael Ian, and Mario Epelbaum (1995), "The Premium for Skills in LDCs: Evidence from Mexico", Working Paper.

Eaton, Jonathan and Akiko Tamura (1994), "Bilateralism and Regionalism in Japanese and US Trade and Direct Foreign Investment Patterns", *Journal of the Japanese and International Economies* 8, 478-510.

Ekholm, Karolina (1995), *Multinational Production and Trade in Technological Knowledge,* Lund Economic Studies, Number 58.

Feenstra, Robert C. and Gordon H. Hanson (1995a), "Foreign Investment, Outsourcing, and Relative Wages", in *Political Economy of Trade Policy: Essays in Honor of Jagdish Bhagwati.* Cambridge: MIT Press.

Feenstra, Robert C. and Gordon H. Hanson (1995b), "Foreign Direct Investment and Relative Wages: Evidence from Mexico's Maquiladoras", Working Paper.

Freeman, Richard B. (1995), "Are Your Wages set in Beijing?", *Journal of Economic Perspectives* 9, 15-32.

Helpman, Elhanan and Paul Krugman (1985), *Market Structure and Foreign Trade.* Cambridge: MIT Press.

Keller, Wolfgang (1994), "Absorptive Capacity: Understanding the Creation and Acquisition of Technology in Development", Working Paper, Yale University.

Markusen, James R. and Anthony J. Venables (1996), The Increased Importance of Multinationals in North American Economic Relationships: A Convergence Hypothesis", in Canzoneri, Matthew W., Wilfred J. Ethier, and Vitoria Grilli (editors), *The New Transatlantic Economy,* London: Cambridge University Press.

Markusen, James R. and Anthony J. Venables (1995 a), "Multinational Firms and the New Trade Theory", working paper.

Markusen, James R. and Anthony J. Venables (1995 b), "The Theory of Endowment, Intra-Industry and Multinational Trade", Working Paper.

Markusen, James R. (1995), "The Boundaries of Multinational Enterprises and the Theory of International Trade", *Journal of Economic Perspectives* 9, 169-189.

Motta, Massimo (1992), "Multinational Firms and the Tariff-Jumping Argument", *European Economic Review* 36, 1557-71.

Richardson, J. David (1995), "Income Inequality and Trade: What to Think, What to Conclude", *Journal of Economic Perspectives* 9, 33-56.

Teece, David (1986), *The Multinational Corporation and the Resource Cost of International Technology Transfer,* Cambridge: Ballinger.

Teece, David (1977), "Technology Transfer By Multinational Firms: The Resource Cost of Transferring Technological Know-How", *Economic Journal* 87, 242-61.

Rutherford, Thomas F. (1994), Applied General-Equilibrium Modelling with MPS/GE as a GAMS subsystem, working paper.

Wood, Adrian (1995), "How Trade Hurt Unskilled Workers", *Journal of Economic Perspectives* 9, 57-80.

14. Foreign Investment as a Vehicle for International Technology Transfer

Magnus Blomström, Ari Kokko*

Economic theory provides two approaches to studying the effects of foreign direct investment (FDI) on host countries. One is rooted in the standard theory of international trade and dates back to MacDougall (1960).This is a partial equilibrium comparative-static approach intended to examine how marginal increments in investment from abroad are distributed. The main prediction of the model is that inflows of foreign capital - whether in the form of FDI or portfolio capital - will raise the marginal product of labor and reduce the marginal product of capital in the host country. In addition, MacDougall suggests that FDI may be connected to other potentially important benefits:

The most important direct gains ... from more rather than less private investment from abroad seem likely to come through higher tax revenue from foreign profits (at least if the higher investment is not induced by lower tax rates), through economies of scale and through external economies generally, especially where (domestic) firms acquire "know-how" or are forced by foreign competition to adopt more efficient methods. (MacDougall, 1960, p. 34)

There is no *a priori* argument, however, as to the relative importance of these various benefits in the MacDougall model.

The other approach departs from the theory of industrial organization, and was pioneered by Hymer (1960). Other important contributions have been made by Buckley and Casson (1976), Caves (1971), Dunning (1973), Kindleberger (1969), and Vernon (1966), among others (for an overview, see Dunning, 1993 or Caves, 1996). The starting point here is the question why firms undertake investment abroad to produce the same goods as they produce at home. The answer has been formulated as follows: "For direct investment to thrive there must be some imperfection in markets for goods or factors, including among the latter technology, or some interference in competition by governments or by firms, which separates markets" (Kindleberger, 1969, p. 13). Thus, to be able to invest in production in foreign markets, a firm must possess some asset (for example, product and/or process technology or management and marketing skills) that can

---

* Stockholm School of Economics, Stockholm

This paper was written as part of a research project on Regionalism and Development managed by the International Trade Division of the World Bank. We would like to thank Adam Jaffee, Robert E. Lipsey, Maurice Schiff, and Alan L. Winters, for valuable comments on earlier versions of the paper, and the World Bank and HSFR for financial support.

be used profitably in the foreign affiliate. Firms investing abroad therefore represent a distinctive kind of enterprise and the distinctive characteristics are pivotal when analyzing the impact of foreign direct investment on host countries. The entry of a multinational corporation (MNC) represents something more than a simple import of capital into a host country, which is generally how the matter is treated in models rooted in traditional trade theory.

This distinction is particularly important for developing countries, where domestic enterprises are likely to be relatively small, weak, and technologically backward. These countries also differ from the developed ones in such aspects as market size, degree of protection, and availability of skills. The entry of MNC subsidiaries into LDCs may therefore have effects, both positive and negative, which are substantially different from those that occur in developed host countries.

Although the traditional trade theory approach and the industrial organization approach are not mutually exclusive, they have so far generally emphasized different aspects of capital movements. Trade theorists have mainly been interested in the direct effects of foreign investment (*direct* as well as portfolio investment) on factor rewards, employment, and capital flows, while those following the industrial organization approach have put more emphasis on indirect effects or externalities. In this study, we will adopt an industrial organization approach, and focus on issues related to the transfer and diffusion of technology and knowledge, as well as the impact of FDI on market structure and competition in host countries. In the next two sections, we will discuss the role of MNCs in the international technology market, and review the empirical evidence on spillovers of technology from MNCs to local host country firms. In section 3, we go on to examine the impact of FDI on the host countries' trade performance. Section 4 examines the effects of FDI on industry structure and competition, while section 5 summarizes and concludes the paper.

14.1 Voluntary and Involuntary Diffusion of MNC Technology

It is well known that MNCs undertake a major part of the world's private research and development (R&D) efforts and produce, own, and control most of the world's advanced technology. It is also known that the multinationals' R&D and technology is heavily concentrated to a few home countries, unlike MNC investment, production, and employment that are more widely spread across both industrialized and developing economies.

A few summary statistics can demonstrate this clustering of technology production. Over four fifths of the global stock of FDI originates from the half dozen home countries that dominate the world's research and technology: the US, the UK, Japan, Germany, Switzerland, and the Netherlands. On average, about a third of the total sales and the total employment of the MNCs based in these countries were accounted for by their foreign affiliates in the early 1990s (see Lipsey, Blomström and Ramstetter, 1995). However, less than 10 per cent of the R&D expenditures of e.g. US manufacturing MNCs were undertaken in their majority-owned foreign affiliates, and more than half of this was recorded by the affiliates located in the UK and Germany. Detailed data on the R&D expenditures of the MNCs based in the other main home countries are not available, but the pattern is likely to be similar and it is not very controversial to conclude that foreign MNCs are the most important sources of modern technology for most economies.

But although we know that MNCs produce and own the bulk of the world's modern technology, it is not obvious exactly how MNC technology spreads across international borders and what role MNCs play in the process. One reason, of course, is that "technology" is an inherently abstract concept, and therefore difficult to observe and evaluate. None of the available measures of technology and technology production - such as R&D expenditures, numbers of new patents, payments for licenses and royalties, stocks of capital equipment, and so forth - cover more than a part of this.

Another reason is that technology is diffused in many different ways. MNC technology can spread to new users through formal market transactions - transfers - or through informal, non-market mediated channels that may be voluntary or involuntary. For each alternative, the role of MNCs can be either active or passive. Table 1 shows some of the possible modes of international technology diffusion, grouped according to the type of transaction and role of the MNC. Foreign direct investment is another potential channel of technology transfer, but we have not included it explicitly in the table: what distinguishes FDI from sales of equipment or licenses to outsiders (or even joint ventures) is that the MNC has chosen to retain the control and ownership of its proprietary technologies within the corporation.

It is an open question which of the transfer modes mentioned in the table is the most important, because it is difficult to compare the technology content of the different transactions, and because there are no comprehensive data available to measure the magnitude of the informal transfers. However, there are some old data provided by the UNCTC (1985) on formal transactions that are useful in outlining the quantitative dimensions, although the figures are neither complete nor extremely accurate. For instance, the industrialized countries imported USD 310 billion worth of machinery and transport equipment in 1980, whereas their yearly payments for technology and technical and management services in the early

Table1. International Diffusion of Technology: Type of Transaction and Role of MNCs

	Role of MNCs	
Type of Transaction	**Active**	**Passive**
Formal	joint ventures, licensing	goods trade
Informal	linkages	trade journals, scientific exchange

Source: Adapted from Fransman (1985).

1980s only amounted to approximately USD 10 billion. Concurrently, the developing countries' imports of machinery and transport equipment from developed countries reached USD 129 billion, and their payments of royalties, fees, and remuneration for technical services totaled about USD 2.5 billion. There are no data on the overall importance of joint ventures, but, for comparison, it can be noted that the industrialized countries' inward stock of foreign investment was estimated at approximately USD 401 billion in 1983, while the corresponding figure for developing countries was USD 138 billion (UNCTC, 1988, p. 25).[1]

There are also some data available to describe the extent of MNC participation in licensing and goods trade. These data are interesting because they confirm that MNCs are the main sources of these types of technology, but also because they indirectly introduce foreign direct investment into the picture. MNCs control technology supplies by virtue of their ownership of proprietary technologies, but they also account for a significant share of the demand, via their foreign affiliates. This is most apparent for the transfers of "disembodied" technology that are captured by data on trade in royalties, licenses, and patent rights. Over 80 per cent of the registered payments to the United States for technology sales during the 1970-1985 period were made by foreign affiliates of US firms (Grosse, 1989). More than 90 per cent of the technology payments from developing countries to Germany in the early 1980s, and over 60 per cent of the payments to Japan, originated from their own foreign affiliates (UNCTC, 1988, p. 177).

The intra-firm character of the technology transfers that take place through trade in capital equipment and other products is less apparent, but still distinguishable. What we know about MNC involvement from statistics on goods trade is that between 70 and 80 per cent of the goods exports of both the US and the UK - the main suppliers of embodied technology together with Japan and Germany - are accounted for by MNCs (UNCTC, 1988, p. 90). Moreover, a significant share of the exports and imports of the major home countries (between one fifth and one third overall, and more for complex and technologically sophisticated goods that supposedly embody more technology) flow between MNC parents and affiliates (UNCTC, 1988, p. 91). A very important part of all formal technology transfers are, therefore, closely tied to FDI.

As already mentioned, there are no comprehensive data on the informal modes of technology diffusion, but it seems that FDI plays an important role also there. For instance, linkage effects can take place between firms in different countries, like when exporters learn from the feedback they receive from their multinational customers abroad (Fransman, 1985), but are perhaps stronger when they arise between local firms and MNC affiliates operating in the same country (as will be seen in the ensuing discussion of spillovers). Similarly, many informal transfers where MNCs have a passive role - those that come about as a result of any kind of personal contact with people who know about MNC technologies - are obviously

[1] Some authors argue, in spite of the lack of more comprehensive evidence, that goods trade is the main channel of technology transfer for most countries. See e.g. Gomulka (1990), p. 161 and Kaplinsky (1990), p. 21.

facilitated by the presence of foreign affiliates. Of course, there are also transfer types that may be entirely independent from FDI. Exports from the MNC's home country may be sufficient to prompt reverse engineering - the practice of taking apart and analyzing products, to learn about the technologies embodied in them - which is often recognized as one of the main sources of involuntary technology dissemination (Zander, 1991, Chapter 5).[2] Other informal transfers, through academic contacts, technical publications, and education abroad can also occur irrespective of the presence of foreign affiliates.

Thus, although FDI was not included explicitly in Table 1, it nevertheless seems that much of the international dissemination of technology is connected to foreign direct investment. A large share of both license sales and sales of technologically advanced products are directed to MNC affiliates, and FDI therefore seems to be more important for the geographical spread of technologies than sales of technology to unrelated parties. In addition, many informal contacts are easier and more important when MNC affiliates are present in the market than when contacts have to be made across international borders. Keeping this in mind, it is not surprising that FDI is probably the transfer mode that has received the most attention (see e.g. Enos, 1989).

However, there are important questions related to the role of FDI as a source of technology for host countries. From the point of view of the host, it is not clear what the actual technological benefits of FDI are and how they come about. A distinguishing characteristic of direct investment, as has already been noted, is that the control and ownership of the technologies used by the affiliates stay in the MNCs' possession. Is there any significant diffusion of technology to new users or is the affiliate able to protect its technology from spreading to outsiders? And if technology is diffused from the MNC affiliates, does it spread through the market or is it diffused informally, and are the multinationals active or passive in this process? Obviously, it is not trivial to pinpoint the correct position of foreign direct investment in Table 1.

14.1.1 Spillovers of MNC Technology

In the debate on the role of MNCs in international technology transfer, it has sometimes been suggested that the most significant channels for the dissemination of modern, advanced technology are external effects or "spillovers" from foreign

[2] One of the few comprehensive quantitative assessments of the importance of imitation and reverse engineering is de Melto et al. (1980). They report that half of a sample of 280 significant innovations commercialized in Canada between 1960 and 1979 could be characterized as "imitations", and that more than half of these resulted from reverse engineering. Mansfield, Schwartz, and Wagner (1981) find that 60 per cent of the patented innovations in their sample were imitated within 4 years. Kim and Kim (1985) also present evidence of imitation and informal technology transfers in 42 South Korean firms. We will not examine these studies in detail because they do not focus on foreign investment, but it should be remembered that MNC presence in the host country probably facilitates imitation of MNC technology.

direct investment, rather than formal technology transfer arrangements (Blomström, 1989). It is argued that when firms establish affiliates abroad and become multinational, they are distinguished from the already established firms in the host country for two reasons. One is that they bring with them some amount of the proprietary technology that constitutes their firm-specific advantage and allows them to compete successfully with local firms who have superior knowledge of local markets, consumer preferences, and business practices. Another reason is that the entry of the MNC affiliate disturbs the existing equilibrium in the market and forces local firms to take action to protect their market shares and profits. Both these changes are likely to cause various types of spillovers that lead to productivity increases in local firms.

Generally, spillovers are said to take place when the MNCs cannot reap all the productivity or efficiency benefits that follow in the host country's local firms as a result of the entry or presence of MNC affiliates. The simplest example of a spillover is perhaps the case where a local firm improves its productivity by copying some technology used by MNC affiliates operating in the local market. Another kind of spillover occurs if the entry of an affiliate leads to more severe competition in the host economy, so that local firms are forced to use existing technology and resources more efficiently; a third type of spillover effect takes place if the competition forces local firms to search for new, more efficient technologies. These effects may take place either in the foreign affiliate's own industry or in other industries, among the affiliate's suppliers or customers.

A first reason to suspect that spillovers are important is that the technologies used by MNC affiliates are not always available in the market (see Blomström and Zejan, 1991). Abstracting from the fact that several means of extracting technology rents may occur simultaneously in reality, we can assume that the MNC has three alternative ways to exploit its technological advantages internationally. The MNC can produce for export in the home country, it can sell its technology to foreigners, or it can establish an affiliate abroad and control foreign production directly.[3]

However, markets for technology are typically imperfect, which makes the transactions costs for sales of technology to outsiders high (Buckley and Casson, 1976; Caves, 1996; Teece, 1981). For instance, it is difficult to judge the value of any specific technology and agree about prices and licensing costs that are acceptable to both parties. Consequently, MNCs often prefer direct investment before licensing, and the preference for FDI may be particularly strong when the newest and most profitable technologies (or those that are very close to the MNCs principal line of business) are exploited. A technology that is exploited through FDI will probably not be licensed to the local competitors in the host country - the

[3] The determinants of the MNCs' choice between exports, licensing, and foreign direct investment are not discussed in this study, but treated in detail by e.g. Blomström and Zejan (1991), Contractor (1984), Davidson (1980, 1983), Davidson and McFetridge (1985), Stobaugh (1988), Telesio (1979), Vernon and Davidson (1979), and Zander (1991).

local firms' only chances to gain access to the technology may lie in reverse engineering or hiring of former MNC employees with special skills, or some other type of spillover. This reason for the importance of spillovers may be most valid for the more developed host countries and industries, because the technical skills required to imitate the newest and most profitable technologies are typically higher (see Blomström, 1991b).

Another reason why spillovers may be significant is that direct contact with users appears to be a principal factor explaining technology diffusion. Before a new process or product innovation is widely spread in the market, potential adopters have limited information about the costs and benefits of the innovation and may therefore associate it with a high degree of risk. As the potential adopters come in contact with existing users, e.g. MNC affiliates, information about the technology is diffused, the uncertainty regarding the pros and cons of the innovation is reduced, and the likelihood of imitation or adoption increases. In this way, the entry of foreign affiliates may demonstrate the existence and profitability of new products and processes, and encourage local firms to adopt some of them: these diffusion processes may even be repeated every time innovations are transferred from the MNC parent to the affiliate. This is an argument for spillovers even when access to new technology is not restricted by proprietary factors, because information about foreign technology is generally more expensive for local firms than for MNC affiliates. In addition, it can be assumed that "contagion" effects are more important for developing host countries, where indigenous skills and information are in shorter supply.[4]

A third reason to expect positive external effects from FDI is related to the typical features of MNCs - scale economies, high initial capital requirements, intensive advertising, and, not least, advanced technology. These are also industry characteristics that signal high barriers to entry, high concentration, and perhaps some inefficiency that follows from low levels of competition. Entry by new domestic firms into such industries in potential host countries is likely to be difficult; MNCs, on the other hand, are both likely to enter just those industries and be well equipped to overcome the entry barriers. They can coordinate their international operations and concentrate specific processes to few locations if scale economies are important entry barriers. If the barriers are made up of high capital costs, MNCs may have larger own funds than local firms, or access to cheaper financing on international markets. Barriers related to product differentiation and technology, finally, are not likely to stop a multinational, since

[4] Mansfield and Romeo (1980) present indirect evidence also for this argument. They claim that the export of technologies from parents to U.S. affiliates abroad speeded up the emergence of competing products or processes in the host countries by an average of 2.5 years in about a third of their cases. In addition, they report that more than half of the managers of a sample of British firms believed that they had introduced some products and processes earlier as a consequence of transfers of technology to U.S. affiliates operating in the United Kingdom. More evidence will be presented later, in the survey of empirical studies of spillovers.

these features often characterize the MNCs themselves. The entry of MNCs into this kind of monopolistic industry is likely to increase the level of competition and force existing firms to become more efficient.

14.2 Empirical Evidence on Spillovers

The earliest discussions of spillovers in the literature on foreign direct investment date back to the early 1960s. The first author to systematically include spillovers (or external effects) among the possible consequences of FDI was MacDougall (1960), who analyzed the general welfare effects of foreign investment. Other early contributions were provided by Corden (1967), who looked at the effects of FDI on optimum tariff policy, and Caves (1971), who examined the industrial pattern and welfare effects of FDI.

The common aim of these studies was to identify the various costs and benefits of FDI, and spillovers were discussed together with several other indirect effects that influence the welfare assessment, such as those arising from the impact of FDI on government revenue, tax policies, terms of trade, and the balance of payments. The fact that spillovers were taken into account was generally motivated by empirical evidence from case studies rather than by comprehensive theoretical arguments - the detailed theoretical models analyzing spillovers did not appear until the late 1970s.[5] Yet, the early analyses made clear that MNCs may improve allocative efficiency by entering into industries with high entry barriers and reducing monopolistic distortions, and induce higher technical efficiency if the increased competitive pressure or some demonstration effect spurs local firms to more efficient use of existing resources. They also proposed that the presence may lead to increases in the rate of technology transfer and diffusion. More specifically, case studies showed that foreign MNCs may:

- contribute to efficiency by breaking supply bottlenecks (but that the effect may become less important as the technology of the host country advances);
- introduce new know-how by demonstrating new technologies and training workers who later take employment in local firms;
- either break down monopolies and stimulate competition and efficiency or create a more monopolistic industry structure, depending on the strength and responses of the local firms;
- transfer techniques for inventory and quality control and standardization to their local suppliers and distribution channels; and,
- force local firms to increase their managerial efforts, or to adopt some of the marketing techniques used by MNCs, either on the local market or internationally.

Although this diverse list gives some clues about the broad range of various spillover effects, it says little about how common or how important they are in

[5] For theoretical studies, see e.g. Findlay, (1978), Koizumi and Kopecky (1977), Das (1987), and Wang and Blomström (1992). Some other early empirical studies are Balasubramanyam (1973), Brash (1966), Deane (1970), Dunning (1958), Forsyth (1972), Gabriel (1967), Rosenbluth (1970), and Safarian (1966).

general. This brings up the question "How can we measure the significance and scope of spillovers?"

It is not difficult to picture an ideal study of productivity spillovers in host countries. To examine how the development of technology and productivity in individual local firms is related to the presence of foreign MNC in the local market, the study would require detailed micro data, both quantitative and qualitative. The study would have to cover several years, to take into account the fact that spillovers are not instantaneous. It should also include a large number of firms and industries, so that inter-industry spillovers could be observed, and so that it would be possible to draw statistically significant conclusions. With this kind of detailed information, it would also be possible to study productivity spillovers in the home countries of MNCs, and examine the presence of market access spillovers. However, to the best of our knowledge, no comprehensive analyses of this character have ever been made - one reason, of course, is the extreme data requirements. Additional empirical evidence on spillovers must therefore be drawn from two other sources.

Firstly, in addition to the few case studies focusing directly on spillovers, there is a large number of detailed case studies discussing other aspects of FDI in different countries and industries, and these studies often contain valuable "circumstantial evidence" of spillovers. For instance, many analyses of the linkages between MNCs and their local suppliers and subcontractors have documented learning and technology transfers that may make up a basis for productivity spillovers or market access spillovers. These studies seldom reveal whether the MNCs are able to extract all the benefits that the new technologies or information generate among their supplier firms, so there is no clear proof of spillovers, but it is reasonable to assume that spillovers are positively related to the extent of linkages. Similarly, there is much written on the relation between MNC entry and presence and market structure in host countries, and this is closely related to the possible effects of FDI on competition in the local markets. There are also studies of demonstration effects, technology diffusion, and labor training in foreign MNCs that are relevant for our purposes.

Secondly, there are a few statistical studies examining the relation between foreign presence in a host country industry and productivity (or productivity growth) in the locally-owned share of the industry or in individual locally-owned firms. These studies typically estimate production functions for locally-owned firms, and include the foreign share of the industry as one of the explanatory variables. They then test whether foreign presence has a significant positive impact on local productivity (or productivity growth) once other firm and industry characteristics have been accounted for. There are no similar statistical studies of productivity spillovers in home countries, but there is one study employing statistical techniques to examine whether firms located near exporting MNCs are more likely than others to become exporters. Although the data used in these analyses is often limited to few variables, aggregated to industry level rather than plant level, and in several cases of a cross-section rather than time-series or panel character, they do provide some important evidence on the presence and pattern of spillover effects.

To provide an overview of the empirical evidence, we have structured the presentation as follows: we begin by distinguishing between spillover effects that are related to backward and forward linkages between MNCs and domestic firms, and go on to discuss MNC training of local employees and demonstration effects. In addition, we review the results of the available statistical studies of spillovers.

14.2.1 Linkages between MNCs and Local Firms

Some of the spillovers from FDI operate via the linkages between the MNC's foreign affiliate and its local suppliers and customers. The spillovers occur when local firms benefit from the MNC affiliate's superior knowledge of product or process technologies or markets, without incurring a cost that exhausts the whole gain from the improvement.[6] Backward linkages arise from the MNC affiliate's relationships with suppliers, while forward linkages stem from contacts with customers.

Backward Linkages

Some of the "complementary activities" that may create spillovers through backward linkages are identified in Lall (1980). In summary, Lall notes that MNCs may contribute to raise the productivity and efficiency in other firms as they:

- help prospective suppliers (domestic as well as foreign) to set up production facilities;
- provide technical assistance or information to raise the quality of suppliers' products or to facilitate innovations;
- provide or assist in purchasing of raw materials and intermediaries;
- provide training and help in management and organization; and
- assist suppliers to diversify by finding additional customers.

In his empirical study, Lall (1980) examines two Indian truck manufacturers (one MNC and one joint venture) and finds significant backward linkages of all five types mentioned above. In particular, he notes that the truck manufacturers had been active in the establishment of supplier firms: of the 36 sampled supplier firms, 16 had been launched by the principals.[7] Behrman and Wallender (1976), who examine the operations of General Motors, ITT, and Pfizer in several host countries, find similar linkages. They emphasize the ongoing character of the

---

[6] However, the existence of linkages does not prove that there are spillovers, but the two are probably closely related. Even if the MNC affiliate charges for the support it provides to their local suppliers and distributors, it is not always able to extract the full value of the resulting productivity increases.

[7] The domestic content in Lall's two cases was extremely high - probably over 90 per cent - and both firms had extensive supplier networks, with 500 and 339 independent suppliers,

contacts and information flows between MNCs and their local suppliers. Evidence on the development of linkages is also provided by e.g. Watanabe (1983a, 1983b) and UNCTC (1981).[8]

Apart from demonstrating various types of linkages that create a potential for spillovers, these studies also suggest that the local content in MNC production is one of the determinants of the strength of linkages. Reuber et al. (1973), in a comprehensive survey of MNC affiliates in developing countries, note that over a third of the total value of goods and services purchased 1970 by all affiliates included in their survey were provided by local firms. However, there were systematic differences in local purchases depending on the affiliates' market orientation, the parent's nationality, and the host country. Local-market oriented affiliates purchased more from local firms than did export-oriented affiliates (perhaps because import licenses are easier to obtain for exporters); European MNCs relied more on local firms than US or Japanese firms (perhaps because they are generally older and have already built up local supplier networks); and affiliates in Latin America and India purchased more local inputs than affiliates in the Far East (probably because of differences in local content requirements). In addition to these factors, it seems that the technical capability of potential local suppliers must be important to take into account.

Moreover, there is a tendency for the share of local inputs to increase over time, also for export-oriented affiliates. McAleese and McDonald (1978), who study Irish manufacturing during the period 1952-1974, show that local purchases of inputs increase as the MNC affiliates mature. Several factors contribute to the gradual development of linkages: further production processing stages are added over time, the autonomous growth of the manufacturing sector brings up new suppliers, and some MNC take deliberate action to attract and develop local suppliers.[9] Hence, it is possible that spillovers also become more common over

respectively. It should be noted that these characteristics already distinguish the Indian experience from others, since an extreme import substitution policy made India a virtually closed economy until the mid-1980s. Hill (1982), who examines the Philippine appliance and motor cycle industries, argues that inter-firm linkages are often significantly weaker than in the Indian case. More liberal import policies reduce local content, the smaller size of of many principal firms makes them incapable of offering technical assistance to suppliers. most markets makes much production economically unviable, and the assembler character Similarly, Lindsey (1989) argues that the positive impact of MNCs on the Philippine economy has been very limited.

[8] Lall (1978) reviews numerous other studies of linkages between MNCs and local firms, and Halbach (1989) summarizes a detailed study of subcontracting and linkages in several South-East Asian industries.

[9] This last point is noted in numerous other studies. Dunning (1958), one of the earliest contributions, maintains that foreign firms are generally engaged in the training of local suppliers. In addition to the comprehensive evidence on local content, Reuber *et al.* (1973) argue that MNCs actively support the establishment of independent local suppliers. Lim and Pang (1982) also underscore this in their study of the Singapore electronics industry:

time, as more and more local firms establish various types of contacts with the foreign MNCs.

In addition to the linkages and spillovers that are result of cooperation between affiliates and local firms, it is also possible that there are effects that occur as suppliers are forced to meet the higher standards of quality, reliability, and speed of delivery of the MNCs. For instance, Brash (1966), in a study of the impact made by General Motors on its Australian local suppliers, emphasizes the importance of the MNC's stricter quality control, which also had an impact on the suppliers' other operations. Katz (1969, p. 154) reports that foreign MNCs operating in Argentina "forced their domestic suppliers to adopt productive processes and techniques used by the suppliers of their main firms in their country of origin". Similarly, Watanabe (1983a) notes complaints from small local producers in the Philippines about the large foreign firms' tough requirements on both product characteristics and prices: in developing countries, in particular, this alone may have an effect on what technologies are used, and perhaps also on the general competitive climate. However, there is very little additional evidence on such "forced linkage effects".

Some less optimistic conclusion on the effects of linkages are suggested by Aitken and Harrison (1991), who examine Venezuelan manufacturing between 1976 and 1989, and conclude that the effect of foreign investment on the productivity of upstream local firms is generally negative. They assert that foreign firms divert demand for domestic inputs to imported inputs, which means that the local supplier firms are not able to benefit from potential economies of scale. Their results differ from most other findings in this respect. One reason is that their study includes also local firms that have not been fortunate enough to establish linkages with foreign affiliates, and because they do not take into account the increase in local content that seems to take place over time. Yet, their conclusions highlight the need for more research where the connection between spillovers and linkages is examined explicitly.

14.2.2 Forward Linkages

There is much less evidence of forward than backward linkages. Only a minority of the firms studied by Reuber et al. (1973) claimed to have contributed significantly to the development of local distributors and sales organizations. However, McAleese and McDonald (1978) report that forward linkages in the Irish economy grew in much the same way as backward linkages. In particular,

---

they point specifically at the role of MNCs in suggesting entrepreneurial possibilities and assisting in the establishment of supplier firms, and their "willingness to bear the initial costs of encouraging and patronizing local suppliers, who in the long run would be cost-competitive" (p. 591). What distinguishes their study is, firstly, that it is concerned with export-oriented TNCs whereas most others look at import-competing industries, and secondly, that they show how the development of linkages in Singapore was relatively rapid during the late 1970s while most other studies seem to suggest a much slower process.

they assert that many MNCs commenced operations with heavy export-orientation, but that the importance of the home market has increased over time.

Blomström (1991a) discusses forward linkages in closer detail, and emphasizes the growing technical complexity in many industries. On the one hand, this could mean that only MNCs can afford the necessary R&D to develop and manufacture modern products; on the other hand, industrial application of e.g. computer-based automation and information technologies might require expertise from the manufacturers. This, he argues, would contribute to increasing the role of MNC-customer contacts, especially in the smaller countries. One of the few empirical works touching upon the issue is the study by Aitken and Harrison (1991) noted above. They conclude that spillovers from forward linkages seem to be important in most industries - in fact, they argue that the downstream effects of foreign investment are generally more beneficial than the upstream effects.

Summarizing, there is much evidence of the existence and potential of backward linkages, and a suspicion about the growing importance of forward linkages as well. Some of the host country characteristics that may influence the extent of linkages - and thereby the extent of spillovers - are market size, local content regulations, and the size and technological capability of local firms. Moreover, linkages are likely to increase over time, as the skill level of local entrepreneurs grows, new suppliers are identified, and local content increases. This constitutes circumstantial evidence for spillovers, but it must also be mentioned that there are hardly any studies where the connection between linkages and spillovers is explicit.

14.2.3 Training of Local Employees in MNC Affiliates

The transfer of technology from MNC parents to affiliates is not only embodied in machinery, equipment, patent rights, and expatriate managers and technicians, but is also realized through the training of the affiliates' local employees. This training affects most levels of employees, from simple manufacturing operatives through supervisors to technically advanced professionals and top-level managers. Types of training range from on-the-job training to seminars and more formal schooling to overseas education, perhaps at the parent company, depending on the skills needed. Although higher positions are often initially reserved for expatriates, the local share typically increases over time. The various skills gained while working for an affiliate may spill over as the employees move to other firms, or set up their own businesses.

The evidence on spillovers from the MNC affiliates' training of local employees is far from complete, and comes mainly from developing country studies. Considering that the public education systems in developing countries are relatively weaker, it is also possible that spillovers from training are relatively more important there. However, there is scattered evidence of effects in the industrialized countries, and then perhaps mainly regarding management skills. It is possible, for instance, that the inter-firm mobility of managers has contributed to spread specific management practices from Japan to the United States and Europe, and, in earlier times, from the US to Europe (Caves, 1996). Moreover, casual observation suggests that the mobility of employees from MNCs in the computer and software industries contributes to spillovers, both within the industry and elsewhere.

Studies in developing countries have recorded spillovers of both technical and management skills. For instance, Gerschenberg (1987) examines MNCs and the training and spread of managerial skills in Kenya. From detailed career data for 72 top and middle level managers in 41 manufacturing firms, he concludes that MNCs offer more training of various sorts to their managers than private local firms do, although not more than joint ventures or public firms. Managers also move from MNCs to other firms and contribute to the diffusion of know-how. Of the managers in private local and public firms who had training from elsewhere, the majority had received it while working for MNCs - joint ventures, on the other hand, seemed to recruit mainly from public firms. Yet, mobility seemed to be lower for managers employed by MNCs than for managers in local firms. This is not surprising remembering the common finding that MNCs pay more for their labor than what local firms do, even taking skill levels into account: in fact, it is not unreasonable to hypothesize that the fear of a "brain-drain" to local firms is one of the reasons behind the higher wages in MNCs. Katz (1987) points out that managers of locally owned firms in Latin America often started their careers and were trained in MNC affiliates.[10]

Chen (1983), in a study of technology transfer to Hong Kong, chooses to emphasize training of operatives. In three out of four sampled industries, the MNCs' incidence of undertaking training and their training expenditures were significantly (several times) higher than those for local firms. Consequently, he concludes that "the major contribution of foreign firms in Hong Kong manufacturing is not so much the production of new techniques and products, but the training of workers at various levels" (p. 61).

Another factor in the dissemination of technology and human capital skills is related to the R&D efforts undertaken by the MNC affiliates. Here, we will only hint at some of the results in a very extensive research field. Firstly, MNCs do undertake R&D in their host countries, although it is strongly concentrated to the home countries. The affiliates' research efforts could be important, and should be compared with the R&D efforts of local firms, rather than with the parents' total

[10] Wasow and Hill (1986) provide similar evidence for the dissemination of management skills in the Philippine insurance industry. Likewise, Yoshihara (1988) underlines the importance of training in foreign companies (and overseas education) for Chinese-owned firms in South-East Asia. Behrman and Wallender (1976) recognize spillovers of both managerial and technical skills. In particular, they note that several of the MNC affiliates' subcontractors had been established by former employees. Hill (1982) also identifies similar cases in the Philippine appliance and motor cycle industries, but argues that they were insignificant. Nevertheless, 12 out of 20 assembler firms had some subcontractors that were established by former employees.

R&D. Doing so, Fairchild and Sosin (1986) conclude that foreign firms in Latin America exhibit more internal local R&D activity than is generally presumed, and that their total expenditures on research are very similar to those of domestic firms. In addition, they have access to the aggregate know-how base of the parent and related affiliates, and sometimes also to the parent's R&D facilities. The affiliates' R&D may therefore be more efficient than that of local firms. Not much is known, however, about what type of R&D is done in affiliates - traditionally, much has been adaptation of products and processes - and even less is known about the mobility of R&D personnel or the effects on the host country's technological capability.[11]

Judging from the aggregate evidence on spillovers from the training of MNC personnel, there seems to be a definite accumulation of human capital skills in the MNCs' employee stock. Some of these skills can be appropriated by local firms when employees move to new jobs, but how much is an open question. The fact that most studies deal with the spread of management skills suggests that they are less firm-specific than technical skills, and can more easily be used in other contexts: the empirical evidence, however, is too limited for any more definite conclusions.

14.2.4 Demonstration Effects

There are a few case studies where pure demonstration effects of FDI on local firms in the host countries of MNCs have been discussed. Riedel (1975) claim that horizontal demonstration effects from the operations of MNCs were an important force behind the development of the manufacturing export sector in Hong Kong in the 1960s. Swan (1973) suggests that multinationals are important not only for the diffusion of the specific technologies they use, but more generally because they strengthen international communications channels, which makes demonstration across international borders possible. Tilton (1971), in a study of the semiconductor industry, points to the importance of new MNCs in introducing US innovations to the European countries. Lake (1979), also examining the semiconductor industry, argues that affiliates of US MNCs have been more active than local firms in the diffusion of new technology in Great Britain. Mansfield and Romeo (1980) show that the technologies transferred to affiliates are younger than those sold to outsiders, and that there are cases where the affiliates' technology imports have induced local competitors to imitate their behavior.

These case studies suggest that demonstration may be an important channel for spillovers. However, there are too few studies to reveal how important the simple demonstration effects are, nor do we know whether they are more important in some countries or industries than in others. One reason is that pure demonstration effects often take place unconsciously: it is seldom documented how and where a firm first learns about a new technology or product that is subsequently adopted.

[11] For some recent studies touching on these issues, see Cantwell (1995), Patel and Pavitt (1994), and Zander (1994).

Another reason is that demonstration effects are often intimately related to competition. Summarizing a comparison of MNC and local technologies, Jenkins (1990, p. 213) notes that "over time, where foreign and local firms are in competition with each other, producing similar products, on the same scale and for the same market, there is a tendency for local firms to adopt similar production techniques to those of the MNCs. Indeed this is part of a general survival strategy, whereby in order to compete successfully with the MNCs local capital attempts to imitate the behavior of the MNCs".

Some case studies at the firm and industry level are also available to describe the combined effects of demonstration and competition from MNCs on local firms. For instance, Langdon (1981), in a study of FDI in the Kenyan soap industry, reports that the entry of foreign MNCs also introduced mechanized production, and local firms found themselves unable to sell handmade soap in the urban markets. Instead, they were forced to introduce mechanized techniques to stay in business. Similarly, foreign entry into the Kenyan footwear industry led to increased competition and changes in the production techniques of local firms (Jenkins, 1990). In the Brazilian textile industry, the establishment of an affiliate by a foreign firm brought in synthetic fibers: the consequent stagnation of demand for cotton textiles led to the disappearance of some local firms, and forced others to seek joint ventures with foreign firms in order to get access to competitive technology (Evans, 1979).

Some authors have actually hypothesized that the most important influences of MNCs on local firms operate through the interaction of demonstration and competition (Blomström, 1986a), and several reasons to expect important effects from competition were noted in the conceptual discussion: most importantly, MNCs are likely to enter into industries where potential local challengers are discouraged by high barriers to entry and where competition between existing local firms may therefore be limited.

In practice, it is difficult to distinguish between effects of demonstration and competition when it comes to imitation or adoption of new technologies, and the most valuable information from case studies may therefore be related to how local firms respond to increased competition in the short run, before imitation takes place. The immediate local reaction may be to merely enforce stricter or more cost-conscious management and motivate employees to work harder, in order to reduce slack or improve X-efficiency. It is possible that this seemingly simple response may make a more substantial contribution to productivity than improvements in resource allocation (see Leibenstein, 1966, 1980). Bergsman (1974), on the basis of a study of industry in six developing countries, argues that X-efficiency is several times as important as allocative efficiency in increasing incomes in these countries. Also Pack (1974), in a study of LDC manufacturing industries, and Page (1980), referring to evidence for three manufacturing industries in Ghana, suggest that factors related to X-efficiency - mainly management and capacity utilization - are more important than changes in resource allocation (via changes in relative factor prices) to improve performance (see also White, 1976).

The potential productivity improvements from these types of reactions are probably larger in the less developed countries than elsewhere, simply because the initial inefficiencies are often larger. On the other hand, local firms in the less developed countries may be too weak to mount a competitive response to foreign entry, whereas the locals in industrialized host countries can often be expected to reply competitively. Various defensive corporate agreements, such as amalgamations among local firms or cooperative ventures with other foreign firms, may improve the local firms' competitiveness, even in developing countries (Blomström, 1986b; Lall, 1979; Evans, 1977), but there are no direct cross-country comparisons available, and there are not enough case studies for more comprehensive conclusions. Exactly what the reaction is - and how important the spillover benefits are - is likely to depend on the initial conditions in the market, and how much of an impact MNC entry makes on concentration and competition. However, the evidence seems to suggest that there is a larger risk that foreign MNCs crowd out local firms in developing countries We will return to this question in section 3 below.

14.2.5 Statistical Testing of Spillovers

Although there is plenty of empirical evidence of spillovers from the studies reviewed above, there are only few direct analyses and tests of the existence and significance of spillovers in a more general setting, presumably because of measurement problems and lack of suitable data.[12] In addition, most of the studies that are available focus on intra-industry effects. An early exception is Katz (1969), who notes that the inflow of foreign capital into the Argentine manufacturing sector in the 1950s had a significant impact on the technologies used by local firms. He asserts that the technical progress did not only take place in the MNCs' own industries, but also in other sectors, because the foreign affiliates forced domestic firms to modernize "by imposing on them minimum standards of quality, delivery dates, prices, etc. in their supplies of parts and raw materials" (Katz, 1969, p. 154).

The earliest statistical analyses of intra-industry spillovers include studies for Australia by Caves (1974), for Canada by Globerman (1979a), and for Mexico by Blomström and Persson (1983).[13] These authors examine the existence of

[12] It should also be noted in this context that both intra-industry and inter-industry R&D spillovers have been identified and estimated, mainly for developed countries, but generally without explicit reference to MNCs and FDI. See e.g. Bernstein (1988, 1989) and Nadiri (1991a). The fact that this kind of spillover seems to take place offers some indirect support to the hypothesis that there are technology spillovers between MNC affiliates and local firms. The conclusion that technological innovations (proxied by R&D measures) in some domestic firms have positive effects on the productivity of other domestic firms is analogous to the situation where technological innovations (proxied by the size of the technology gap or the amount of technology imports) in foreign affiliates have positive effects on the productivity of local firms.

[13] See also Blomström (1989). Moreover, Chen (1983) presents a detailed discussion and

spillovers by testing whether foreign presence - expressed in terms of the foreign share of each industry's employment or value added - has any impact on labor productivity in local firms in a production function framework. Foreign presence is simply included among other firm and industry characteristics as an explanatory variable in a multiple regression. All three studies conclude that spillovers are significant at this aggregate level, although they cannot say anything about how spillovers take place.

Some more recent studies also present results that are consistent with these early analyses. Blomström and Wolff (1994) ask whether the spillovers in the Mexican manufacturing sector were large enough to help Mexican firms converge toward US productivity levels during the period 1965-1982. Their answer is affirmative: foreign presence seems to have a significant positive impact on the rates of growth of local productivity. Nadiri (1991b), in a study of the impact of US direct investment in plant and equipment on the manufacturing sectors in France, Germany, Japan, and the UK between 1968 and 1988, comes to similar conclusions. Increases in the capital stock owned by US multinationals seem to stimulate new domestic investment in plant and equipment, and it appears that there is also a positive impact of FDI on the growth of total factor productivity in the host countries' manufacturing sectors.

There are also some studies suggesting that the effects of foreign presence are not always beneficial for local firms. For instance, Haddad and Harrison (1991, 1993), in a test of the spillover hypothesis for Moroccan manufacturing during the period 1985-1989, conclude that spillovers do not take place in all industrial sectors. Like Blomström (1986a), they find that foreign presence lowers the average dispersion of a sector's productivity, but they also observe that the effect is more significant in sectors with simpler technology. This is interpreted to mean that foreign presence forces local firms to become more productive in sectors where best practice technology lies within their capability, but that there are no significant transfers of modern technology. Furthermore, they find no significant effects of foreign presence on the rate of productivity growth of local firms, and interpret this as additional support to the conclusion that technology spillovers do not occur.

Aitken and Harrison (1991) use plant-level data for Venezuelan manufacturing between 1976 and 1989 to test the impact of foreign presence on total factor productivity growth. They conclude that domestic firms exhibited higher productivity in sectors with a larger foreign share, but argue that it may be wrong to conclude that spillovers have taken place if MNC affiliates systematically locate in the more productive sectors. In addition, they are also able to perform some more detailed tests of regional differences in spillovers. Examining the

some statistical evidence of spillovers in the major manufacturing industries in Hong Kong, although he does not examine the whole manufacturing sector. More specifically, he shows that foreign firms have been more active than local firms in importing new technologies to Hong Kong, and that the rates of technology diffusion have been higher in the industries where foreign firms hold larger market shares.

geographical dispersion of foreign investment, they suggest that the positive impact of FDI accrued mainly to the domestic firms located close to the MNC affiliates. However, effects seem to vary between industries. Aitken and Harrison (1991) is also one of the few studies, apart from Katz (1969), where inter-industry spillovers from foreign investment are discussed explicitly. As noted earlier, they assert that forward linkages generally brought positive spillover effects, but that backward linkages appeared to be less beneficial because of the foreign firms' high import propensities (although there were differences between industrial sectors).

Cantwell (1989), who investigates the responses of local firms to the increase in competition caused by the entry of US multinationals into European markets between 1955 and 1975, also argues that positive technology spillovers did not occur in all industries. His analysis differs notably from the other studies discussed in this section - he does not focus on productivity, but rather on changes in the market shares of foreign and local firms - but his conclusions are interesting. He asserts that "the technological capacity of indigenous firms ... was the major factor in determining the success of the European corporate response" (p. 86) to the US challenge, and that the size of the national market was an additional determinant. More specifically, Cantwell suggests that the entry of US affiliates provided a highly beneficial competitive spur in the industries where local firms had some traditional technological strength, whereas local firms in other industries - especially in countries where markets were too small to allow both kinds of firms to operate at efficient scale - were forced out of business or pushed to market segments that were ignored by the foreign MNCs.

Recently, some authors have also explicitly discussed the apparent contradictions between the earlier statistical spillover studies. In line with Cantwell (1989), Kokko (1994) argues that spillovers should perhaps not be expected in all kinds of industries. In particular, foreign MNCs may sometimes operate in "enclaves", where neither products nor technologies have much in common with those of local firms. In such circumstances, there may be little scope for learning, and spillovers may not materialize. Conversely, when foreign affiliates and local firms are in more direct competition with each other, spillovers are more likely. Examining data for Mexican manufacturing, he finds no signs of spillovers in industries where the foreign affiliates have much higher productivity and larger market shares than local firms; in industries without these enclave characteristics, on the other hand, there appears to be a positive relation between foreign presence and local productivity. Kokko, Tansini and Zejan (1996a) present similar findings for the Uruguayan manufacturing sector.

Another possible explanation for the divergent findings from the existing statistical tests is proposed by Kokko (1997), who focuses on effects of competition in Mexican manufacturing. The earlier studies have tested the hypothesis that productivity spillovers are strictly proportional to foreign presence, but Kokko (1997) argues that this is not always the case. Spillovers from competition, in particular, are not determined by foreign presence alone, but rather by the simultaneous interactions between foreign and local firms. Hence, it is possible that the spillovers are larger in case where a few foreign MNC stir up a previously protected market than in a situation where foreign affiliates hold large market shares, but refrain from competing hard with local firms. In fact, sometimes large foreign presence may even be a sign of a weak local industry, where local firms have not been able to absorb any productivity spillovers at all and have therefore been forced to yield market shares to the foreign MNCs.

These recent analyses point to the significance of local conditions in host countries as determinants of the magnitude and scope of spillovers. A high level of local competence and a competitive environment, both contribute to raise the absorptive capacity of the host country. In addition to explaining some of the differences between countries and industries when it comes to productivity benefits from FDI, they also highlight a possible role for economic policy in host countries. So far, foreign MNCs have typically been controlled through various types of performance and technology transfer requirements, but it appears that policies supporting a more competitive environment are useful alternatives for countries aiming to maximize the productivity benefits from FDI.[14]

14.3 Trade Effects

To enter a foreign market and to become a successful exporter, a company must not only be a competent manufacturer, but it will also need to manage the international marketing, distribution, and servicing of its products - tasks that are typically connected with high fixed costs. Few local firms, particularly those in developing countries, have the skills and resources to take on all these challenges on their own (see further Keesing and Lall, 1992). A MNC parent or affiliate is likely to be in a better position to establish export operations, since it can benefit from the existing international network of the entire corporation. The contacts with other parts of the corporation provide both knowledge of international market conditions and access to foreign marketing and distribution networks. Moreover, MNCs are often larger than local firms and may be able to afford the high fixed costs for the development of transport, communications, and financial services that are needed to support export activities.

Examining the impact of FDI on the host countries' trade performance, it is useful to distinguish between direct and indirect effects. Regarding the direct effects on host country exports, we follow Helleiner (1973), and divide the exports and MNC activities into four different categories according to production characteristics:

- local raw material processing,
- conversion of import-substituting industry to exporting,
- new labor-intensive final product exports, and

[14] For instance, Blomström, Kokko and Zejan (1994) and Kokko and Blomström (1995) show that the technology imports of foreign affiliates are partly determined by the competition in the host country market.

• labor-intensive processes and component specialization within vertically integrated international industries.

In the first category, i.e. processing of locally produced raw materials, MNCs may have better export potential than indigenous firms in developing countries, because of their business contacts abroad, their marketing skills, superior technology both in product and processes, and their greater know-how in general. Particularly for the poorest developing countries, where most of these assets are lacking, foreign firms may be one of the few available alternatives, at least for the time being, if they want to increase their exports.

The same MNC advantages should also be important when countries try to convert import-substituting industries to exporting (for empirical evidence, see Blomström and Lipsey, 1993). If this avenue for the expansion of manufactured exports is chosen, one should also consider the suggestion that multinational corporations may be a significant factor in speeding up free trade, not least among the developing countries. For instance, it has been noted that MNCs have been firm supporters of common markets, customs unions, and free trade areas in the developing countries, since these enable them to rationalize small-scale facilities and develop exports (see e.g. Helleiner, 1973 and Dunning, 1993). On the other hand, there are reasons to believe that the MNCs are eager to support protection in host developing countries if the reason for the investment from the very beginning was the profits provided by protected market. However, with the decreasing appeal of import-substituting strategies in general in developing countries, the anti-trade influence of MNCs is expected to be weak.

Multinational corporations are also believed to have more general interests in trade liberalization in their home countries, particularly in their own products, and it is often said that these firms have become a significant source of political pressure for freer trade. Given the currently widespread perception that there is a hardening of commercial policies in the developed countries, it is important to analyze what role the MNCs may play here. Are these firms in a position to facilitate exports from the developing to the developed countries, for instance, in labor-intensive final products, and, if so, will they do it on a general basis or will they lobby only for their own exports?

Some efforts have been made to test the hypothesis that protection is lower on products that are of special interest to the MNCs. For instance, Lavergne (1983) and Helleiner (1977) tested whether United States MNCs acted as a free-trade force in support of imports of their majority-owned foreign affiliates, but no conclusive evidence was found. In this context, it can be noted that Bhagwati (1988) argues that it is not meaningful to look for pro-trade influences of multinational corporations in cases of successful opposition to particular protectionist pressures. He asserts that MNCs are in opposition to protectionism in general in their home countries to avoid possible outbreaks of protectionism elsewhere, so that they act as an important free-trade force, not in specific sectors, but at higher levels.

When it comes to exports of new labor-intensive final products, such as textiles and other consumer goods, history seems to tell that there are many opportunities for developing countries to become significant exporters even without the help from foreign firms. Such a conclusion would, however, understate the importance and contribution of multinational corporations in these exports. Firms in developing countries seeking to expand their exports to developed countries face immense difficulties in setting up a distribution network, keeping in close touch with the rapid change in consumer tastes, mastering the technicalities of industrial norms and safety standards, and building up a new product image. In many cases, the design, packaging, distribution, and servicing of the products are as important as being able to produce them at (or below) ruling prices in world markets, and the lack of such skills constitutes a key entry barrier to markets for developing country exporters.

The story behind the success of many developing country firms entering world markets in light consumer goods is that foreign firms help them by providing the links to the final buyers (Blomström, 1990). Usually, firms in developed countries seek out manufacturers in developing countries and bring them under contract as suppliers. The former may be import-wholesalers in narrow product lines, large department store chains or, as in East Asia, Japanese trading companies dealing in a wide variety of products. Many of these firms are true MNCs with outlets in several countries, and by not taking their activities into account, we significantly overstate the opportunities for developing countries to enter markets for labor-intensive final products.

Hence, although the local suppliers of MNCs do not always export under their own name, they benefit from access to foreign markets. This may allow them to expand output and achieve economies of scale. It is also likely that the linkages with export oriented MNCs provide knowledge about product and process technologies and foreign market conditions - for instance, foreign preferences regarding design, packaging, and product quality - and if this information can be used profitably in the supplier companies' other operations, there may be important external effects from contacts with foreign MNCs (see Keesing and Lall, 1992).

Exports of labor-intensive components within vertically integrated industries are almost by definition dependent on the participation of MNCs. Generally, we think of these exports as intra-firm trade, but a great part of them are arm's-length transactions between MNCs and indigenous LDC firms (although the MNCs usually take responsibility for the marketing and distribution of the products). However, since this type of production is often dependent on imports of raw materials and intermediate goods, it is not obvious that the net earnings of foreign exchange will be significant. The main benefits from export processing are instead related to increased employment, skills, wages, and taxes, at least in the short run (see Kobba, 1986, for an analysis of the effects of such foreign direct investment in Tunisia).

Countries that choose to specialize in labor-intensive processes and components production for MNCs also have to take into account that these affiliates are relatively "footloose", with little physical capital nailed down to hinder movement to the most favorable environment (see e.g. Flamm, 1984). Decisions taken by the parents may lead to sudden changes in the production of their affiliates in different countries, without necessarily taking into account the interest of host countries. This may happen as a result of changes in the costs of production, the perception of risks, or the policy environment in different host countries (UNCTC, 1985).

In addition to the export influences that require some type of linkage between MNCs and local firms, there may also be several indirect effects that benefit local export performance. In the simplest case, local firms may learn how to succeed in foreign markets simply by copying MNCs, although more tangible externalities are usually needed. For instance, MNCs may have affiliated firms in the prospective export market who can lobby for trade liberalization, and local firms may benefit from any reductions of trade barriers that are achieved. There may be spare capacity in the distribution or marketing facilities created by MNCs, which local firms may use at or slightly above marginal cost. The MNCs may also train their local staff in export management, and these skills may spill over to local firms if the MNCs employees change jobs. Other channels for the diffusion of information on foreign market conditions are trade associations and other industry organizations, of which MNCs are often prominent members. This kind of "market access spillovers" may be most important where the indigenous resources are weakest, i.e. in developing countries.

In one of the few available statistical analyses of export externalities, Aitken, Hanson and Harrison (1994) hypothesize that one firm's export activities may reduce the costs for foreign market access of other potential exporters that are located nearby. Testing a logit-specification for over 2,000 Mexican manufacturing plants during the period 1986-1990, they find that locating near an exporting MNC raises the probability of exporting for an individual firm, but that there is no corresponding effect from locating near locally-owned exporters. Hence, Aitken, Hanson and Harrison (1994, p. 25) conclude that "Foreign-owned enterprises are a natural conduit for information about foreign markets and technology, and a natural channel through which domestic firms can distribute their goods. To the extent that foreign investors directly or indirectly provide information and distribution services, their activities enhance the export prospects of local firms." Although their study cannot say anything about what the channels for market access spillovers are, they are able to demonstrate that the effects are significant at the national level.

Similarly, Kokko, Tansini and Zejan (1996b) find that the presence of foreign MNCs raises the likelihood that local Uruguayan firms are involved in exporting. However, this effect only applies for exports to the world market, while exports to the neighboring countries Argentina and Brazil appear not to be influenced by foreign presence. Moreover, only foreign MNCs established in Uruguay during the relatively outward-oriented period since 1973 - but not MNCs established during the preceding import-substituting period - seem to have any influence on the probability that local firms are engaged in exporting.

14.4 Competitive and Anti-Competitive Effects

It was argued earlier that MNCs may improve industrial efficiency and resource allocation in their host countries by entering into industries where high entry barriers reduce the degree of domestic competition. The entry of MNCs into these monopolistic industries is likely to raise the level of competition and force existing firms to become more efficient. Foreign entry may, of course, also lead to a fall in the number of firms in the industry if the least efficient local companies are forced out of business. This raises the fear that foreign MNCs may outcompete all local firms and establish monopolies that are even worse than the domestic oligopolies they replace: in addition to restricting competition, there is a risk that MNC monopolies may also repatriate profits and avoid taxation through transfer pricing.

However, it is likely that competition generally becomes more fierce, because the MNC affiliates' strategies typically stir up the established patterns of "gentlemanly competition". Hence, it has been said that "whatever the market structure that results from the influence of direct investment, it can be argued that entry by a foreign subsidiary is likely to produce more active rivalrous behavior and improvement in market performance than would a domestic entry at the same initial scale" (Caves, 1971, p. 15). Another point to note is that the increase in competition is often more effective in inducing technological change and productivity improvements than profit incentives, since "threats of deterioration or actual deterioration from some previous state are more powerful attention-focusing devices than are vague possibilities for improvements" (Rosenberg, 1976, p. 124).

Yet, the uncertainty regarding the generality of competition effects motivates an examination of some studies of FDI and industry structure in host countries. One central problem here is the question whether MNC entry and presence explain industry structure, or whether industry structure determines if MNCs will enter or not. This is an important question, since we have argued that one of the reasons to expect significant effects of foreign presence is the improvement in efficiency and resource allocation that may follow from MNC entry into monopolistic host country industries. Another problem is that there is some confusion regarding effects that are endemic to MNCs and those that are only speeded up by MNC presence. Few authors have been able to make a proper distinction between these two effects, but it may not be a crucial issue in the present context. What matters is the impact made by MNCs, and not the question of whether it is caused by foreign ownership or some other of the MNCs' characteristics. Yet another (perhaps more important) complication is that there is no simple relationship between competition and efficiency, on the one hand, and concentration, on the other hand. This will warrant some further comments later.

Moving to the empirical findings, it is clear that the overwhelming majority of studies are able to establish a positive correlation between foreign entry and presence and seller concentration in host country industries (see e.g. Dunning, 1993 and Caves, 1996, for surveys). However, the causal links are more difficult to establish. One finding is that the correlation disappears once other determinants of concentration are taken into account, and that MNCs do not cause concentration but are drawn to concentrated industries (Fishwick, 1981; Globerman, 1979b). Knickerbocker (1976) shows that entries by MNCs into the US market in the 1960s led to lower concentration, and that the same pattern was evident also for Canada, Italy, France, and West Germany. Commenting on these and other studies, Caves (1996, p. 89) concludes that the "correlations do not themselves prove that any direct causal relationships exist between foreign investment and concentration". It should, however, be noted that most of the studies look at effects of MNC entry, and it is possible that the concentration-reducing impact does not hold for already established affiliates, who may instead be interested in building barriers to entry.

Regarding the studies of developing countries, most authors have not been able to - or have not even tried to - determine whether the high degrees of concentration in the industries where foreign affiliates are present have been caused by MNCs or whether MNCs have just been attracted to these industries by good profit opportunities. Two (seemingly contradictory) exceptions are Evans (1977), who claims that MNCs tended to reduce concentration in the oligopolistic Brazilian pharmaceutical industry, and Newfarmer (1979), who argues for the opposite effect - caused by interlocking directorates, collusion, cross-subsidization, and other "oligopolistic tactics" - in the Brazilian electrical equipment industry.

Lall (1978) hypothesizes that it is plausible that MNCs speed up the natural concentration process in LDCs, or that the weakness of local competitors allows MNCs to achieve a higher degree of market dominance than in developed countries. Lall (1979) proceeds to argue that the level of concentration probably falls in the short run following MNC entry, as the affiliate adds to the number of firms in the industry, but that this may be reversed in the long run. The MNCs may buy out local firms or force them out of business, their success may force local firms to fusions and amalgamations, or they may be more skilled as lobbyists than others, thus adding to entry barriers and protection. Looking at the effects of MNCs on concentration in 46 Malaysian industries, he asserts that the presence of foreign firms on balance increased concentration. This was brought about both by the MNCs' impact on general industry characteristics - such as higher initial capital requirements, capital intensity, and advertising intensity - and by some apparently independent effect of foreign presence, perhaps related to "predatory" conduct, changes in technology and marketing practices, or gains of policy concessions from the government. Similar results were reported for Mexico in Blomström (1986b). Thus, the evidence seems to suggest that there is a larger risk that MNCs crowd out local firms in LDCs than in developed countries.

The assumption implicit in much of the discussion above is that competition improves efficiency and welfare, but there are cases where it must not necessarily be that way. Firstly, economies of scale are important determinants of industrial productivity. To the extent that foreign entry increases concentration in relatively small national industries, resource allocation and efficiency may well improve from the increase in average firm size. Whether this effect is stronger than that from the presumably reduced competition depends on market characteristics and trade policy. For instance, a fall in the number of competitors from thirty to twenty must not necessarily harm the competitive environment, but a reduction from three to two certainly will. Similarly, increased concentration is likely to have more harmful effects in protected industries than in import-competing or export-oriented industries.[15]

In fact, free trade and imports may well be good substitutes for large numbers of domestic competitors: Scandinavian, and particularly Swedish, industrial policies have for a long time built on this assertion (Hjalmarsson, 1991), although the competition between the few remaining large firms has also been important (Porter, 1990; Sölvell, Zander, and Porter, 1991). The conclusion by Chen (1983, p. 90) from his study of Hong Kong manufacturing, where all industries are either export-oriented or import-competing, is consistent with these arguments: "There are indications that the presence of foreign investment in an industry may have the effect of eliminating wasteful competition ... [without introducing] damaging monopolistic elements into the industry".

Secondly, focusing more closely on technology, there is the classic "Schumpeterian Dilemma" of weighing the static allocative efficiency of competitive markets against the supposed dynamic efficiency of monopolistic and oligopolistic firms. The rate of technical progress can perhaps be higher in concentrated markets, since firms there have internally generated profits to use for R&D, and are generally larger and more able to enjoy economies of scale in R&D. It is also possible that market structure has some impact on what the R&D efforts aim to achieve.

In fact, empirical studies seem to show that market structure affects both the rate and type of technical progress. Looking at the overall rate of technical change, Kamien and Schwartz (1982) summarize a survey of research in industrialized countries by concluding that neither perfect competition nor perfect monopoly, but rather mildly oligopolistic markets, are most conducive to technical progress. Moreover, Katz (1984) and Teitel (1984) in studies of Latin America, and Lall (1980) for India, show that technical change in industries with limited competition largely aims to overcome supply bottlenecks, e.g. by substituting imported raw materials and components, while change in more competitive industries is characterized by cost-reducing and quality-improving innovations. In these cases, however, limited competition is intimately tied to import-substitution rather than concentration, although there is a certain overlap.

Summarizing the evidence on the relation between MNC entry and presence and industry structure, it seems that MNCs enter mainly into industries where barriers

[15] The Peruvian automotive industry in the late 1960s and early 1970s (like many other industries in countries with extreme import-substitution policies) provides a striking example of the fact that low concentration does not necessarily equal high efficiency. At that time, "13 firms, each with some foreign ownership, were assembling 18 brands and over 25 models of automotive vehicles, mostly passenger cars. Facing a limited local market, none of these firms was able to use more than 30 per cent of its installed capacity" (UNCTC, 1981, p. 19).

to entry and concentration are relatively high, and initially add to the number of firms in the market. In the long run, MNCs may contribute to some increase in concentration, but efficiency may still benefit, particularly if protection does not guarantee an easy life also for the MNC affiliate. Most of the evidence, however, is related to MNC entry rather than to MNC presence - the dynamic aspects of MNCs and competition in host country markets are not well researched. Moreover, much of the evidence refers to effects in developed countries, and it is not possible to disregard the risk that MNC entry into developing countries replaces local production and forces local firms out of business, rather than forcing them to become more efficient.

14.5 Concluding Remarks

This paper has reviewed the evidence on host country effects of foreign direct investment. The focus of the discussion has been on the transfer and diffusion of technology from foreign multinationals to the host country, since MNCs own and control the bulk of the world's commercial technology. In addition, we have examined the impact of foreign MNCs for the trade performance of their host countries, and the effects on competition and industry structure in the host countries.

A tentative conclusion of the review is that foreign direct investment may promote economic development by contributing to productivity growth and exports in their host countries. However, the exact nature of the relation between foreign MNCs and their host economies seems to vary between industries and countries. It is reasonable to assume that the characteristics of the host country's industry and policy environment are important determinants of the net benefits of FDI. This paper has not explicitly dealt with questions related to host country policies vis-à-vis foreign MNCs, but the findings of the paper highlight the need for future research in this area.

References

Aitken, B., G.D. Hanson and A. Harrison (1994), "Spillover, Foreign Investment, and Export Behavior", NBER Working Paper No. 4967, December.

Aitken, B. and A. Harrison (1991), "Are There Spillovers From Foreign Direct Investment? Evidence from Panel Data for Venezuela", mimeo, MIT and the World Bank, November.

Balasubramanyam, U.N. (1973), *International Transfer of Technology to India*, New York; Praeger.

Behrman, J. and H. Wallender (1976), *Transfer of Manufacturing Technology within Multinational Enterprises*, Cambridge, Mass.; Ballinger.

Bergsman, J. (1974), "Commercial Policy, Allocative Efficiency and 'X-efficiency'", *Quarterly Journal of Economics*, Vol. 86, 409-433.

Bernstein, J.I. (1988), "Cost of Production, Intra- and Interindustry R&D Spillovers: Canadian Evidence", *Canadian Journal of Economics*, Vol. 21, 324-347.

Bernstein, J.I. (1989), "The Structure of Canadian Interindustry R&D Spillovers, and the Rates of Return to R&D", *Journal of Industrial Economics*, Vol. 37, 315-328.

Bhagwati, J. (1988), *Protectionism*, Cambridge; MIT Press.

Blomström, M. (1986a), "Foreign Investment and Productive Efficiency: The Case of Mexico", *Journal of Industrial Economics*, Vol. 15, 97-110.

Blomström, M. (1986b), "Multinationals and Market Structure in Mexico", *World Development*, Vol. 14, 523-530.

Blomström, M. (1989), *Foreign Investment and Spillovers: A Study of Technology Transfer to Mexico*, London; Routledge

Blomström, M. (1990), *Transnational Corporations and Manufacturing Exports from Developing Countries*, New York; United Nations.

Blomström, M. (1991a), "Host Country Benefits of Foreign Investment", in D.G. McFetridge, ed., *Foreign Investment, Technology and Economic Growth*, Toronto and London; Toronto University Press.

Blomström, M. (1991b), "Technology Transfer through Various Forms of International Investment: Some Evidence from Swedish MNCs", in H.W. Singer, N. Hatti and R. Tandon, eds., *Joint Ventures and Collaborations*, New Dehli; Indus Publishing Company.

Blomström, M., A. Kokko and M. Zejan (1994), "Host Country Competition and Technology Transfer by Multinationals", *Weltwirtschaftliches Archiv*, Band 130, 521-533.

Blomström, M. and R.E. Lipsey (1993), "Foreign Firms and Structural Adjustment in Latin America: Lessons from the Debt Crisis", in G. Hansson, ed., *Trade, Growth and Development: The Role of Politics and Institutions*, London and New York; Routledge.

Blomström, M. and H. Persson (1983), "Foreign Investment and Spillover Efficiency in an Underdeveloped Economy: Evidence from the Mexican Manufacturing Industry", *World Development*, Vol. 11, 493-501.

Blomström, M. and E. Wolff (1994), "Multinational Corporations and Productivity Convergence in Mexico", in W. Baumol, R. Nelson and E. Wolff, eds., *Convergence of Productivity: Cross-National Studies and Historical Evidence*, Oxford; Oxford University Press.

Blomström, M. and M. Zejan (1991), "Why Do Multinational Firms Seek Out Joint Ventures?", *Journal of International Development*, Vol. 3, 53-63.

Brash, D.T. (1966), *American Investment in Australian Industry*, Cambridge, Mass.; Harvard University Press.

Buckley, P.J. and M. Casson (1976), *The Future of the Multinational Enterprise*, London; Macmillan.

Cantwell, J. (1989), *Technological Innovation and Multinational Corporations*, Oxford; Basil Blackwell.

Cantwell, J. (1995), "The Globalization of Technology: What Remains of the Product Cycle Model?", *Cambridge Journal of Economics*, Vol. 19, 155-174.

Caves, R.E. (1971), "International Corporations: The Industrial Economics of Foreign Investment", *Economica*, Vol. 38, 1-27.

Caves, R.E. (1974), "Multinational Firms, Competition and Productivity in Host-Country Markets", *Economica*, Vol. 41, 176-193.

Caves, R.E. (1996), *Multinational Enterprise and Economic Analysis*, Second Edition. Cambridge; Cambridge University Press.

Chen, E.K.Y. (1983), *Multinational Corporations, Technology and Employment*, London; Macmillan.

Contractor, F. (1984), "Choosing Between Direct Investment and Licensing: Theoretical Considerations and Empirical Tests", *Journal of International Business Studies*, Vol. 15, 167-188.

Corden, W.M. (1967), "Protection and Foreign Investment", *Economic Record*, Vol. 43, 209-232.

Das, S. (1987), "Externalities, and Technology Transfers through Multinational Corporations: A Theoretical Analysis", *Journal of International Economics*, Vol. 22, 171-182.

Davidson, W.H. (1980), *Experience Effects in International Investment and Technology Transfer*, Ann Arbor; U.M.I. Research Press.

Davidson, W.H. (1983), "Structure and Performance in International Technology Transfer", *Journal of Management Studies*, Vol. 20, 453-465.

Davidson, W.H. and D.G. McFetridge (1985), "Key Characteristics in the Choice of International Technology Transfer", *Journal of International Business Studies*, Vol. 15, 5-21.

Deane, R.S. (1970), *Foreign Investment in New Zealand Manufacturing*, Wellington; Sweet and Maxwell.

Dunning, J. (1958), *American Investment in British Manufacturing*, London; George Allen & Unwin.

Dunning, J. (1973), "The Determinants of International Production", *Oxford Economic Papers*, Vol. 25, 289-336.

Dunning, J. (1993), *Multinational Enterprises and the Global Economy*, Reading; Addison-Wesley Publ. Co.

Enos, J.L (1989), "Transfer of Technology", *Asia-Pacific Economic Literature*, Vol. 3, 3-37.

Evans, P.B. (1977), "Direct Investment and Industrial Concentration", *Journal of Development Studies*, Vol. 13, 373-385.

Evans, P. (1979), *Dependent Development: The Alliance of Multinational, State and Local Capital in Brazil*, Princeton; Princeton University Press.

Fairchild, L. and K. Sosin (1986), "Evaluating Differences in Technological Activity between Transnational and Domestic Firms in Latin America", *Journal of Development Studies*, Vol. 22, 697-708.

Findlay, R. (1978), "Relative Backwardness, Direct Foreign Investment, and the Transfer of Technology: A Simple Dynamic Model", *Quarterly Journal of Economics*, Vol. 92, 1-16.

Fishwick, F. (1981), "Multinational Companies and Economic Concentration in Europe", mimeo, Paris; Institute for Research and Information on Multinationals.

Flamm, K. (1984), "The Volatility of Offshore Investment", *Journal of Development Economics*, Vol. 16, 231-248.

Forsyth, D. (1972), *U.S. Investment in Scotland*, New York; Praeger.

Fransman, M. (1985), "Conceptualising Technical Change in the Third World in the 1980s: An Interpretive Survey", *Journal of Development Studies*, Vol. 21, 572-652.

Gabriel, P.P. (1967), *The International Transfer of Corporate Skills: Manager Contracts in Less Developed Countries*, Cambridge, Mass.; Harvard University Press.

Gerschenberg, I. (1987), "The Training and Spread of Managerial Know-How. A Comparative Analysis of Multinational and Other Firms in Kenya", *World Development*, Vol. 15, 931-939.

Globerman, S. (1979a), "Foreign Direct Investment and 'Spillover' Efficiency Benefits in Canadian Manufacturing Industries", *Canadian Journal of Economics*, Vol. 12, 42-56.

Globerman, S. (1979b), "A Note on Foreign Ownership and Market Structure in the United Kingdom", *Applied Economics*, Vol. 11, 35-42.

Gomulka, S. (1990), *The Theory of Technological Change and Economic Growth*, London and New York; Routledge.

Grosse, R. (1989), *Multinationals in Latin America*, London and New York; Routledge.

Haddad, M. and A. Harrison (1991), "Are there Positive Spillovers from Direct Foreign Investment? Evidence from Panel Data for Morocco", mimeo, Harvard University and the World Bank, September.

Haddad, M. and A. Harrison (1993), "Are there Positive Spillovers from Direct Foreign Investment? Evidence from Panel Data for Morocco", *Journal of Development Economics*, Vol 42, 51-74.

Halbach, A.J. (1989), "Multinational Enterprises and Subcontracting in the Third World: A Study of Inter-Industrial Linkages", Multinational Enterprises Programme, Working Paper No. 58. Geneva; International Labour Office.

Helleiner, G.K. (1973), "Manufactured Exports from Less-Developed Countries and Multinational Firms", *Economic Journal*, Vol. 83,

Helleiner, G.K. (1977), "Transnational Enterprises and the New Political Economy of United States Trade Policy", *Oxford Economic* Papers, Vol. 16,

Hill, H. (1982), "Vertical Inter-Firm Linkages in LDCs: A Note on the Philippines", *Oxford Bulletin of Economics and Statistics*, Vol. 44, 261-271.

Hjalmarsson, L. (1991), "The Scandinavian Model of Industry Policy", in M. Blomström and P. Meller, eds., *Diverging Paths: Comparing a Century of Scandinavian and Latin American Economic Development*, Washington, D.C. and Baltimore; Inter-American Development Bank and Johns Hopkins University Press.

Hymer, S. (1960), "The International Operations of National Firms: A Study of Direct Investment", unpublished PhD thesis, Massachusetts Institute of Technology.

Jenkins, R. (1990), "Comparing Foreign Subsidiaries and Local Firms in LDCs: Theoretical Issues and Empirical Evidence", *Journal of Development Studies*, Vol. 26, 205-228.

Kamien, M.I. and N.L. Schwartz (1982), *Market Structure and Innovation*, Cambridge; Cambridge University Press.

Kaplinsky, R. (1990), "Technology Transfer, Adaptation, and Generation", in M. Chatterji, ed., *Technology Transfer in the Developing Countries*, London; Macmillan.

Katz, J.M. (1969), *Production Functions, Foreign Investment and Growth*, Amsterdam; North Holland.

Katz, J.M. (1984), "Domestic, Technological Innovations and Dynamic Comparative Advantage: Further Reflections on a Comparative Case Study Program", *Journal of Development Economics*, Vol. 16, 13-38.

Katz, J.M. (1987), *Technology Creation in Latin American Manufacturing Industries*, New York; St. Martin's Press.

Keesing, D.B. and S. Lall (1992), "Marketing Manufactured Exports from Developing Countries: Learning Sequences and Public Support", in G. Helleiner, ed., *Trade Policy, Industrialization, and Development: New Perspectives*, Oxford; Clarendon Press.

Kim, L. and Y. Kim (1985), "Innovation in a Newly Industrializing Country: A Multiple Discriminant Analysis", *Management Science*, Vol. 31, 312-322.

Kindleberger, C.P. (1969), *American Business Abroad*, New Haven; Yale University Press.

Knickerbocker, F.T. (1976), "Market Structure and Market Power Consequences of Foreign Direct Investment by Multinational Companies", Occasional Paper No. 8, Washington, D.C.; Center for Multinational Studies.

Kobba, F. (1986), *Foreign Direct Investment in Tunisia*, PhD Thesis, Stockholm; University of Stockholm.

Koizumi, T. and K.J. Kopecky (1977), "Economic Growth, Capital Movements and the International Transfer of Technical Knowledge", *Journal of International Economics*, Vol. 7, 45-65.

Kokko, A. (1994), "Technology, Market Characteristics, and Spillovers", *Journal of Development Economics*, Vol. 43, 279-293.

Kokko, A. (1997), "Productivity Spillovers from Competition between Local Firms and Foreign Affiliates", *Journal of International Development*, forthcoming.

Kokko, A. and M. Blomström (1995), "Policies to Encourage Inflows of Technology through Foreign Multinationals", *World Development*, Vol. 23, 459-468.

Kokko, A., R. Tansini and M. Zejan (1996a), "Local Technological Capability and Spillovers from FDI in the Uruguayan Manufacturing Sector", *Journal of Development Studies*, Vol. 34, 602-611.

Kokko, A., R. Tansini and M. Zejan (1996b), "Trade Regimes and Effects of FDI: Evidence from Uruguay", paper presented at workshop on *Multinationals, Trade and Economic Geography*, IUI, Stockholm, May 23-24.

Lake, A.W. (1979), "Technology Creation and Technology Transfer by Multinational Firms", in R.G. Hawkins, ed., *Research in International Business and Finance, Vol. 1: The Economic Effects of Multinational Corporations*, Greenwich; JAI Press.

Lall, S. (1978), "Transnationals, Domestic Enterprises and Industrial Structure in LDCs: A Survey", *Oxford Economic Papers*, Vol. 30, 217-248.

Lall, S. (1979), "Multinationals and Market Structure in an Open Developing Economy: The Case of Malaysia", *Weltwirtschaftliches Archiv*, Band 115, 325-350.

Lall, S. (1980), "Vertical Interfirm Linkages in LDCs: An Empirical Study", *Oxford Bulletin of Economics and Statistics*, Vol. 42, 203-226.

Langdon, S. (1981), *Multinational Corporations in the Political Economy of Kenya*, London; Macmillan.

Lavergne, R. (1983), The *Political Economy of U.S. Tariffs: An Empirical Investigation*, New York; Academic.

Leibenstein, H. (1966), "Allocative Efficiency vs. X-Efficiency", *American Economic Review*, Vol. 56, June, 392-415.

Leibenstein, H. (1980), "X-Efficiency, Intrafirm Behavior, and Growth", in S. Maital and N. Meltz, eds., *Lagging Productivity Growth*, Cambridge, Mass.; Ballinger.

Lim, L.Y.C. and E.F. Pang (1982), "Vertical Linkages and Multinational Enterprises in Developing Countries", *World Development*, Vol. 10, 585-595.

Lindsey, C.W. (1989), "Commodities, Technology, and Trade: Transnational Corporations and Philippine Economic Development"", *Philippine Review of Economics and Business*, Vol. 26, 67-108.

Lipsey, R.E., M. Blomström and E. Ramstetter (1995), "Internationalized Production in World Output", NBER Working Paper No. 5385, December.

MacDougall, G.D.A. (1960), "The Benefits and Costs of Private Investment from Abroad: A Theoretical Approach", *Economic Record*, Vol. 36, 13-35.

Mansfield, E. and A. Romeo (1980), "Technology Transfer to Overseas Subsidiaries by U.S.-based Firms", *Quarterly Journal of Economics*, Vol. 95, 737-750.

Mansfield, E., M. Schwartz, and S. Wagner (1981), "Imitation Costs and Patents: An Empirical Study", *Economic Journal*, Vol. 91, 907-918.

McAleese, D. and D. McDonald (1978), "Employment Growth and Development of Linkages in Foreign-Owned and Domestic Manufacturing Enterprises", *Oxford Bulletin of Economics and Statistics*, Vol. 40, 321-339.

de Melto, D. *et al* (1980), "Innovation and Technological Change in Five Canadian Industries", Economic Council of Canada, Discussion Paper No. 176.

Nadiri, M.I. (1991a), "Innovations and Technological Spillovers", mimeo, New York University and NBER, September.

Nadiri, M.I. (1991b), "U.S. Direct Investment and the Production Structure of the Manufacturing Sector in France, Germany, Japan, and the U.K.", mimeo, New York University and NBER, December.

Newfarmer, R.S. (1979), "Oligopolistic Tactics to Control Markets and the Growth of TNCs in Brazil's Electrical Industry", *Journal of Development Studies*, Vol. 15, 108-140.

Pack, H. (1974), "The Employment-Output Tradeoff in LDCs - A Microeconomic Approach", *Oxford Economic Papers*, Vol. 26, 388-404.

Page, J.M. (1980), "Technical Efficiency and Economic Performance: Some Evidence from Ghana", *Oxford Economic Papers*, Vol. 32, 319-339.

Patel, P. and K. Pavitt (1994), "Technological Competencies in the World's Largest Firms: Characteristics, Constraints and Scope for Managerial Choice", mimeo.

Porter, M.E. (1990), The Competitive Advantage of Nations, New York; The Free Press.

Reuber, G.L., with H. Crookell, M. Emerson, and G. Gallais-Hamonno (1973), *Private Foreign Investment in Development*, Oxford; Clarendon Press.

Riedel, J. (1975), "The Nature and Determinants of Export-Oriented Direct Foreign Investment in a Developing Country: A Case Study of Taiwan", *Weltwirtschaftliches Archiv*, Band 111, 505-528.

Rosenberg, N. (1976), *Perspectives on Technology*, New York; Cambridge University Press.

Rosenbluth, G. (1970), "The Relation between Foreign Control and Concentration in Canadian Industry", *Canadian Journal of Economics*, Vol. 3, 14-38.

Safarian, A.E. (1966), Foreign Ownership of Canadian Industry, Toronto; McGraw-Hill.

Stobaugh, R. (1988), *Innovation and Competition*, Cambridge, Mass.; Harvard Business School Press.

Swan, P.L. (1973), "The International Diffusion of an Innovation", *Journal of Industrial Economics*, Vol. 22, 61-69.

Sölvell, Ö., I. Zander and M. Porter (1991), *Advantage Sweden*, Stockholm; Norstedts.

Teece, D.J. (1981), "The Market for Know-How and the Efficient International Transfer of Technology", *The Annals of the American Academy of Political and Social Science*, Vol. 458, 81-96.

Teitel, S. (1984), "Technology Creation in Semi-Industrial Economies", *Journal of Development Economics*, Vol. 6, 39-61.

Telesio, P. (1979), *Technology Licensing and Multinational Enterprises*, New York; Praeger Publishers.

Tilton, J.E. (1971), *The International Diffusion of Technology: The Case of Semiconductors*, Washington; Brookings Institution.

UNCTC (1981), *Transnational Corporation Linkages in Developing Countries: The Case of Backward Linkages via Subcontracting*, New York; United Nations Centre on Transnational Corporations.

UNCTC (1985), *Transnational Corporations in World Development: Third Survey*, London; Graham & Trotman Limited (for the United Nations Centre on Transnational Corporations).

UNCTC (1988), *Transnational Corporations in World Development: Trends and Prospects*, New York; United Nations Centre on Transnational Corporations.

Vernon, R. (1966), "International Investment and International Trade in the Product Cycle", *Quarterly Journal of Economics*, Vol. 80, 190-207.

Vernon, R. and W.H. Davidson (1979), *Foreign Production of Technology-Intensive Products by U.S.-Based Multinational Enterprises*, Washington, DC; National Science Foundation.

Wang, Y. and M. Blomström (1992), "Foreign Investment and Technology Transfer: A Simple Model", *European Economic Review*, Vol. 36, 137-155.

Wasow, B. and H. Hill (1986), *The Insurance Industry in Economic Development*, New York; New York University Press.

Watanabe, S. (1983a), "Technical Co-operation between Large and Small Firms in the Filipino Automobile Industry", in S. Watanabe, ed., *Technology Marketing and*

Industrialization: Linkages between Small and Large Enterprises, New Delhi; Macmillan.

Watanabe, S. (1983b), "Technological Linkages through Subcontracting in Mexican Industries", in S. Watanabe, ed., *Technology Marketing and Industrialization: Linkages between Small and Large Enterprises*, New Delhi; Macmillan.

White, L.J. (1976), "Appropriate Technology, X-inefficiency and the Competitive Environment: Some Evidence From Pakistan", *Quarterly Journal of Economics*, Vol. 90, 575-589.

Yoshihara, (1988), *The Rise of Ersatz Capital in South-East Asia*, Oxford; Oxford University Press.

Zander, U. (1991), *Exploiting a Technological Edge - Voluntary and Involuntary Dissemination of Technology*, Stockholm; Institute of International Business, Stockholm School of Economics.

Zander, I. (1994), *The Tortoise Evolution of the Multinational Corporation - Foreign Technological Activity in Swedish Multinational Firms 1890-1990*, Stockholm; Institute of International Business, Stockholm School of Economics.

Acknowledgements

Chapter 3, *Opening the Black Box of Innovation*, by Philippe Aghion and Jean Tirole is reprinted from The European Economic Review, 38, 1994, pp.701-710, with kind permission from Elsevier Science - NL, Sara Burgerhartstraat 25, 1055 KV Amsterdam, The Netherlands.

Chapter 6, *Universities as a Source of Commercial Technology: A Detailed Analysis of University Patenting 1965-1988*, by Rebecca Henderson, Adam B. Jaffe and Manuel Trajtenberg is reprinted from The Review of Economics and Statistics, 80:1, 1998, pp. 119-127, with kind permission from the President and Fellows of Harvard College and the Massachusetts Institute of Technology.

MIX
Papier aus verantwortungsvollen Quellen
Paper from responsible sources
FSC® C105338

If you have any concerns about our products,
you can contact us on
ProductSafety@springernature.com

In case Publisher is established outside the EU,
the EU authorized representative is:
Springer Nature Customer Service Center GmbH
Europaplatz 3, 69115 Heidelberg, Germany

Printed by Libri Plureos GmbH
in Hamburg, Germany